AF386602

Springer Optimization and Its Applications

Volume 234

Aims and Scope

Optimization has continued to expand in all directions at an astonishing rate. New algorithmic and theoretical techniques are continually developing and the diffusion into other disciplines is proceeding at a rapid pace, with a spot light on machine learning, artificial intelligence, and quantum computing. Our knowledge of all aspects of the field has grown even more profound. At the same time, one of the most striking trends in optimization is the constantly increasing emphasis on the interdisciplinary nature of the field. Optimization has been a basic tool in areas not limited to applied mathematics, engineering, medicine, economics, computer science, operations research, and other sciences.

The series **Springer Optimization and Its Applications (SOIA)** aims to publish state-of-the-art expository works (monographs, contributed volumes, textbooks, handbooks) that focus on theory, methods, and applications of optimization. Topics covered include, but are not limited to, nonlinear optimization, combinatorial optimization, continuous optimization, stochastic optimization, Bayesian optimization, optimal control, discrete optimization, multi-objective optimization, and more. New to the series portfolio include Works at the intersection of optimization and machine learning, artificial intelligence, and quantum computing.

Volumes from this series are indexed by Web of Science, zbMATH, Mathematical Reviews, and SCOPUS.

Weili Wu · Zhao Zhang · Ding-Zhu Du

Computational Aspects of Social Networks

 Springer

Weili Wu
Department of Computer Science
University of Texas at Dallas
Richardson, TX, USA

Ding-Zhu Du
Department of Computer Science
University of Texas at Dallas
Richardson, TX, USA

Zhao Zhang
School of Mathematics
Zhejiang Normal University
Jinhua, Zhejiang, China

ISSN 1931-6828 ISSN 1931-6836 (electronic)
Springer Optimization and Its Applications
ISBN 978-3-032-14832-2 ISBN 978-3-032-14833-9 (eBook)
https://doi.org/10.1007/978-3-032-14833-9

Focus on how to be social, not on how to do social.

—Jay Baer, Convince and Convert

When you say it, it is marketing. When they say it, it is social proof.

—Andy Crestodina

Privacy is dead, and social media hold the smoking gun.

—Pete Cashmore, Mashable CEO

Preface

An important class of graph data is the social data, generated from the online social network (OSN). The social data have been growing rapidly over the internet through various OSNs, including online websites, such as Facebook, LinkedIn, ResearchGate, and messengers, such as Skype. The widespread use of them leads to an increasing interest in efficiently and effectively discovering important, useful, and implicit information. Efficient techniques for algorithm design and analysis in social data are crucial to applications across many domains, including public safety, environment management, election, viral marketing, and misinformation clarification. During last eight years, to meet the need in the development of this technology, we established three courses on computational social networks, *Introduction to Computational Aspects of Social Networks*, *Combinatorial Optimization in Social Networks*, and *Advances in Computational Aspects of Social Networks*, and taught them as regular courses for graduate and senior undergraduate students, summer graduate courses, or short graduate courses in University of Texas at Dallas, University of Chinese Academy of Sciences, Zhejiang Normal University, Beijing Jiaotong University, Ocean University of China, Beijing University, Shandong University, Lanzhou University, Hebei Normal University, East China University of Science and Technology, and Hong Kong Polytechnic University. This book is based on lecture notes for these three courses, together with selected materials from papers published in IEEE/ACM Transactions and conferences, especially in *IEEE Transactions on Computational Social Systems*, *IEEE/ACM Transactions on Networking*, and *IEEE INFORCOM*.

There exists a very large number of research publications in the literature on computational aspects of social networks. This book does not intend to cover all of them. In fact, it selects only a portion of materials with the computational aspect, the algorithmic aspect, and the optimization aspect. Meanwhile, touch a few works in empirical study.

Intended audience The aim of the book is to teach a course for mainly graduate and senior undergraduate students in computer science and applied mathematics. The book contains four types of building blocks, fundamental knowledge about computational aspects of social networks (Chaps. 1, 2, 4, 7), combinatorial optimization methods in social networks (Chaps. 2–6), machine learning in social networks (Chaps. 1, 10, 11), advanced topics in computational aspects of social networks (Chaps. 7–10). Since intended audiences are in computer science and applied mathematics, we assume that the reader already has fundamental knowledge in programming skills, as well as algorithm design and analysis.

Richardson, USA
Jinhua, China
Richardson, USA
January 2025

Weili Wu
Zhao Zhang
Ding-Zhu Du

Acknowledgments

In preparation of our lecture notes and current book, we received many helps and supports from our colleagues and students, including Xujin Chen, My Thai, Deying Li, Hongwei Du, Zhenhua Duan, Cong Tian, Hejiao Huang, Qizhi Fang, Bin Liu, Qingqin Nong, Dachuan Xu, Wenguo Yang, Tiande Guo, Jianming Zhu, Jing Yuan, Shaojie Tang, Xianyue Li, Yapu Zhang, Guanmo Tong, Lingchen Kong, Nahua Xiu, Jianxiong Guo, Tiantian Chen, Shuning Gong, Zaixin Lu, Yuanjun Bi, Lidong Wu, Lidan Fan, Yuqing Zhu, Ruidong Yan, Qiufen Ni, Yi Li, Rong Jin, Ke Su, Dongyu Mao, Yisheng Zhou, Jialing Dai. We really appreciate very much for their suggestions and comments.

Moreover, we are grateful for supports of Profs. Andy Yao, Francis Yao, Xiaohua Jia, Jiannong Cao, Xiaodong Hu, Xiwen Lu, Wenan Zang, Xiaoming Yuan, and Xiaofeng Gao. A lot of work for this book were done when we visited at Institute for Interdisciplinary Information Sciences, Tsinghua University, City University of Hong Kong, Hong Kong Polytechnic University, Institute of Applied Mathematics, Chinese Academy of Sciences, East China University of Science and Technology, Hong Kong University, and Shanghai Jiaotong University. Finally, we would like to acknowledge the support in part by NSF of USA under grants 1747818 and 1907472, by NSF of China under grants U20A2068.

<table>
<tr><td>Richardson, USA</td><td>Weili Wu</td></tr>
<tr><td>Jinhua, China</td><td>Zhao Zhang</td></tr>
<tr><td>Richardson, USA</td><td>Ding-Zhu Du</td></tr>
<tr><td>January 2025</td><td></td></tr>
</table>

Contents

Chapter 1
Introduction and Community

Social networks aren't about Web sites. They're about experiences.
— Mike DiLorenzo

In this chapter, we introduce fundamental concepts, such as social network, social influence, and community, as well as community detection problem.

1.1 What Is Social Network?

A *network* is also called a graph, consisting of a set of nodes and a set of edges. Each edge is a pair of nodes, also called a *link*. There are two types of networks, undirected and directed. In a directed network, each edge is an ordered pair of nodes (u, v) which is directed from u to v. A directed edge is also called an *arc*. In an undirected network, each edge is an unordered pair of nodes $\{u, v\}$ without direction.

The social network is a special type of network. In a social network, nodes usually represent social actors and edges usually represent social relations. For example, consider all persons in the world as nodes and an edge exists between two persons if and only if the two persons are friends. This network is called the *friendship* network. The friendship network has been studied in social science for a long time. Social scientists found a very interesting property of the friendship network, called the *six degree of separation*. Figure 1.1 shows an example. In this example, Dr. Lidong Wu has distance four away from Bill Gates. All authors of this book are friends of Lidong Wu. This means that if you know one of them, then you have distance at most six from Bill Gates.

The six degree of separation was discovered by several experiments. One of them was designed by Stanley Milgram (1933–1984) in 1967. He randomly selected a group of people in Nebraska and asked them to send letters based on the following rule: If he/she knows a stock broker in Boston, then send a letter to the broker;

© The Author(s), under exclusive license to Springer Nature Switzerland AG 2026
W. Wu et al., *Computational Aspects of Social Networks*, Springer Optimization and Its Applications 234, https://doi.org/10.1007/978-3-032-14833-9_1

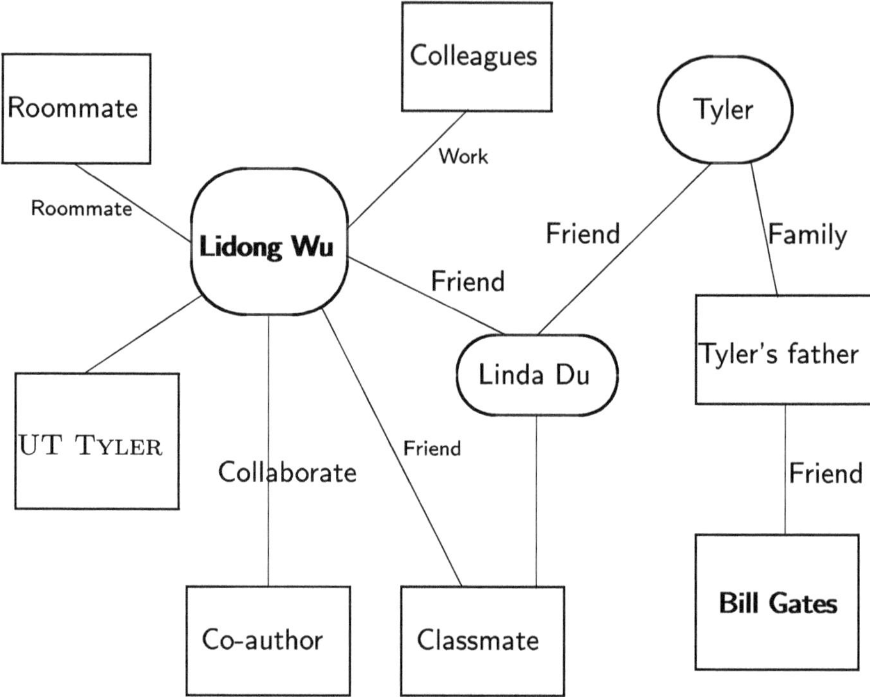

Fig. 1.1 Six degree of separation

otherwise, send letters to his/her friends and ask their friends to put name on the letter and to pass the letter in the same way. After a period of time, Milgram collected letters from those stock brokers in Boston and found that almost all letters passed at most six hands before reaching the brokers. This means that in the friendship network, every two persons have distance at most six. We may say that the property holds in probability sense. The reader may find some exceptional cases in which this property does not hold. For example, a person, such as Robinson, is isolated in an island away from the rest of the world. Then this person may have a much larger distance from others.

The six degree of separation indicates that although the friendship network has a large size, every pair of nodes only has a small distance. Networks of such a type are called *small world networks*, which is a large class of social networks. Let us show more examples.

Consider all authors, each of them has at least one publication. A link exists between two authors if and only if they are co-authors for a publication. This network is called the *co-authorship network*. The table in Fig. 1.2 shows the number of authors with certain distance from a well-known mathematician Paul Erdös. Such a distance is called the *Erdös number*. A person with distance n from Paul Erdös is said to have Erdös number n.

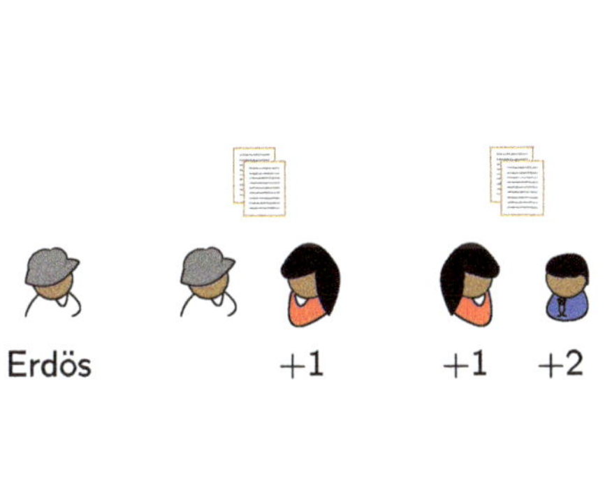

Fig. 1.2 Erdös number

Two authors in UTD have Erdös number two because they are co-authors of Ronald Graham who is a co-author of Erdös. This implies that their students have Erdös number at most three (Could you calculate your Erdös number?). The table shows that nobody has Erdös number thirteen, which means that in the co-authorship network, the maximum distance is thirteen. Hence, it is a small world network.

Consider a network which shows connections made by an airline company, such as Delta airline. The maximum distance in this network is also small. How should we know? Let us think about this question. Suppose that you want to travel from one city to another city and obtain a travel schedule from a travel agency. If the schedule asks you to transfer more than five times from one flight to another flight, are you satisfied? Of course no. Thus, the maximum distance in a airline network could not be high and hence the network is a small world network.

Social networks have a lot of applications. The friendship network is the foundation for Facebook and LinkedIn and the co-authorship network is the foundation for ResearchGate. Those computer networks which represent social networks are called *online social networks*. Online social networks have a property in common, that is, they are dynamic, growing rapidly. The airline network does not have this property and hence nobody uses it to build an online social network. On the other hand, the airline network does have an important application, search for cheap air-ticket.

Actually, the cheapest ticket problem can be formulated as the shortest path problem in the airline network. Let each air-flight be represented by a directed edge connecting two airports (Fig. 1.3).

The edge weight is the price to take this flight. Each airport is represented by two sets, a set of flight start-points and a set of flight end-points. If there is a possibility to transfer from a flight to another flight at an airport, then at this airport, put an edge from the end-point of the first flight to the start-point of the second flight, also put edge-weight to be the transfer cost (Fig. 1.4).

Fig. 1.3 Each flight is represented by a directed edge

Fig. 1.4 Each airport
contains a set of ending
points and a set of starting
points. Each edge between
them represents a possible
transfer from one air-flight to
another one

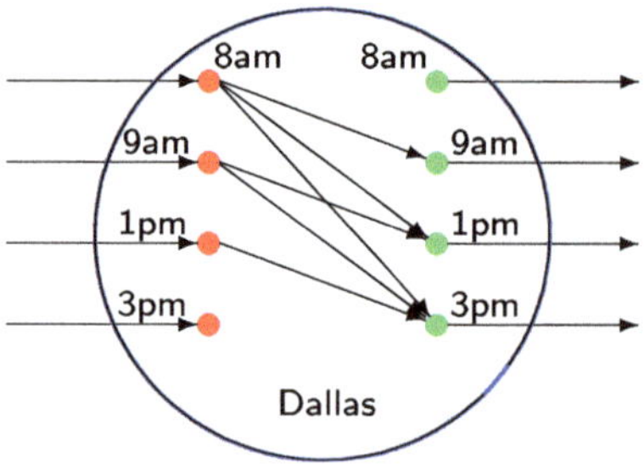

Fig. 1.5 Add a virtual
starting point and a virtual
ending point in each airport

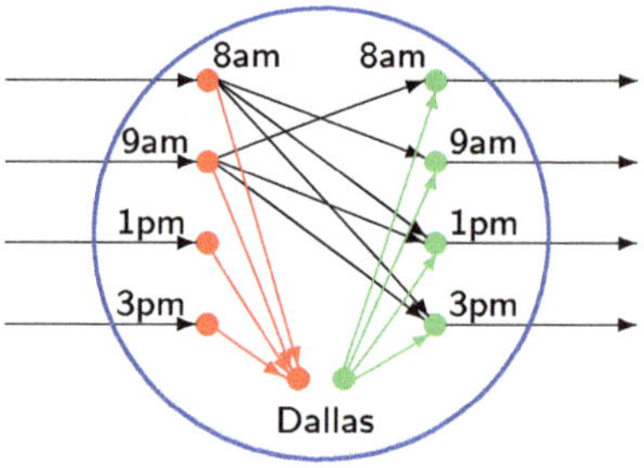

At each airport, create one virtual start-point with virtual edges connecting it
to all start-points of flights, and also create one virtual end-point with virtual edges
connecting all end-points of flights to it. All virtual edges have zero-weight (Fig. 1.5).
Now, the cheapest ticket from an airport to another airport is equivalent to the shortest
path from the virtual start-point of the former airport to the virtual end-point of the
latter airport.

1.2   Social Influence

In the study of social networks, social influence is an important subject. In fact, each
social actor always has an influence on his/her neighbors through social relations.
For example, when you are happy because of something, you would like to tell
your friends so that your friends may also feel happy for you. Such a happiness is

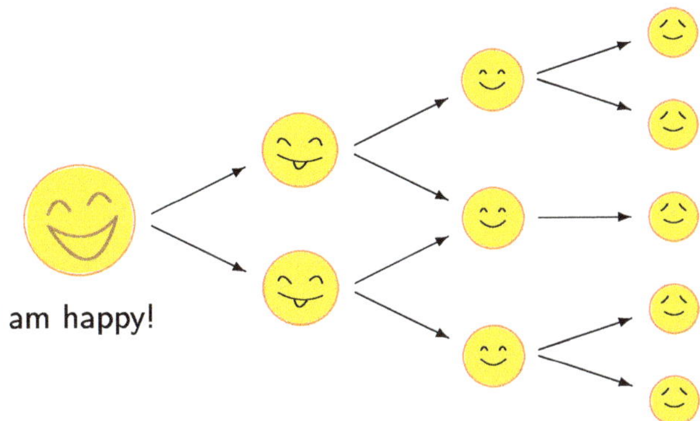

Fig. 1.6 Three degree of influence

transformable, that is, your friends' friends may also feel happy for you in a certain level. Of course, the influence could be weaker and weaker with the increase of distance. A question is how far a social influence can reach. With a high probability, cannot be as far as six. Otherwise, everybody can influence the whole world, which is incredible. The researchers found by experiments that most social influence would be dismissed after three steps of spreading. This property of the friendship network is called the *three degree of influence* (Fig. 1.6).

There are many applications of social influence, such as political election, viral marketing, and rumor blocking. Those applications generate many optimization problems. Those problems are main topics that we will study later.

1.3 Information Diffusion Model

To study social influence, we have to make clear of information diffusion process. When a social network is considered, its information diffusion model is often determined and assumed to be known. There are many information diffusion models. The following three are used quite frequently.

- the deterministic model,
- the independent cascade model, and
- the linear threshold model.

Let us explain them in detail as follows.

In the deterministic model, every node has two states, active and inactive (or infected and uninfected). The information diffusion process usually consists of a finite number of steps. Before a diffusion process starts, all nodes are in inactive state. To initiate the process, several nodes are chosen as *seeds*, which are activated, i.e., changed to active state. In each follow-up step, active nodes will make their neighbors (or out-neighbors in a directed network) activated. This process continues until no more inactive node could become active. The number of active nodes at the end is called the *influence spread*.

In the independent cascade (IC) model, every node also has two states, active and inactive. For generality, let us describe the model on a directed network $G = (V, E)$. Every link (u, v) is associated with a number $p_{uv} \in [0, 1]$. The diffusion process also consists of a finite number of steps. Before a diffusion process starts, all nodes are in inactive state. In initiate step (i.e., step 0), several nodes are chosen as seeds which are changed to active. A node is *freshly active* in a step if it becomes active in the latest step. For example, in step 1, all seeds are freshly active which are also active in step 2, but they are not freshly active in step 2. In step k for $k \geq 1$, the following rules are applied:

- Only freshly-active nodes have power to influence their out-neighbors.
- Let A_{uv} be the event that a freshly-active node u succeeded in influencing its inactive out-neighbor v. The event A_{uv} occurs with probability p_{uv}.
- Suppose v has h active in-neighbors $u_1, u_2, \ldots, u_h$. Then events $A_{u_1 v}, A_{u_2 v}, \ldots, A_{u_h v}$ are independent. Therefore, v becomes active with probability $1 - (1 - p_{u_1 v})(1 - p_{u_2 v}) \ldots (1 - p_{u_h v})$.

Note that in the third condition, the statement is only "active," not "freshly active." Actually, events $A_{u_1 v}, A_{u_2 v}, \ldots, A_{u_h v}$ may happen in different steps. For example, suppose u_1 is freshly active in step 1 and u_2 is freshly active in step 2. Then v may be activated by u_1 at step 1 with probability $p_{u_1 v}$ and by u_2 at step 2 with probability $(1 - p_{u_1 v}) p_{u_2 v}$. Putting together, the probability that v is activated by either u_1 or u_2 is

$$p_{u_1 v} + (1 - p_{u_1 v}) p_{u_2 v} = 1 - (1 - p_{u_1 v})(1 - p_{u_2 v}).$$

The process stops at the step when no more inactive nodes could be activated. Since the independent cascade model is a random process, the number of active nodes at the end of the process is a random variable. Its expectation is called the *influence spread*. An example of information diffusion in the IC model is shown in Fig. 1.7.

The linear threshold (LT) model is also a probabilistic one. Consider a directed network $G = (V, E)$. Each link (u, v) is associated with a number $b_{uv} \in [0, 1]$. Let $N^-(v)$ denote the set of in-neighbors of v, i.e., $N^-(v) = \{u \mid (u, v) \in E\}$. The link labels $\{b_{uv}\}$ are required to satisfy $\sum_{u \in N^-(v)} b_{uv} \leq 1$. Every node has two states, active and inactive. Before a diffusion process starts, all nodes are in inactive state, and each node v is required to select a number θ_v randomly and uniformly from $[0, 1]$, which is called the *threshold* at node v. In initiate step (i.e., step 0), several nodes are chosen as seeds which are changed to active. In step k for $k \geq 1$, every inactive node evaluates the value of a linear function

$$\sum_{u \in N^-(v)} b_{uv} x_u$$

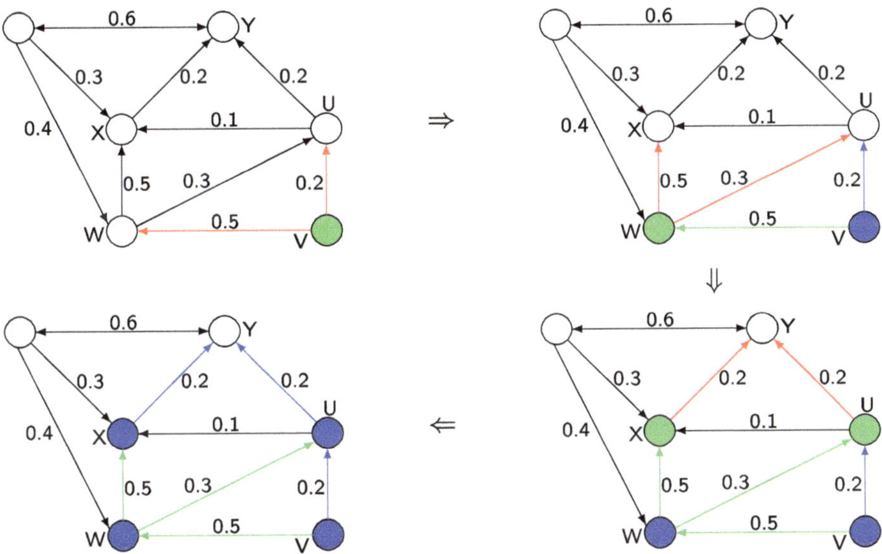

Fig. 1.7 An example of information diffusion process in the IC model. Green nodes are freshly-active nodes. Blue nodes are active (but not freshly-active) nodes. Red lines indicate attempts to active out-neighbors. If an attempt succeeds, then the line becomes green, otherwise, the line becomes blue

where

$$x_u = \begin{cases} 1 & \text{if } u \text{ is active,} \\ 0 & \text{otherwise.} \end{cases}$$

An inactive node v is activated if and only if $\sum_{u \in N^-(v)} b_{uv} x_u \geq \theta_v$. The process halts at the step when no inactive node could be further activated. Since the threshold θ_v is randomly chosen, the process is probabilistic and the number of active nodes at the end of the process is a random variable. The influence spread is also defined to be the expected number of active nodes at the end of the process. An example of information diffusion in the LT model is shown in Fig. 1.8.

1.4 Community

Why the friendship network has two surprising properties, the six degree of separation and the three degree of influence? An explanation comes from the network structure, called community structure. A *community* is a group of persons who have certain interests in common. For example, all persons with interest in computer science form a community and all persons with interest in mathematics form another community. News about computer science would spread very fast in computer science community, but not in mathematics community.

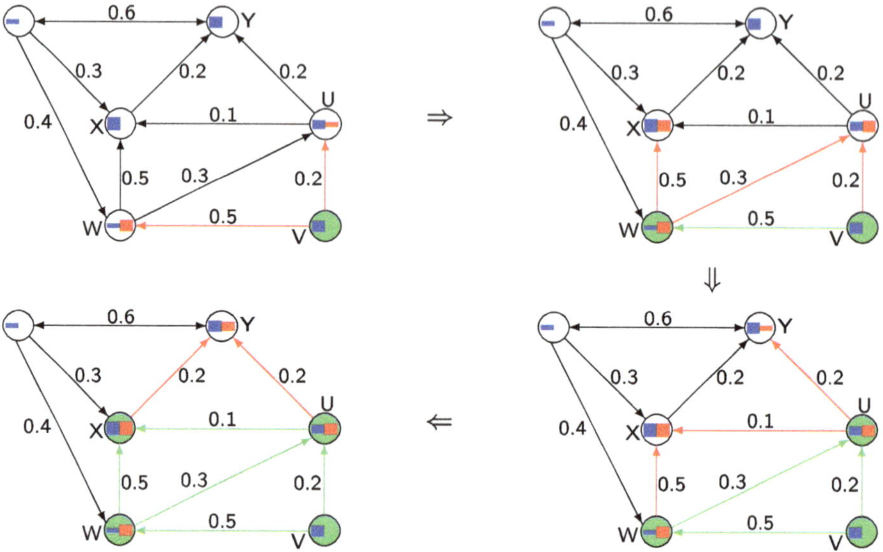

Fig. 1.8 An example of information diffusion process in the LT model. Green nodes are active nodes. Red lines indicate attempts of active nodes to activate their out-neighbors. Blue squares are the thresholds. Red squares are accumulated influence. When the accumulated influence is at least the threshold, then the node is activated and becomes green

Fig. 1.9 Why the friendship network has three degree of influence? The green part is the leading group

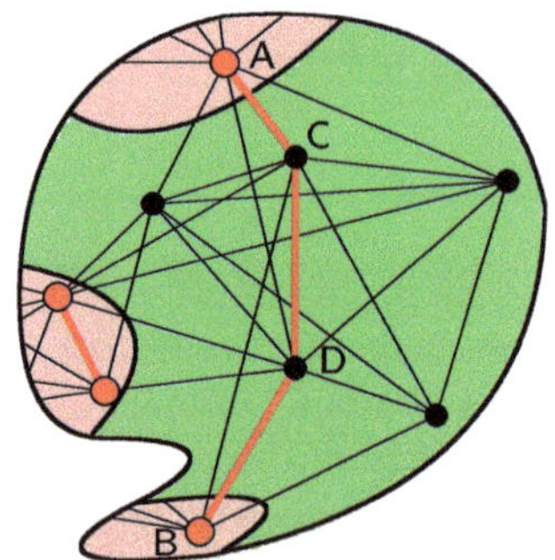

Within the same community, any two persons may have distance at most three. This is because each community in the friendship network contains a leading group. The leading group consists of those persons who are key persons knowing each other. Every member in the community knows at least one person in the leading group. This special property of community establishes the three degree of influence (Fig. 1.9).

Now, consider a person A in the computer science community and a person B in the mathematics community. Select a person C in both the computer science community and the mathematics community. Such a person C exists since a lot of professors in theoretical computer science have interest in both areas. From the above discussion, we see that the distance between A and C is at most three and the distance

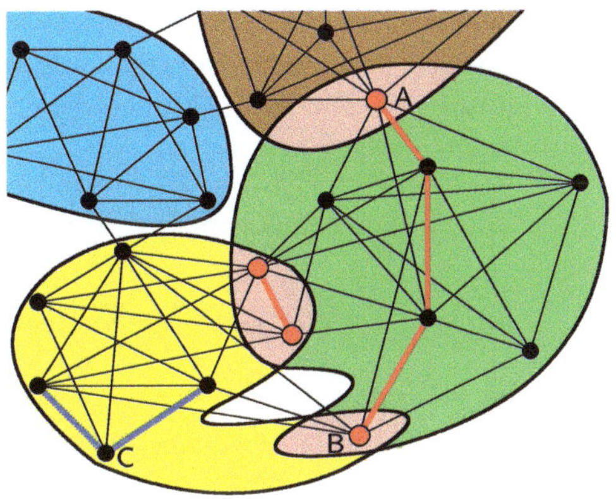

Fig. 1.10 Why the friendship network has six degree of separation? With high probability that any two communities are overlapping

between B and C is also at most three. Therefore, the distance between A and B is at most six. This example shows that the distance between two persons in two different communities is at most six if these two communities are overlapping. Actually, in the friendship network, it is true with high probability that any two communities are overlapping (Fig. 1.10).

1.5 Community Detection

Community structure plays an important role in the study of problems on social networks. Therefore, discover communities for a given social network becomes an attractive research problem, called *community detection*. However, community detection is a quite tough problem. Although a large amount of efforts have been made on this problem. It is still considered as a not well-solved problem. Actually, in a general social network, how should a community be defined? This is already a hard problem.

There are some efforts which tried to build an axiom system for the definition of community. However, those efforts are not widely accepted in research community. Right now, based on different considerations, many methods have been proposed for community detection. Almost all of them formulated detection into optimization problems and these problems are hard to solve. Let us look at a few of them in the following.

1.5.1 Modularity Maximization

This method was proposed by Newman in 2006 [305]. The work has received about 8000 citations now. This method is based on maximization a function, called *modularity function*, defined as follows.

Consider a network $G = (V, E)$ with adjacency matrix (a_{ij}). Given a partition C of V, let us call each component of C as a community. Define

$$Q(C) = \frac{1}{2|E|} \sum_{i,j \in V} \left[a_{ij} - \frac{k_i k_j}{2|E|} \right] \delta_{C_i, C_j}$$

where k_i is the degree of node i, C_i is the community containing node i, and δ_{C_i, C_j} is the Kronecher delta symbol, i.e.,

$$\delta_{C_i, C_j} = \begin{cases} 1 & \text{if } C_i = C_j, \\ 0 & \text{otherwise.} \end{cases}$$

$Q(C)$ can be rewritten as a difference of two terms:

$$Q(C) = \frac{1}{2|E|} \sum_{i,j \in V} a_{ij} \delta_{C_i, C_j} - \sum_{i,j \in V} \frac{k_i}{2|E|} \cdot \frac{k_j}{2|E|} \delta_{C_i, C_j}$$

where the first term is the fraction of edges within communities. If an edge is distributed at random, then it has endpoint i with probability $\frac{k_i}{2|E|}$ and has endpoint j with probability $\frac{k_j}{2|E|}$. Hence, it lies at (i, j) with probability $\frac{k_i}{2|E|} \cdot \frac{k_j}{2|E|}$. So the second term is the expected fraction of edges within communities when edges are distributed at random.

The modularity maximization is NP-hard [29]. There are many heuristics in the literature. Let us look at one of them. For this purpose, we derive another expression of modularity function as follows.

Theorem 1.5.1 *Suppose $C = (C_1, \ldots, C_m)$. Then*

$$Q(C) = \sum_{s=1}^{m} \left[\frac{L(C_s, C_s)}{L(V, V)} - \left(\frac{L(C_s, V)}{L(V, V)} \right)^2 \right]$$

where $L(U, W) = \sum_{i \in U, j \in W} a_{ij}$.

Proof By a direct calculation,

$$
Q(C) = \frac{1}{2|E|} \sum_{i,j\in V} \left[a_{ij} - \frac{k_i k_j}{2|E|} \right] \delta_{C_i, C_j}
$$

$$
= \frac{1}{2|E|} \sum_{s=1}^{m} \sum_{i,j\in C_s} \left[a_{ij} - \frac{k_i k_j}{2|E|} \right]
$$

$$
= \frac{1}{2|E|} \sum_{s=1}^{m} \left[L(C_s, C_s) - \frac{\left(\sum_{i\in C_s} k_i\right)\left(\sum_{j\in C_s} k_j\right)}{2|E|} \right]
$$

$$
= \sum_{s=1}^{m} \left[\frac{L(C_s, C_s)}{L(V, V)} - \left(\frac{L(C_s, V)}{L(V, V)}\right)^2 \right].
$$

The theorem is proved. $\qquad\square$

Corollary 1.5.2 *When communities C_i and C_j are merged, the increment of Q is*

$$
\Delta_{C_i \cup C_j} Q(C) = \frac{2}{L(V, V)^2}[L(C_i, C_j)L(V, V) - L(C_i, V)L(C_j, V)].
$$

Proof By Theorem 1.5.1,

$$
\Delta_{C_i \cup C_j} Q(C)
$$

$$
= \left[\frac{L(C_i \cup C_j, C_i \cup C_j)}{L(V, V)} - \left(\frac{L(C_i \cup C_j, V)}{L(V, V)}\right)^2 \right]
$$

$$
- \left[\frac{L(C_i, C_i)}{L(V, V)} - \left(\frac{L(C_i, V)}{L(V, V)}\right)^2 \right] - \left[\frac{L(C_j, C_j)}{L(V, V)} - \left(\frac{L(C_j, V)}{L(V, V)}\right)^2 \right]
$$

$$
= \left[\frac{L(C_i, C_i) + L(C_j, C_j) + 2L(C_i, C_j)}{L(V, V)} - \left(\frac{L(C_i, V) + L(C_j, V)}{L(V, V)}\right)^2 \right]
$$

$$
- \left[\frac{L(C_i, C_i)}{L(V, V)} - \left(\frac{L(C_i, V)}{L(V, V)}\right)^2 \right] - \left[\frac{L(C_j, C_j)}{L(V, V)} - \left(\frac{L(C_j, V)}{L(V, V)}\right)^2 \right]
$$

$$
= \frac{2}{L(V, V)^2}[L(C_i, C_j)L(V, V) - L(C_i, V)L(C_j, V)].
$$

The corollary is proved. $\qquad\square$

Denote $\tilde{\Delta}_{C_i \cup C_j} Q(C) = L(C_i, C_j)L(V, V) - L(C_i, V)L(C_j, V)$. The following is a greedy algorithm for modularity maximization.

Algorithm 1 Greedy Algorithm for Modularity Maximization.

Input: A network $G = (V, E)$.
Output: A community partition U^{k^*}.
1: $U^1 \leftarrow \{\{v\} \mid v \in V\}$
2: **for** $k = 1$ to $n - 1$ **do**
3:    choose C_i and C_j from U^k to maximize $\tilde{\Delta}_{C_i \cup C_j} Q(U^k)$
4:    $U^{k+1} \leftarrow (U^k \setminus \{C_i, C_j\}) \cup \{C_i \cup C_j\}$
5:    $k^* = \text{argmax}_{1 \leq k \leq n} Q(U^k)$
6: **end for**
7: **return** U^{k^*}.

1.5.2   Connection-Based Detection

Members within a community should have a closer relationship compared with members in different communities. Since edges represent social relations, there exist more edges within each community and less edges between different communities. There are several ways to make such an idea precise.

Connection-Based Condition 1. Consider a network $G = (V, E)$ with adjacency matrix (a_{ij}). A node subset U is a community in weak sense if

$$L(U, U) > L(U, \bar{U}),$$

where $\bar{U} = V \setminus U$ is the complement of U.

This condition was proposed by Radicchi et al. [338] in 2004, which means that each community has more connections inside than outside.

Connection-Based Condition 2. Consider a network $G = (V, E)$ with adjacency matrix (a_{ij}). A partition $C = (C_1, \ldots, C_m)$ of V is considered as a community partition in the most weak sense if for any s $(1 \leq s \leq m)$,

$$L(C_s, C_s) > \max_{t:t \neq s} L(C_s, C_t).$$

This condition was proposed by Hu et al. [190] in 2008, which means that each community has more connections inside than to any other community.

Connection-Based Condition 3. Consider a network $G = (V, E)$ with adjacency matrix (a_{ij}). A node subset U is a community if for any $x \in U$,

$$L(x, U) > L(x, \bar{U}).$$

This condition means that each node in a community has more connections inside than outside.

Connection-Based Condition 4. Consider a network $G = (V, E)$ with adjacency matrix (a_{ij}). A partition $C = (C_1, \ldots, C_m)$ of V is considered as a community partition if for any s $(1 \leq s \leq m)$ and for any $x \in C_s$,

$$L(x, C_s) > \max_{t: t \neq s} L(x, C_t).$$

This condition means that each node in a community has more connections inside than to any other community.

Among the above four conditions, Condition 3 is the strongest one, which can imply the others. Condition 2 is the weakest one, which can be implied by any of the others.

Based on each condition, we can form an optimization problem as follows.

Problem 1.5.3 (*Maximum Community Partition in Terms of Condition k*) Given a network $G = (V, E)$, find a community partition satisfying condition k described in the above, such that the number of parts is maximum.

Lu et al. [272] found that this class of problems is hard.

Theorem 1.5.4 (Lu et al. [272]) *For $k = 1, 2, 3, 4$, the maximum community partition problem in terms of condition k is NP-hard.*

A 0-1 variable x is said to be an *indicator* of an event if the event occurs if and only if $x = 1$. Let x_{ik} be an indicator for the event that node v_i belongs to the kth community C_k, z_{lk} be an indicator for the event that edge e_l is within community C_k, and y_k be an indicator for the event that community C_k is not empty (that is C_k exists). Zhang et al. [484] formulate the maximum community partition problem in terms of Condition 1 into a 0-1 LP (linear programming).

$$\max \sum_{k=1}^{n} y_k \tag{1.1}$$

$$\text{s.t.} \sum_{k=1}^{n} x_{ik} = 1, \text{ for } i = 1, 2, \ldots, n, \tag{1.2}$$

$$z_{lk} \leq x_{ik}, z_{lk} \leq x_{jk}, x_{ik} + x_{jk} - 1 \leq z_{lk},$$
$$\text{for } e_l = (i, j) \in E, k = 1, 2, \ldots, n, \tag{1.3}$$

$$4 \sum_{l=1}^{|E|} z_{lk} \geq \sum_{j=1}^{n} \sum_{i=1}^{n} x_{ik} a_{ij} + y_k \text{ for } k = 1, 2, \ldots, n, \tag{1.4}$$

$$\frac{1}{n} \sum_{i=1}^{n} x_{ik} \leq y_k \leq \sum_{i=1}^{n} x_{ik} \text{ for } k = 1, 2, \ldots, n, \tag{1.5}$$

$$x_{ik}, z_{lk}, y_k \in \{0, 1\} \text{ for } i = 1, 2, \ldots, n, l = 1, 2, \ldots, |E|, k = 1, 2, \ldots, n.$$

Let us make some explanation for the above constraints in the 0-1 LP.

Constraint (1.2) means that every node should belong to exactly one community. In Constraint (1.3), $z_{lk} \leq x_{ik}$ means that if edge e_l lies in C_k, so is node i; $z_{lk} \leq x_{jk}$ has the same meaning; and $x_{ik} + x_{jk} - 1 \leq z_{lk}$ means that if edge e_l does not lie within C_k, then either $i \notin C_k$ or $j \notin C_k$. Constraint (1.4) means that Condition 1 holds since Condition 1 is equivalent to

$$2 \sum_{l=1}^{|E|} z_{lk} \geq \sum_{j=1}^{n} \sum_{i=1}^{n} x_{ik} a_{ij} - 2 \sum_{l=1}^{|E|} z_{lk} + y_k$$

when $y_k = 1$ (i.e., C_k exists). Constraint (1.5) means that $C_k \neq \emptyset$ if and only if $y_k = 1$.

This 0-1 LP provides a path to design heuristics.

1.5.3 Influence-Based Detection

Connection-based community detection actually uses the deterministic information diffusion model in which a connection is established as long as an edge exists. In a probabilistic information diffusion model, such as independent cascade (IC) model or linear threshold (LT) model, existence of an edge is not sufficient to establish a connection or influence. Moreover, an influence may occur not only directly through an edge, but also through a simple path. In the IC model, consider a simple path $(v_0, v_1, \ldots, v_k)$. The probability that v_k is influenced by v_0 is $p_{v_0 v_1} p_{v_1 v_2} \cdots p_{v_{k-1} v_k}$ under the assumption that events of influence through the edges in the path are independent.

In the LT model, what is the probability of influence through an edge? Notice that an edge is labeled by its weight. So, the answer is not direct. However, the weight can also be treated as a probability similar to that in the IC model. Actually, in the next chapter, we will show that the LT model is equivalent to a mutually-exclusive cascade (MC) model as follows: In the MC model, everything is the same as the IC model except one. Recall the following condition in the IC model:

- Suppose v has h active in-neighbors $u_1, u_2, \ldots, u_h$. Then events $A_{u_1 v}, A_{u_2 v}, \ldots, A_{u_h v}$ are independent. Therefore, v becomes active with probability $1 - (1 - p_{u_1 v})(1 - p_{u_2 v}) \cdots (1 - p_{u_h v})$.

In the MC model, this condition is replaced by the following:

- Suppose v has h active in-neighbors $u_1, u_2, \ldots, u_h$. Then events $A_{u_1 v}, A_{u_2 v}, \ldots, A_{u_h v}$ are *mutually-exclusive*. Therefore, v becomes active with probability $p_{u_1 v} + p_{u_2 v} + \cdots + p_{u_h v}$.

In the MC model which is equivalent to the LT model, for each edge (u, v), the label p_{uv} in the MC model is equal to the label b_{uv} in the LT model. Therefore, we can treat b_{uv} as p_{uv}. This also explains why we ask $\sum_{u \in N^-(v)} b_{uv} \leq 1$ in the definition of the LT model.

Now, consider community detection based on influence. For each connection-based community condition defined previously and each probabilistic model, we may introduce an influence-based community condition. The condition will compare inside influence with outside influence. In order to prevent the formulated problem from being too complicated to be solved, Lu et al. [153] consider a simple case that only inside influence is counted and outside influence is ignored.

Consider a social network $G = (V, E)$ with diffusion model m. Given a partition $C = (C_1, \ldots, C_k)$ of V, define $\sigma_m[C_i] = \sum_{u,w \in C_i, u \neq w} p_{C_i}(u, w)$, where $p_{C_i}(u, w)$ is the probability that w accepts influence of u through C_i. As an example, we calculate $\sigma_m[C_i]$ in Fig. 1.11, where C_i has five nodes. Denote by $\bar{\sigma}_m(w) = \sum_{u \in C \setminus \{w\}} p_{C_i}(u, w)$. The calculated results are shown as follows.

For the IC model,

$$\bar{\sigma}_{IC}(1) = 3 \times 1 + (1 - 0.75^3) = 3.578125,$$
$$\bar{\sigma}_{IC}(2) = \sigma_{IC}(3) = \sigma_{IC}(4) = 0.25,$$
$$\bar{\sigma}_{IC}(5) = 0,$$
$$\sigma_{IC}[\{1, 2, 3, 4, 5\}] = 4.328125.$$

For the LT model.

$$\bar{\sigma}_{LT}(1) = 3 \times 1 + 3 \times 0.25 = 3.75,$$
$$\bar{\sigma}_{LT}(2) = \sigma_{LT}(3) = \sigma_{LT}(4) = 0.25,$$
$$\bar{\sigma}_{LT}(5) = 0,$$
$$\sigma_{LT}[\{1, 2, 3, 4, 5\}] = 4.5.$$

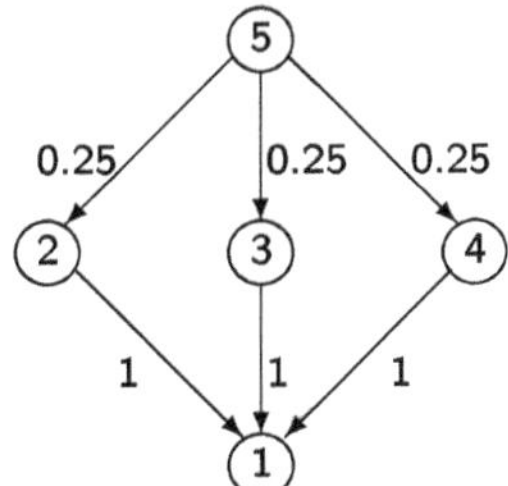

Fig. 1.11 An example for the calculation of $\sigma_{IC}[C_i]$ and $\sigma_{LT}[C_i]$

The influence-based community partition problem with diffusion model m is defined as

$$\max \quad \sigma_m[C_1] + \cdots + \sigma_m[C_k] \tag{1.6}$$
$$\text{subject to} \quad C = (C_1, \ldots, C_k) \text{ does not have empty component.}$$

For a set function $f : 2^V \to R$ and two sets $A \in 2^V$ and $x \in V$, denote

$$\Delta_x f(A) = f(A \cup \{x\}) - f(A)$$

the marginal profit of x over A. A set function $f : 2^V \to R$ is *supermodular* if for any $A, B \in 2^V$ with $A \subseteq B$, and any $x \in V \setminus B$,

$$\Delta_x f(A) \leq \Delta_x f(B). \tag{1.7}$$

This is also equivalent to

$$f(A) + f(B) \leq f(A \cup B) + f(A \cap B) \tag{1.8}$$

for any $A, B \in 2^V$ with $A \subseteq B$, that is,

$$f(A \cup D) - f(A) \leq f(B \cup D) - f(B)$$

for any $A, B, D \in 2^V$ with $A \subseteq B$.

Lemma 1.5.5 *Function $\sigma_{LT}[C]$ is supermodular.*

Proof For any node sets $A, B \subseteq V$ with $A \subseteq B$, and any node $u \in V \setminus B$,

$$\Delta_u \sigma_{LT}(A) = \sum_{j \in A} p_A(u, j) + \sum_{i \in A} p_A(i, u) + \sum_{i, j \in A: i \neq j} \{p_{A \cup \{u\}}(i, j) - p_A(i, j)\},$$

$$\Delta_u \sigma_{LT}(B) = \sum_{j \in B} p_B(u, j) + \sum_{i \in B} p_B(i, u) + \sum_{i, j \in B: i \neq j} \{p_{B \cup \{u\}}(i, j) - p_B(i, j)\}.$$

Since $A \subset B$, we have

$$\sum_{j \in A} p_A(u, j) \leq \sum_{j \in B} p_B(u, j),$$
$$\sum_{i \in A} p_A(i, u) \leq \sum_{i \in B} p_B(i, u).$$

Furthermore,

$$p_{A \cup \{u\}}(i, j) - p_A(i, j) \leq p_{B \cup \{u\}} - p_B(i, j),$$

which holds because $p_{A\cup\{u\}}(i, j) - p_A(i, j)$ is the sum of probabilities over all paths from i to j passing u within A, which is certainly no bigger than the sum of probabilities over all paths from i to j passing u within B. Then $\Delta_u \sigma_{LT}(A) \leq \Delta_u \sigma_{LT}(B)$, and the lemma follows from the definition of supermodular in (1.7). $\qquad\square$

Based on the property described in Lemma 1.5.5, we can design an algorithm for optimal community bisection, i.e., finding an optimal solution to Problem (1.6) when $k = 2$. To do so, let us first show the following result.

Consider a partition $\mathcal{P}$ of V. Let us sort $\mathcal{P}$ into $\mathcal{S}$ by Algorithm 2.

Algorithm 2 Sorting Procedure.

Input: A community partition $\mathcal{P}$.
Output: An ordering of communities in $cal\,P$.
1: choose S_1 from $\mathcal{P}$ arbitrarily, set $\mathcal{S} \leftarrow \{S_1\}$ and $U \leftarrow S_1$;
2: **for** $i = 2$ to $|\mathcal{P}|$ **do**
3:    $S_i = \text{argmax}_{S \in \mathcal{P}\setminus \mathcal{S}}\{\sigma_{LT}[U \cup S] - \sigma_{LT}[S]\}$
4:    $\mathcal{S} \leftarrow \mathcal{S} \cup \{S_i\}$;
5:    $U \leftarrow U \cup S_i$;
6: **end for**
7: **return** $\mathcal{S}$.

The ordered collection $\mathcal{S}$ has the following property.

Lemma 1.5.6 *Denote $\mathcal{S}_i = \{S_1, \ldots, S_i\}$. Then for any subset $\mathcal{A} \subseteq \mathcal{S}_{|\mathcal{P}|-2}$*

$$\sigma_{LT}[S_{|\mathcal{P}|}] + \sigma_{LT}[V - S_{|\mathcal{P}|}] \geq \sigma_{LT}[(\cup_{S\in\mathcal{A}}S) \cup S_{|\mathcal{P}|}] + \sigma_{LT}[V - (\cup_{S\in\mathcal{A}}S) \cup S_{|\mathcal{P}|}].$$

This means that the best bisection for separating $S_{|\mathcal{P}|}$ and $\mathcal{S}_{|\mathcal{P}|-1}$ is $(V - S_{|\mathcal{P}|}, S_{|\mathcal{P}|})$.

Proof Denote $\sigma[A] = \sigma_{LT}[A]$ and $v(\mathcal{A}) = \cup_{S\in\mathcal{A}}S$. We prove by induction on $k = 2, \ldots, |\mathcal{P}|$ that for any $\mathcal{A} \subseteq \mathcal{S}_{k-1}$ and any $j \geq k+1$,

$$\sigma[v(\mathcal{S}_k)] - \sigma[v(\mathcal{S}_k \setminus \mathcal{A})] \geq \sigma[S_j \cup v(\mathcal{A})] - \sigma[S_j]. \tag{1.9}$$

For $k = |\mathcal{P}|$, inequality (1.9) is exactly what the lemma claims.

For $k = 2$, $\mathcal{A} \subseteq \mathcal{S}_1$ implies that $\mathcal{A} = \emptyset$ or $\mathcal{S}_1$. When $\mathcal{A} = \emptyset$, both sides of (1.9) equal 0. When $\mathcal{A} = \mathcal{S}_1$, for any $j \geq 2$, (1.9) becomes

$$\sigma[v(\mathcal{S}_2)] - \sigma[v(\mathcal{S}_2 \setminus \mathcal{S}_1)] \geq \sigma[S_j \cup S_1] - \sigma[S_j],$$

that is,

$$\sigma[S_1 \cup S_2] - \sigma[S_2] \geq \sigma[S_1 \cup S_j] - \sigma[S_j].$$

This inequality holds by the choice of S_2. The induction basis is proved. Next, we prove the induction step by considering two cases.

Case 1. $\mathcal{S}_{k-1} \notin \mathcal{A}$. By induction hypothesis,

$$\sigma[v(\mathcal{S}_{k-1})] - \sigma[v(\mathcal{S}_{k-1} \setminus \mathcal{A})] \geq \sigma[S_j \cup v(\mathcal{A})] - \sigma[S_j].$$

Note that $v(\mathcal{S}_k \setminus \mathcal{A}) \cap v(\mathcal{S}_{k-1}) = v(\mathcal{S}_{k-1} \setminus \mathcal{A})$ and $v(\mathcal{S}_k \setminus \mathcal{A}) \cup v(\mathcal{S}_{k-1}) = v(\mathcal{S}_k)$. By supermodularity of $\sigma[\cdot]$,

$$\sigma[v(\mathcal{S}_k)] - \sigma[v(\mathcal{S}_k \setminus \mathcal{A})] \geq \sigma[v(\mathcal{S}_{k-1})] - \sigma[v(\mathcal{S}_{k-1} \setminus \mathcal{A})].$$

Therefore, (1.9) hold for k.

Case 2. $\mathcal{S}_{k-1} \in \mathcal{A}$. By induction hypothesis,

$$\sigma[v(\mathcal{S}_{k-1})] - \sigma[v(\mathcal{S}_{k-1} \setminus (\mathcal{A}_{k-1} \setminus \mathcal{A}))] \geq \sigma[S_k \cup v(\mathcal{S}_{k-1} \setminus \mathcal{A})] - \sigma[S_k],$$

i.e.,

$$\sigma[v(\mathcal{S}_{k-1})] - \sigma[v(\mathcal{A})] \geq \sigma[v(\mathcal{S}_k \setminus \mathcal{A})] - \sigma[S_k].$$

Note that $(v(\mathcal{A}) \cup S_j) \cap v(\mathcal{S}_{k-1} = v(\mathcal{A})$ and $(v(\mathcal{A}) \cup S_j) \cup v(\mathcal{S}_{k-1} = v(\mathcal{S}_{k-1}) \cup S_j$. By supermodularity,

$$\sigma[v(\mathcal{S}_{k-1}) \cup S_j] - \sigma[v(\mathcal{S}_{k-1}] \geq \sigma[v(\mathcal{A} \cup S_j] - \sigma[\mathcal{A}].$$

Therefore,

$$\sigma[S_j \cup v(\mathcal{S}_{k-1})] + \sigma[S_k] \geq \sigma[v(\mathcal{S}_k \setminus \mathcal{A}] + \sigma[v(\mathcal{A}) \cup S_j].$$

Now, to show (1.9), it suffices to show

$$\sigma[v(\mathcal{S}_k)] + \sigma[S_j] \geq \sigma[S_j \cup v(\mathcal{S}_{k-1})] + \sigma[S_k],$$

i.e.,

$$\sigma[v(\mathcal{S}_k)] - \sigma[S_k] \geq \sigma[S_j \cup v(\mathcal{S}_{k-1})] - \sigma[S_j].$$

This can be obtained by the choice of S_k. $\square$

The Algorithm 3 produces an optimal bisection.

Algorithm 3 Optimal Algorithm for Problem (1.6) with $k = 2$.

Input: A social network $G = (V, E)$ with LT model.
Output: A community partition $\mathcal{P}_{max}$.
1: $\mathcal{P} \leftarrow \{\{u\} \mid u \in V\}$;
2: $\mathcal{P}_{max} \leftarrow \mathcal{P}$;
3: **for** $k = 1$ to $|V| - 2$ **do**
4:     use Sort Procedure to sort $\mathcal{P}$ into $\mathcal{S}$;
5:     **if** $\sigma_{LT}[\mathcal{P}_{max}] < \sigma_{LT}[(v(\mathcal{S}_{|\mathcal{P}|-1}), S_{|\mathcal{P}|})]$ **then**
6:         $\mathcal{P}_{max} \leftarrow (v(\mathcal{S}_{|\mathcal{P}|-1}), S_{|\mathcal{P}|})$;
7:     **end if**
8:     $\mathcal{P} \leftarrow (\mathcal{P} \setminus \{S_{|\mathcal{P}|-1}, S_{|\mathcal{P}|}\} \cup \{S_{|\mathcal{P}|-1} \cup S_{|\mathcal{P}|}\}$;
9: **end for**
10: **return** $\mathcal{P}_{max}$.

Theorem 1.5.7 *Problem (1.6) is polynomial-time solvable for $k = 2$.*

Proof The algorithm first considers the case that $S_{|\mathcal{P}|-1}$ and $S_{|\mathcal{P}|}$ are separated by the bisection and compares it with the best bisection kept in $\mathcal{P}_{max}$. Then it considers the case that $S_{|\mathcal{P}|-1}$ and $S_{|\mathcal{P}|}$ are not separated by the bisection. Thus, the algorithm selected the best bisection from all possibilities. $\square$

1.5.4 Overlapping and Hierarchy

In previous subsections, we only deal with community partition, in which V is divided into *disjoint* parts. However, in the introduction of community, we mentioned that communities are often overlapping. Overlapping community detection is a mathematically very hard task. Therefore, there are only quite small number of publications in the study of them and those works are not strong theoretically [74, 230, 361]. Community partition might be considered as the first step simplification of general community detection problem.

In many social networks, a hierarchy structure may exist in communities. For example, computer science community contains smaller communities, such as theoretical computer science community, data science community, and computer network community. Discovering communities' hierarchy structure is also a very hard task, which has also attracted a few research efforts [361], however not theoretically strong.

The above study focuses on single layer networks. Efforts have also be made on multi-layer networks [218, 377, 425] and signed social networks [251, 265].

1.6 Heuristic and Approximation

In the previous section, we have presented several algorithms. They provide either polynomial-time solutions or heuristics if the problem is NP-hard, i.e., unlikely to have a polynomial-time solution. However, in this course, we may talk more about approximation. An algorithm for an NP-hard optimization problem is called an *approximation* if its performance can be theoretical guaranteed. Otherwise, the algorithm is called a *heuristic*. Let opt and apx denote the objective values of an optimal solution and an approximate solution on the same input of an optimization problem, respectively. The algorithm is called a ρ-approximation if

$$opt \leq apx \leq \rho \cdot opt \quad \text{for minimization problem}$$
$$opt \geq apx \geq \rho \cdot opt \quad \text{for maximization problem.}$$

Here, ρ is called the *approximation ratio* or *performance ratio*.

In the study of social networks, we would meet a lot of hard problems which are unlikely to have a polynomial-time algorithm. Therefore, researchers resort to either heuristics or approximations. Actually, a large quantities of existing algorithms are heuristics. Only a small number of algorithms are approximations. Techniques for designing approximation algorithms are favorite for theoretical researchers. Two important ones are the maximization and the minimization of monotone nondecreasing submodular set functions, which are both generalizations of the set cover problem.

Let $\mathcal{C}$ be a collection of subsets of a finite set X. Given an integer $k > 0$, find k subsets from $\mathcal{C}$ to cover the maximum number of elements in X. This is called the *maximum set coverage* problem, which is well-known to be NP-hard. The following is a greedy approximation.

Algorithm 4 Greedy Approximation for Maximum Set Coverage.

Input: A collection $\mathcal{C}$ of subsets of a finite set X, and an integer $k > 0$.
Output: k subsets.

1: **repeat**
2: pick an unselected subset from $\mathcal{C}$ to cover the maximum number of uncovered elements, and mark elements in the selected subset as covered
3: **until** k subsets are selected
4: **return** k picked subsets.

Theorem 1.6.1 *The above greedy algorithm is a $(1 - e^{-1})$-approximation for the maximum set coverage problem.*

Proof Suppose that when the ith subset is selected, it covers x_i uncovered elements. Denote $z_i = opt - x_1 - \cdots - x_i$, where opt is the number of elements covered by an optimal solution. Note that the optimal solution must cover at least z_i elements which are not covered by the first i selected subsets. Therefore, there must exist a subset in the optimal solution covering at least z_i/k uncovered elements. Hence, $x_{i+1} \geq z_i/k$. Thus, $z_{i+1} \leq z_i - z_i/k = z_i(1 - 1/k) \leq z_i e^{-1/k}$, where the last inequality uses the fact $1 + x \leq e^x$. This implies that $z_k \leq z_0 e^{-1} = opt \cdot e^{-1}$, i.e., $x_1 + \cdots + x_k \geq (1 - e^{-1}) \cdot opt$. $\qquad\square$

To generalize the above result, consider a set function f over 2^X where X is a finite set. f is called *submodular* if for any subsets A, B of X,

$$f(A) + f(B) \geq f(A \cup B) + f(A \cap B).$$

f is said to be *monotone nondecreasing* if for any subsets A, $B \subseteq X$ with $A \subset B$,

$$f(A) \leq f(B).$$

The submodular function f on 2^X has the following important properties:
1. For any subsets A, $B \subseteq X$ with $A \subset B$, and for any $x \in X \setminus B$,

$$\Delta_x f(A) \geq \Delta_x f(B)$$

where $\Delta_x f(A) = f(A \cup \{x\}) - f(A)$.
2. For any subsets A, D of X,

$$\Delta_D f(A) \leq \sum_{x \in D} \Delta_x f(A), \tag{1.10}$$

where $\Delta_D f(A) = f(A \cup D) - f(A)$.

Suppose f is a monotone nondecreasing, submodular function on 2^X. Furthermore, suppose $f(\emptyset) = 0$. Such a function f is called a *polymatroid function*. Given a positive integer k, consider a maximization problem

$$\max \ f(A)$$
$$\text{subject to} \ \ |A| \leq k,$$
$$A \in 2^X.$$

The following is a greedy algorithm for this problem.

Algorithm 5 Greedy Algorithm for Submodular Maximization.

Input: A polymotroid function f over 2^X and a positive integer k.
Output: A subset of k elements in X.
1: $A_0 \leftarrow \emptyset$;
2: **for** $i = 1$ to k **do**
3:    $v \in X \setminus A_{i-1}$ to maximize $\Delta_v f(A_{i-1})$ and
4:    $A_i \leftarrow A_{i-1} \cup \{v\}$;
5: **end for**
6: **return** A_k.

The following result is well-known, which indicates that this greedy algorithm has a good performance ratio.

Theorem 1.6.2 (Nemhauser and Wolsey [301, 302]) *Suppose A_k is produced by the above algorithm. Then $f(A_k) \geq (1 - e^{-1}) \cdot opt$ where opt is the optimal value.*

Proof Suppose x_1, x_2, ..., x_k are obtained by the above greedy algorithm. Denote $A_i = \{x_1, x_2, \ldots, x_i\}$ for $i = 1, \ldots, k$ and $A_0 = \emptyset$. Suppose $A^* = \{u_1, u_2, \ldots, u_k\}$ is an optimal solution. Then for $i = 0, 1, \ldots, k - 1$, we have

$$
\begin{aligned}
f(A^*) &\leq f(A^* \cup A_i) \\
&= f(A_i) + \Delta_{u_1} f(A_i) + \cdots + \Delta_{u_k} f(A_i \cup \{u_1, u_2, \ldots, u_{k-1}\}) \\
&\leq f(A_i) + \Delta_{u_1} f(A_i) + \cdots + \Delta_{u_k} f(A_i) \\
&\leq f(A_i) + k \cdot \Delta_{x_{i+1}} f(A_i),
\end{aligned}
\tag{1.11}
$$

where the first inequality is due to the monotonicity of f, the second inequality is due to the submodularity of f, and the third inequality is due to the greedy rule in the algorithm.

Denote $a_i = f(A^*) - f(A_i)$. Then it follows from (1.11) that $a_i \leq k(a_i - a_{i+1})$. Hence

$$
a_{i+1} \leq a_i(1 - 1/k) \leq a_i \cdot e^{-1/k}.
$$

Iteratively using this inequality, we have $a_k \leq a_0 e^{-1}$. Note that $a_0 = f(A^*) - f(\emptyset) = f(A^*) = opt$. Therefore, $f(A_k) \geq (1 - e^{-1})opt$. $\square$

Next, we study the minimum set cover problem: Given a collection $\mathcal{C}$ of subsets of a finite set X, find a minimum subcollection of $\mathcal{C}$ covering all elements of X. Such a subcollection is called a *set cover*. Clearly, the problem has a solution if and only if the given collection $\mathcal{C}$ is a set cover. The minimum set cover problem is also NP-hard and has a greedy approximation.

Algorithm 6 Greedy Approximation for Minimum Set Cover.

Input: A collection $\mathcal{C}$ of subsets of a finite set X.
Output: A set cover.

1: **repeat**
2:    pick an unselected subset from $\mathcal{C}$ to cover the maximum number of uncovered elements, and mark elements in the selected subset as covered
3: **until** all elements are covered
4: **return** the collection of selected subsets.

Theorem 1.6.3 *The above greedy algorithm is a* $(1 + \ln n)$*-approximation for the minimum set cover problem.*

Proof Let *opt* denote the number of subsets in an optimal solution. Notice that these *opt* subsets cover all elements and hence cover all uncovered elements at any step. Thus, each selected subset could cover at least $1/opt$ fraction of uncovered elements. Therefore, after t steps, at most $(1 - 1/opt)^t$ fraction of elements are not covered. For $t = \lfloor (1 + \ln n) \rfloor opt$, we have $(1 - 1/opt)^t n \le e^{-(1/opt)t} n < 1$, which implies that after t steps, all elements are covered. Hence $apx \le t \le (1 + \ln n)opt$. $\square$

A generalization of the minimum set cover is the minimum submodular set cover problem: Let f be a polymatroid function over 2^X where X is a finite set, and c be a nonnegative cost function on X. Consider the minimization problem:

$$\min \ c(A) = \sum_{x \in A} c(x) \tag{1.12}$$
$$\text{subject to} \ \ f(A) = f(X),$$
$$A \in 2^X.$$

This problem has a greedy approximation as follows.

Algorithm 7 Greedy Algorithm for Minimum Submodular Set Cover.

Input: A polymotroid function f over 2^X and a nonnegative cost function c on X.
Output: A subset A of X such that $f(A) = f(X)$.

1: $A \leftarrow \emptyset$;
2: **while** $f(A) < f(X)$ **do**
3:    choose $x \in X \setminus A$ to maximize $\Delta_x f(A)/c(x)$ and
4:    $A \leftarrow A \cup \{x\}$;
5: **end while**
6: **return** A.

Theorem 1.6.4 (Wolsey   [435])   *The   above   greedy   algorithm   is   an* $H(\gamma)$*-approximation for the minimum submodular cover problem where*

$$\gamma = \max_{x \in X} f(\{x\}).$$

Proof We claim that the minimum submodular cover problem can be formulated as the following integer LP.

$$\min \ \sum_{v \in X} c(v) x_v \tag{1.13}$$

$$\text{s.t.} \ \sum_{v \in X - S} \Delta_v f(S) x_v \geq \Delta_{X-S} f(S) \text{ for all } S \in 2^X,$$

$$x_v \in \{0, 1\} \text{ for } v \in X.$$

To show the claim, we first prove that for any set $A \in 2^X$ satisfing $f(A) = f(X)$, its indicator vector $\mathbf{1}_A$ is a feasible solution of LP (1.13), where $\mathbf{1}_A = (x_v)_{v \in X}$ is defined by

$$x_v = \begin{cases} 1 \text{ if } v \in A, \\ 0 \text{ otherwise.} \end{cases}$$

In fact, for any $S \in 2^X$,

$$\sum_{v \in X - S} \Delta_v f(S) x_v = \sum_{v \in A \setminus S} \Delta_v f(S)$$
$$\geq \Delta_{A \setminus S} f(S)$$
$$= f(A) - f(S)$$
$$= f(X) - f(S)$$
$$= \Delta_{X-S} f(S),$$

where the inequality comes from (1.10).

Next, we show that if x is a feasible solution of LP (1.13), then $A = \{v \mid x_v = 1\}$ satisfies $f(A) = f(X)$. In fact, the inequality constraint for $S = A$ is

$$\sum_{v \in X - A} \Delta_v f(A) x_v \geq \Delta_{X-A} f(A),$$

that is,

$$0 \geq f(X) - f(A).$$

So, $f(A) \geq f(X)$. Since f is monotone nondecreasing, we must have $f(X) = f(A)$.

To sum up, the constraints of (1.13) and (1.12) are the same. We shall use dual fitting method to analyze the performance of the greedy algorithm.

The dual LP of the relaxed LP of (1.13) is

$$\max \ \sum_{S \in 2^X} \Delta_{X-S} f(S) y_S \tag{1.14}$$

$$\text{s.t.} \ \sum_{S:v \notin S} \Delta_v f(S) y_S \le c(v) \text{ for } v \in X,$$

$$y_S \ge 0 \text{ for } S \in 2^X.$$

Suppose the above greedy algorithm selects $x_1, \ldots, x_g$ in turn. Denote $A_0 = \emptyset$ and $A_k = \{x_1, \ldots, x_k\}$ for $k = 1, \ldots, g$. Then A_g is the output. Denote $r_k = \Delta_{x_k} f(A_{k-1})$ and $c_i = c(x_i)$. Define a set of dual variables

$$y_S = \begin{cases} \frac{1}{H(\gamma)} \left(\frac{c_{k+1}}{r_{k+1}} - \frac{c_k}{r_k} \right) & \text{if } S = A_k, 0 \le k \le g-1, \\ 0 & \text{otherwise.} \end{cases}$$

where c_0 / r_0 is viewed as 0.

It can be shown that $\{y_S\}$ is a feasible solution to (1.14). By the greedy rule and the submodularity of f, we have

$$\frac{c_k}{r_k} \le \frac{c_{k+1}}{\Delta_{v_{k+1}} f(A_{k-1})} \le \frac{c_{k+1}}{\Delta_{v_{k+1}} f(A_k)} = \frac{c_{k+1}}{r_{k+1}}.$$

Hence $y_S \ge 0$ for any $S \in 2^X$. Next, consider any $v \in X$, we have

$$\sum_{S:v \notin S} \Delta_v f(S) y_S = \frac{1}{H(\gamma)} \sum_{k=0}^{k_v-1} \Delta_v f(A_k) \left(\frac{c_{k+1}}{r_{k+1}} - \frac{c_k}{r_k} \right), \tag{1.15}$$

where $k_v = i$ if $v = x_i$ and $k_v = g$ if $v \notin A_g$. The summation term in the right-hand side can be rewritten as

$$\sum_{k=0}^{k_v-1} \Delta_v f(A_k) \left(\frac{c_{k+1}}{r_{k+1}} - \frac{c_k}{r_k} \right)$$

$$= \sum_{k=1}^{k_v-1} \frac{c_k}{r_k} \left(\Delta_v f(A_{k-1}) - \Delta_v f(A_k) \right) + \frac{c_{k_v}}{r_{k_v}} \Delta_v f(A_{k_v-1})$$

$$= \sum_{k=1}^{k_v} \frac{c_k}{r_k} \left(\Delta_v f(A_{k-1}) - \Delta_v f(A_k) \right), \tag{1.16}$$

where $\Delta_v f(A_{k_v}) = 0$ is used in absorbing the second term into the summation. For any $k = 0, 1, \ldots, k_v - 1$, since $v \notin A_k$, we have

$$\frac{c_k}{r_k} \leq \frac{c(v)}{\Delta_v f(A_k)}$$

by the greedy rule. Furthermore, by the submodularity of f, we have $\Delta_v f(A_{k-1}) - \Delta_v f(A_k) \geq 0$. Hence (1.16) becomes

$$\sum_{k=0}^{k_v-1} \Delta_v f(A_k) \left(\frac{c_{k+1}}{r_{k+1}} - \frac{c_k}{r_k} \right) \leq \sum_{k=1}^{k_v} c(v) \frac{\Delta_v f(A_{k-1}) - \Delta_v f(A_k)}{\Delta_v f(A_k)}$$

$$\leq c(v) H(\Delta_v f(A_0)) \leq c(v) H(\gamma).$$

Combining this with (1.15), the constraint of dual LP (1.14) is satisfied. Since $f(A_g) = f(X)$, it can be calculated that

$$c(A_g) = \sum_{k=1}^{g} c_k = \sum_{k=1}^{g} \frac{c_k}{r_k} \Delta_{x_k} f(A_{k-1}) = \sum_{k=1}^{g} \frac{c_k}{r_k} (f(A_k) - f(A_{k-1}))$$

$$= \sum_{k=1}^{g-1} f(A_k) \left(\frac{c_k}{r_k} - \frac{c_{k+1}}{r_{k+1}} \right) + \frac{c_g}{r_g} f(A_g)$$

$$= \sum_{k=1}^{g-1} f(A_k) \left(\frac{c_k}{r_k} - \frac{c_{k+1}}{r_{k+1}} \right) + \sum_{k=1}^{g-1} \left(\frac{c_{k+1}}{r_{k+1}} - \frac{c_k}{r_k} \right) f(X)$$

$$= \sum_{k=1}^{g-1} \left(\frac{c_{k+1}}{r_{k+1}} - \frac{c_k}{r_k} \right) (f(X) - f(A_k))$$

$$= H(\gamma) \sum_{k=1}^{g-1} \Delta_{X-A_k} f(A_k) \cdot y_{A_k}$$

$$= H(\gamma) \sum_{S \in 2^X} \Delta_{X-S} f(S) \cdot y_S$$

$$\leq H(\gamma) opt,$$

where opt is the optimal value of the integer LP (1.13), which is an upper bound for the objective value of $\{y_S\}$ in the dual LP (1.14). $\qquad \square$

1.7　Power Law Structure

Let us first explain what is a power law graph and why we study the power law graph.

A graph G belongs to class $P(\alpha, \gamma)$ if

$$|\{v \in V(G) \mid \text{degree}(v) = k\}| = \left\lfloor \frac{e^\alpha}{k^\gamma} \right\rfloor$$

for $k = 1, 2, \ldots$. If we use the real number instead of rounding down the integer, then we can say that the number of nodes with degree k is $\frac{e^\alpha}{k^\gamma}$, i.e., it follows the power law distribution $f(x) \sim x^{-\gamma}$. Thus, those graphs are called *power law graphs*.

Actually, **in the study of power law graphs, one usually use the real number instead of rounding down the integer without effecting the conclusion. In fact, the error terms are sufficiently small in mathematical proofs and entire discussion. In this section, we follow this simplification.** For example, we say the following.

The maximum node degree of every graph in $P(\alpha, \gamma)$ is $e^{\alpha/\gamma}$. The number of nodes in each graph in $P(\alpha, \gamma)$ is

$$n = \sum_{k=1}^{e^{alpha/\gamma}} \frac{e^\alpha}{k^\gamma} \approx \begin{cases} \zeta(\gamma)e^\alpha & \text{if } \gamma > 1 \\ \alpha e^\alpha & \text{if } \gamma = 1 \\ \frac{e^{\alpha/\gamma}}{1-\gamma} & \text{if } \gamma < 1 \end{cases}$$

and the number of edges in each graph in $P(\alpha, \gamma)$ is

$$m = \frac{1}{2} \sum_{k=1}^{e^{\alpha/\gamma}} k \cdot \frac{e^\alpha}{k^\gamma} \approx \begin{cases} \frac{1}{2} \cdot \zeta(\gamma - 1)e^\alpha & \text{if } \gamma > 2 \\ \frac{1}{4} \cdot \alpha e^\alpha & \text{if } \gamma = 2 \\ \frac{1}{2} \cdot \frac{e^{2\alpha/\gamma}}{2-\gamma} & \text{if } \gamma < 2 \end{cases}$$

where $\zeta(\gamma) = \sum_{i=1}^{\infty} \frac{1}{i^\gamma}$.

In the real world, many online social networks have the power law structure. For example, scientific collaboration networks have power law structures with $2.1 < \gamma < 2.45$ [11, 12], the World Wide Web has a directed power law structure with $\gamma = 2.1$ (in-degree) and $\gamma = 2.45$ (out-degree) [3], and Internet at router and intra-domain level has the power law structure $\gamma = 2, 48$ [99].

An NP-hard optimization problem in graphs could find the following.

- The problem is still NP-hard in power law graphs.
- The problem has approximation solutions with better performance ratio in power law graphs.

Why the problem remains NP-hard in power law graphs? The main reason is that usually the problem remains NP-hard in graphs with certain fixed node-degree. Then, such a graph can be embedded into a power law graph in which the rest part has a fixed structure and only keeps a window for the graph with fixed node-degree.

For example, consider the minimum vertex-cover problem in graphs. It is known that the problem remains NP-hard in cubic graphs (note: a cubic graph is a graph with degree three.) Now, we reduce it to the problem in power law graphs by embedding the cubic graph into the power law graph. Given a cubic graph G, select α such that

$$|V(G)| = \frac{e^\alpha}{3^\gamma}.$$

Construct a power law graph $H \in P(\alpha, \gamma)$ such that H is the union of G and G' where G' contains nodes with degree $k \neq 3$ and has a structure with an easily-determined minimum vertex-cover.

Why the approximation solution for an NP-hard problem can be easily obtained in power law graphs? The reason is as follows.

- More nodes with low degree.
- Less nodes with high degree.
- Size of optimal solution is often determined by the number of nodes with low degree.

For example, consider the modularity maximization problem. No approximation algorithm has been found to have theoretical guaranteed performance. However, for power law graphs, we can have a polynomial-time $O(1)$-approximation as shown in Algorithm 8.

Algorithm 8 Low-Degree Following Algorithm

Input: A power-law graph $G = (V, E) \in P(\alpha, \gamma)$.
Output: A community partition.
1: $L \leftarrow \emptyset, M \leftarrow \emptyset, O \leftarrow \emptyset$;
2: $p_i \leftarrow, \forall i = 1..n$;
3: **for** $d = 1$ to n **do**
4:     **for** each $i \in V$ **do**
5:         **if** $deg(i) \leq d$ and $i \notin L \cup M$ **then**
6:             **if** $N(i) \not\subset M$ **then**
7:                 select $j \in N(i) \setminus M$ and
8:                 let $M \leftarrow M \cup \{i\}, L \leftarrow L \cup \{j\}, p_i \leftarrow j$
9:             **else**
10:                 select $t \in N(i)$ and set $O \leftarrow O \cup \{i\}, p_i \leftarrow t$;
11:             **end if**
12:         **end if**
13:         $C_d = \{C(i) \mid i \in V - (M \cup O)\}$
14:             where $C(i) = \{i\} \cup \{j \in M \mid p_i = i\} \cup \{t \in O \mid p_{p_t} = i\}$;
15:     **end for**
16:     $d_0 = \text{argmax}_{1 \leq d \leq n} Q(C_d)$;
17: **end for**
18: **return** C_{d_0}.

Before analyzing this algorithm, let us first establish a property of the modularity function.

Lemma 1.7.1 *For any network $G = (V, E)$ and any partition C of V, $Q(C) \leq 1$.*

Proof Suppose $C = (V_1, V_2, \ldots, V_k)$

$$Q(C) = \sum_{s=1}^{k} \left[\frac{L(V_s, V_s)}{L(V, V)} - \left(\frac{L(V_s, V_s) + L(V_s, \bar{V}_s)}{L(V, V)} \right) \right]$$

$$\geq \frac{\sum_{s=1}^{k} L(V_s, V_s)}{L(V, V)}$$

$$\geq 1.$$

$\square$

Theorem 1.7.2 (Dinh and Thai [75]) *For power law graphs with $\gamma > 2$ and for any $\varepsilon > 0$, Low-Degree Following Algorithm (Algorithm 8) can generate a community partition C such that*

$$Q(C) \geq \frac{\zeta(\gamma)}{\zeta(\gamma)} - \varepsilon.$$

Proof In the algorithm, we choose d_0 to reach the largest modularity. In the proof, we will show that there exists a d_0 which gives a community partition with modularity at least

$$\frac{\zeta(\gamma)}{\zeta(\gamma)} - \varepsilon.$$

To start the proof, let the following denote the community partition generated by the algorithm.

$$C = (V_1, V_2, \ldots, V_h) = \{C(i) \mid V \setminus (M \cup O)\}$$

where

$$C(i) = \{i\} \cup \{j \in M \mid p_j = i\} \cup \{t \in O \mid p_{p_t} = i\}.$$

Note that for any $i \in M \cup O$, i and p_i are in the same community. Hence, we have a lower bound for the positive part of the modularity,

$$\sum_{s=1}^{h} L(V_s, V_s) \geq 2|M \cup O| \geq e^{\alpha} \sum_{k=1}^{d_0} k^{-\gamma} \geq e^{\alpha}(\zeta(\gamma) - \varepsilon$$

for sufficiently large d_0, independent from α.

To establish an upper bound of the modularity, we have that for $i \in L$ with $V_s = C(i)$,

$$L(V_s, V_s) + L(V_s, \bar{V}_s) \leq \text{degree}(i)\text{degree}(i) \cdot d_0 + \text{degree}(i) \cdot d_0^2$$

$$\leq 2\text{degree} \cdot d_0^2.$$

Therefore,

$$\sum_{s=1}^{h} (L(V_s, V_s) + L(V_s, \bar{V}_s))^2 \leq \sum_{i \in V \setminus (M \cup O)} (2\text{degree}(i) \cdot d_0^2)^2$$

$$\leq \sum_{k=1}^{e^{\alpha/\gamma}} \frac{e^{\alpha}}{k^{\gamma}} \cdot (2kd_0^2)^2$$

$$\leq e^{\alpha/\gamma} \cdot e^{\alpha} (2d_0^2)^2.$$

Putting above bounds together, we obtain

$$Q(C) == \sum_{s=1}^{h} \left[\frac{L(V_s, V_s)}{L(V, V)} - \left(\frac{L(V_s, V_s) + L(V_s, \bar{V}_s)}{L(V, V)} \right) \right]$$

$$\geq \frac{\zeta(\gamma) - \varepsilon}{\zeta(\gamma - 1)} - \frac{(2d_0^2)^2}{\zeta(\gamma - 1)e^{\alpha(1 - 1/\gamma)}}$$

$$\geq \frac{\zeta(\gamma)}{\zeta(\gamma - 1)} - \varepsilon$$

for sufficiently large α. (Note: d_0 depends on γ. α is independent from γ.) $\qquad\square$

1.8　Dunbar's Number

In performance evaluation of algorithms, empirical analysis is another important method in addition to the theoretical one. In this section, we present an example.

In previous sections, we introduced several algorithms for community detection. Each algorithm is designed based on a measure of community quality. Different measures induce different algorithms, which may fit different applications of communities. In this section, we would like to mention an additional measure, the community size.

Dunbar [94] suggests that both in traditional social networks and in social networks based on the Internet, any community should have a size of not greater than 150 regardless of the method used for community detection. This is because in a large community of more than 150 members, connections are weak among its members. Therefore, 150 is called the Dunbar number.

Actually, Dunbar number is the limit of stable online community size, which is calculated as a direct function of relative neocortex size. The neocortical processing capacity limits the number of individuals with whom a stable interpersonal relationship can be maintained, and hence, in turn limits the size of the community.

In addition, communities of sizes 1, 2, and 3 are too small to be considered a group. Therefore, the desirable community size is 4 to 150. In [413] it is observed that by scrutinizing the top 5000 highest-quality communities provided in DBLP, a coauthorship network [14], there are 4951 of sizes 4–150, that is, 99% of the highest-quality communities in DBLP network conform to Dunbar's number. This is strong evidence of Dunbar's number.

Dunbar [94] also provides a series of numbers with the corresponding closeness of a person to the people he interacts with, which is summarized in the following table of the social circle.

Size	Closeness
4–5	Support clique
6–15	Sympathy group
16–50	Close friends
51–150	Casual friends
151–500	Acquaintance
501–1500	Just a face

This means that each person can have about five people in their support group of closest friends, about 15 people in a sympathy group who are close enough to exchange their confidence, about 50 close friends, about 150 casual friends, about 500 acquaintances, and totally, any given person can identify about 1500 faces.

In [413], an expensive empirical study is conducted to systematically evaluate and compare various community detection algorithms (listed below) taking into account the size of the identified communities.

- Infomap [348],
- Multilevel [20],
- Fastgreedy [303],
- Eigenvector [304],
- Label Propagation Algorithm (LPA) [339],
- DEMON (an overlapping community detection algorithm) [65], and
- CAA (clique-based algorithm) [413].

Those algorithms, except the last one, have open source implementations, and hence are easily to reproducible for comparison purposes.

The comparison is done on two data sets.

- The first data set is Twitter user follower topology, collected from Twitter over a 3-month period in summer 2013. It consists of 318,233 Twitter users with 3,545,258 directed edges.
- The second one is DBLP data set in [460], which is a co-authorship network consisting of totally 317,080 nodes and 1,049,866 undirected edges.

The performance evaluation metrics used in the comparison are as follows.

- Community Size Distribution: The communities are divided, by their size, into four groups of [4–50], [51–150], [151–500], and [501+]. Then, compare the quality of communities in each group with other criteria.

- Community Coverage: The desirable community coverage is to measure the number of users assigned to communities of size 4–150. The experimental results show that Infomap performs the best in terms of this metric. In fact, it assigns 62% of all users in the Twitter network and 98% of all users in DBLP network into communities of size 4–150.
- Extended Modularity: This metric is defined in [361] to measure the goodness of overlapping communities in the following.

$$EQ = \frac{1}{2m} \sum_i \sum_{v \in C_i, w \in C_i} \frac{1}{O_v O_w} \left[A_{vw} - \frac{k_v k_w}{2m} \right] \tag{1.17}$$

where C_i is the ith community, O_v is the number of communities to which v belongs, A is the adjacency matrix and $A_{ve} = 1$ if there exists an edge between v and w, and $A_{vw} = 0$, otherwise. k_v and k_w are the node degrees of v and w, respectively. $(k_v k_W / 2m)$ describe the expected number of edges between v and w. m is the total number of edges in the whole topology. The extended modularity range is $[-1, 1]$. If there is no overlap between the communities, then EQ becomes the traditional one. In (1.17), the first summation is for C_i over all communities. In [413], the C_i is over all communities in each group, i.e., four groups of [4–50], [51–150], [151–500], and [501+]. Through comparison on those partial extended modularity values, it indicates in [413] that Infomap and DEMON outperform others.
- Triangle Participation (TPR): In [460], it is found that TPR is a good metric when looking for ground-truth communities, where TPR is defined as the number of nodes in a community that forms a triad, divided by the total number of nodes in the community. DEMON scores a perfect value of 1.0 due to its design. Among others, CAA achieves significantly higher TPR on both networks, with a score of close to 0.9 on the Twitter network and a score of above 0.9 TPR on the DBLP network.
- Conductance: In [237] it is defined that the conductance of a community is the ratio of the number of edges on the boundary of the community to the sum of degrees of nodes in the community. In [413], the average conductance of the communities is calculated in different size groups. Then, it is found that both CAA and DEMON perform poorly in conductance.
- Internal Density: The internal density $f(S)$ of a community S is defined in [237] as follows.

$$f(S) = \frac{m_s}{n_s(n_s - 1)/2} \tag{1.18}$$

where m_s is the number of edges in the community S and n_s is the number of nodes in the community S. Experiments in [413] show that smaller community size typically indicates higher internal density value. Moreover, all algorithms show similar performance and CAA consistently perform slightly better.

- Transitivity: It is defined in [137] to be

$$\frac{3 \times \text{number of triangles}}{\text{number of triads}}$$

which measures the ties between individuals. The average transitivity score in each group is computed in [413]. It is found that CAA and DEMON perform similar in DBLP network, and DEMON performs slightly better in Twitter network.

All above analyses are attached with data and figures which can be brought here due to copyright. The reader may find them in [413]. The conclusion in [413] is that in general Infomap and CAA are able to discover communities of desirable sizes with better performance.

Exercises

1. Consider a set function f defined on subsets of X. Show that f is submodular if and only if for any two subsets A and B of X with $A \subset B$, and any $x \in X \setminus B$,

$$\Delta_x f(A) \geq \Delta_x f(B).$$

2. Consider a set function f defined on subsets of X. Show that f is monotone nondecreasing and submodular if and only if for any two subsets A and B of X with $A \subset B$, and any $x \in X \setminus A$,

$$\Delta_x f(A) \geq \Delta_x f(B).$$

3. A set function f is *modular* if both f and $-f$ are submodular. For any subset A of a finite set X, let $f(A) = |A|$. Show that $f(A)$ is modular.

4. Suppose f is a monotone nondecreasing and submodular function on 2^V. Show that for any positive constant $c > 0$, $\min(f(A), c)$ is also a monotone nondecreasing and submodular function. Furthermore, if f is a polymatroid function, then $\min(f(A), c)$ is also a polymatroid function. (A set function f is a polymatriod function if f is monotone nondecreasing, submodular, and $f(\emptyset) = 0$.)

5. Let f be a submodular function. Show that for any two subsets,

$$f(A \cup B) - f(B) \leq \sum_{x \in A \setminus B} [f(B \cup \{x\}) - f(B)].$$

6. Suppose f is a submodular function on 2^V. Prove or disprove that for any positive constant $c > 0$, $\min(f(A), c)$ is also a submodular function.

7. Let f be a submodular function. Show that for any two sets A and B,

$$\sum_{x \in A \setminus B} [f(A) - f(A \setminus \{x\})] \leq f(A) - f(A \cap B).$$

8. Consider a directed network $G = (V, E)$ with adjacency matrix (a_{ij}). For any partition C of V, define the modularity function of C by

$$Q(C) = \frac{1}{|E|} \sum_{i,j \in V} \left[a_{ij} - \frac{k_i^{out} k_j^{in}}{|E|} \right] \delta_{C_i, C_j}$$

where k_i^{out} is the out-degree of node i, k_j^{in} is the in-degree of node j, C_i is the community containing i, and δ_{C_i, C_j} is the Kronecher delta symbol. Show that

$$Q(C) = \sum_{C_s \in C} \left[\frac{L(C_s, C_s)}{L(V, V)} - \frac{L(C_s, V)L(V, C_s)}{L(V, V)^2} \right].$$

Design a greedy algorithm for modularity maximization in directed networks.
9. Please formulate the maximum community partition 2 into a 0-1 LP to together with an explanation.
10. (**Power-Law Graphs**) A graph G belongs to power-law graph class $P(\alpha, \gamma)$ if

$$|\{v \in V(G) \mid \deg(v) = k\}| = \lfloor \frac{e^\alpha}{k^\gamma} \rfloor.$$

Consider a graph $G = (V, E)$ with adjacency matrix (a_{ij}). Suppose $G \in P(\alpha, \gamma)$ for $\gamma > 2$. Show that (a) the maximum degree in G is $\lfloor e^{\alpha/\gamma} \rfloor$, (b) $|V(G) \leq \zeta(\gamma)e^\alpha$, and (c) $|E(G)| \leq \frac{1}{2} \cdot \zeta(\gamma - 1)e^\alpha$ where $\zeta(\gamma) = \sum_{i=1}^{\infty} \frac{1}{i^\gamma}$.
11. Based on an observation on power-law graphs that lower-degree nodes follow higher-degree nodes, an algorithm is designed for the modularity maximization as shown in Algorithm 8.
    Show that for power law graphs with $\gamma > 2$, the low-degree following algorithm is a $O(1)$-approximation for the modularity maximization problem without using simplification of using the real number instead of rounding down to the integer.
12. Show that in a social network with IC model, finding the largest influence path from a node to another node is equivalent to finding the shortest path for properly defined distance.
13. Consider a set X of n-dimensional vectors. For any subset A of X, $f(A)$ is defined to be the maximum number of independent vectors in A. Show that $f(x)$ is submodular.
14. Consider a graph $G = (V, E)$. For any edge subset $A \subseteq E$, define $f(A)$ to be the maximum number of edges in A which does not contain a cycle. Show that $f(A)$ is submodular.

15. Consider a graph $G = (V, E)$. For any node subset $A \subseteq V$, define $f(A)$ to be the number of connected components in subgraph with node set V and edges each with at least one endpoint in A. Show that $|V| - f(A)$ is submodular.

16. Consider a graph $G = (V, E)$. For any node subset A, $f(A)$ is defined to be the number of connected components in the subgraph induced by A. Show that $|V| - f(A)$ is not submodular.

17. Please analyze the proof of Theorem 1.6.1, where the submodularity of a set function is implicitly used.

18. Show that following problem has a greedy $(1 - e^{-1})$-approximation: Given a collection $\mathcal{C}$ of subsets of a finite set X and an integer $k > 0$, find a subset A of X with $|A| = k$ to maximize the number of subsets in $\mathcal{C}$ hit by A where a subset S is said to be hit by A if $A \cap S \neq \emptyset$.

19. (Positive Influence Dominating Set [417]) Given a social network $G = (V, E)$. A subset D of V is called a *positive influence dominating set* if every node u is either in D or adjacent to at least $deg(u)/2$ nodes in D where $deg(u)$ is the degree of u. Show that finding the minimum positive influence dominating set is APX-hard and it is a submodular cover problem.

20. (Connected Positive Influence Dominating Set [87, 465]) If furthermore, a positive influence dominating set induced a connected subgraph, then it is called a connected positive influence dominating set. Please determine whether finding the minimum connected positive influence dominating set for a given graph is a minimum submodular cover problem, or not.

21. (Partial Positive Influence Dominating Set [341]) Given a graph $G = (V, E)$ and a number $p \in (0, 1]$. A node subset P is a *partial positive influence dominating set* if P positive influence dominates at least p percent of all nodes. Design a polynomial-time approximation algorithm for finding the minimum partial positive influence dominating set, to have performance ratio $\gamma H(\Delta)$ where $H(\cdot)$ is the Harmonic number, $\gamma = 1/(1 - (1 - p)(\Delta/\delta))$, and Δ and δ are the maximum degree and the minimum degree of the graph G, respectively.

22. ([341]) Show that finding the minimum partial positive influence dominating set in power-law graph can have a polynomial-time constant-approximation.

Historical Notes

In this chapter, the main subject is community detection. Community is a natural structure existing in various networks [116, 137, 286]. Identifying such a structure is often an important step for understanding the network deeply. Its early work can be traced to [433]. However, it is a difficult task due to no widely accepted quality measure existing in the literature [4]. The community detection is usually to be formulated to minimization or maximization problems based on various structure metrics [66, 137, 256, 272, 306, 323, 444, 502] and consideration on density [206, 288]. Comparison on some of them are made in [116, 177, 237, 413, 421, 464]. Especially, in [413], six algorithms, respectively in [20, 65, 303, 304, 339, 348], are compared through sizes of produced community. The importance of community size is noticed by Dunbar in [94]. In [413], the comparison is made on co-authorship network DBLP and Twitter follower graph. They observed that "larger communities achieve higher modularity score which is consistent with the resolution limit of modularity as indicated in [117]." Another criterion for quality of community is proposed in [459]. Tong et al. [391] proposed a sequence of cut-related formulations and obtained several theoretical results based on existing theory for some classical problems, assignment [389], network flow [115, 123, 332], and multi-cut [140, 172, 207]. They also accumulated a list of popular quality criteria for the community [237]: Conductance [208], expansion [338], cut ratio [116], normalized cut [367], average-ODF, internal density [338]. Using those criteria, they compared algorithms for six formulations on Zachary karate club network [479], Girvan-Newman benchmark network [137], and the artificial networks generated by LFR benchmark [227]. Those networks were also used in [91, 93, 257] for testing. There are still many other criteria, such as out-degree [114] and the fraction of the edge volume [206]. Actually, some ways other than optimization can also be employed for community detection, such as random walk [334], fuzzy membership degree [344], and edge prediction [447]. There are a huge number of publications about community detection in the literature. What we mentioned in above is only a small portion of them. In fact, from the following fact, we may image how large this number is. Currently, the paper [305] has received nearly 10,000 citations. In Sect. 1.5, we presented a few examples to show some typical algorithms. They may have new progress on community partition [317, 318, 485] in the literature. Very different from others, an Axiomatic Approach for community detection is presented in [23].

Many complex networks, such as online social networks and biological networks, have the power law structure [11, 62]. Community detection is an important task in the study of such networks [118, 137]. Community structure provides insight into how network function and topology affect each other [491]. Knowing such structure would be helpful in the study of some problems on those networks [76, 194, 308, 320, 325, 411, 472, 506]. Special techniques can be found in [364, 365, 481] for the study of power law graphs on several issues, such as complexity, approximation, and robustness.

There is a popular way to design a heuristic that is to simplify the input graph into a tree. Usually, the problem on trees can be solved easily. In this approach, the shortest path tree plays an important role. In [58], one may find an interesting work on estimation of the shortest path in social networks.

Chapter 2
Social Influence

In this chapter, we discuss an important problem, influence maximization, and related issues.

2.1 Influence Maximization

Suppose a company wants to find a representative for a product. Clearly, the company would like to find a person with the highest social influence. Suppose the company wants to find two representatives. Then the two persons with the highest and the second highest social influence may not be a good choice if they influence almost the same group of people. In general, if the company wants to find k representatives, then the situation becomes more complicated. It becomes an interesting problem actively discussed in the study of social networks.

Problem 2.1.1 (*Influence Maximization*) Given a social network $G = (V, E)$ with an information diffusion model m and an integer $k > 0$, find locations for a set S of k seeds to maximize the influence spread.

For simplicity, we first consider the problem in the deterministic information diffusion model. In this case, the influence spread is the number of nodes which are reachable from the seed set.

Theorem 2.1.2 *The influence maximization problem is NP-hard under the deterministic model.*

© The Author(s), under exclusive license to Springer Nature Switzerland AG 2026
W. Wu et al., *Computational Aspects of Social Networks*, Springer Optimization and Its Applications 234, https://doi.org/10.1007/978-3-032-14833-9_2

Proof Consider a well-known NP-hard problem, the minimum set cover problem. Its decision version is NP-complete, which is described as follows.

Problem 2.1.3 (*Set Cover*) Given a collection $\mathcal{C}$ of subsets of a finite set X, and an integer $k > 0$, determine whether $\mathcal{C}$ contains k subsets covering all elements in X.

We will construct a polynomial-time many-one reduction from the set cover problem to the decision version of the influence maximization problem, which is described as follows: Given a social network $G = (V, E)$ with an information diffusion model m, an integer $k > 0$, and an integer $h > 0$, determine whether there exist k locations for seeds to make the influence spread bigger than or equal to h.

To do so, construct a bipartite graph G with two node sets $\mathcal{C}$ and X and directed edge set

$$\{(A, x) \mid x \in A \text{ for } x \in X, A \in \mathcal{C}\}.$$

Clearly, if G is considered as a social network with deterministic diffusion model, then all seeds should be selected from node set $\mathcal{C}$. Let $h = |X| + k$. Then the influence spread is at least h if and only if the selected k seeds could cover X. Therefore, the construction maps the input of the set cover problem to the input of the decision version of the influence maximization problem such that the output is "yes" for the set cover problem if and only if the output is "yes" for the decision version of the influence maximization problem. $\qquad\square$

However, the following lemma shows that the influence maximization problem may have a good approximate solution.

Lemma 2.1.4 *The influence spread $\sigma_D(S)$ under deterministic model is a monotone nondecreasing submodular function with respect to seed set S.*

Proof Consider a social network $G = (V, E)$ and two node subsets A and B with $A \subseteq B$. Let $I(A)$ denote the set of nodes that can be reached from A through a path in G. Then $\sigma_D(A) = |I(A)|$. Thus, for any node $x \in V \setminus A$,

$$\sigma_D(A \cup \{x\}) - \sigma_D(A) = |I(A \cup \{x\})| - |I(A)| = |I(A \cup \{x\}) \setminus I(A)|.$$

Similarly,

$$\sigma_D(B \cup \{x\}) - \sigma_D(B) = |I(B \cup \{x\}) \setminus I(B)|.$$

Notice that $I(A \cup \{x\}) \setminus I(A)$ is the set of nodes which can be reached from x through a path, but cannot be reached from A through any path, and $I(B \cup \{x\}) \setminus I(B)$ is the set of nodes which can be reached from x through a path, but cannot be reached from B through any path. Since $A \subseteq B$, we have

$$I(B \cup \{x\}) \setminus I(B) \subseteq I(A \cup \{x\}) \setminus I(A).$$

Therefore

$$\sigma_D(A \cup \{x\}) - \sigma_D(A) \geq \sigma_D(B \cup \{x\}) - \sigma_D(B).$$

Hence, $\sigma_D(S)$ is monotone nondecreasing and submodular. $\square$

Now, we can employ a greedy algorithm to obtain the following by Theorem 1.6.2.

Theorem 2.1.5 *There exists a polynomial-time* $(1 - e^{-1})$*-approximation for the influence maximization problem under deterministic diffusion model.*

We now consider the IC model. In this case, the influence spread $\sigma_{IC}(S)$ is the expected number of active nodes at the end of information diffusion process. Notice that the deterministic model can be seen as a special case of the IC model, in which every edge (u, v) has label $p_{uv} = 1$. Therefore, the influence maximization problem is also NP-hard under the IC model. Moreover, we can obtain the following.

Lemma 2.1.6 *The influence spread* $\sigma_{IC}(S)$ *is monotone nondecreasing and submodular.*

Proof For any subgraph H of G, denote by $\Pr[H]$ the probability that all edges in H are connected and all edges not in H are disconnected. Here, by saying that an edge is connected, we mean that the information can pass through the edge. Let $\sigma_D^H(S)$ denote the influence spread in subgraph H under deterministic model. Then, we have

$$\sigma_{IC}(S) = \sum_{H \subseteq G} \Pr[H]\sigma_D^H(S).$$

By Lemma 2.1.4, $\sigma_D^H(S)$ is monotone nondecreasing and submodular. Thus, as a linear combination of $\{\sigma_D^H(S)\}$'s, $\sigma_{IC}(S)$ is also monotone nondecreasing and submodular. $\square$

Using a similar argument, we can prove that the influence spread is monotone nondecreasing and submodular under the MC model. However, the deterministic diffusion model is not a special case of the MC model. Therefore, the NP-hardness of the influence maximization problem has to be established by a different proof. Since MC model and LT model are equivalent (see Theorem 2.1.9), we establish the NP-hardness for LT in the following.

Theorem 2.1.7 *The influence maximization problem is* NP*-hard in the LT model.*

Proof Consider the NP-complete vertex-cover problem as follows.

Problem 2.1.8 (*Vertex-Cover*) Given a graph $G = (V, E)$ and an integer $k > 0$, determine whether there is a subset C of at most k vertices such that every edge has at least one endpoint in C. Such a vertex subset C is called a *vertex cover*.

For any instance of the vertex-cover problem, (G, k), construct a directed graph G' by replacing each edge $\{u, v\}$ with two arcs (u, v) and (v, u). Add for each arc (u, v), assign a weight $b_{uv} = 1/indeg(v)$ where $indeg(v)$ is the in-degree of v, i.e., $indeg(v) = |N^-(v)|$.

If G has a vertex cover C of size at most k, then every vertex in C can be placed with a seed so that for every vertex v which is not a seed, $N^-(v) \subseteq C$. This means v can be activated with probability one, and hence we obtain $\sigma_{LT}(C) = |V| = n$.

If G does not have a vertex cover of size at most k, then no matter how to place k seeds, there exist two arcs (u, v) and (v, u) such that there is no seed at u and v. This implies that u and v are inactive at the initial step. At any latter step, since u being inactive implies that v cannot be activated with probability one and v being inactive implies that u cannot be activated with probability one. Therefore, to get u and v activated, at least one of them must be activated with probability less than one. Hence for any seed set S, we have $\sigma_{LT}(S) < n$. $\square$

To extend the result from the MC model to the LT model, we now show the equivalence between the MC model and the LT model.

Theorem 2.1.9 *Let p_{uv} and b_{uv} be labels of arc (u, v) in the MC model and the LT model, respectively. If $p_{uv} = b_{uv}$ for any arc (u, v), then for any seed set S, $\sigma_{LT}(S) = \sigma_{MC}(S)$.*

Proof It is sufficient to show that starting from a same seed set, LT process and MC process have the same distribution over sets of active nodes at the completion time. For this purpose, we show by induction on k that at any step k, the active sets have the same distribution in the LT process and the MC process.

Let $A_0 = S$ and A_k denote the set of active nodes at step k. Suppose that A_k has the same distribution in the LT process and the MC process. To show that A_{k+1} has the same distribution in the two processes, it suffices to prove that the probability that a node u is inactive at step k, but becomes active at step $k + 1$, i.e.,

$$\Pr[u \in A_{k+1} \mid u \notin A_k]$$

is the same in the LT process and the MC process. This is because

$$\begin{aligned}
\Pr[u \in A_{k+1}] \\
= \Pr[u \in A_k] + \Pr[u \in A_{k+1} \setminus A_k] \\
= \Pr[u \in A_k] + \Pr[(u \in A_{k+1}) \wedge (u \notin A_k)] \\
= \Pr[u \in A_k] + \Pr[u \in A_{k+1} \mid u \notin A_k] \cdot \Pr[u \notin A_k].
\end{aligned}$$

First, consider the LT process. Let $N^-(u)$ denote the set of in-neighbors of u. Since u is inactive at step k, θ_u must lie in interval $(\sum_{w \in N^-(u) \cap A_{k-1}} b_{wu}, 1]$, which occurs with probability

$$1 - \sum_{w \in N^-(u) \cap A_{k-1}} b_{wu}.$$

Meanwhile, u becomes active at step $k + 1$, i.e., θ_u lies in interval

$$\left(\sum_{w \in N^-(u) \cap A_{k-1}} b_{wu}, \sum_{w \in N^-(u) \cap A_k} b_{wu} \right].$$

This occurs with probability

$$\sum_{w \in N^-(u) \cap A_k} b_{wu} - \sum_{w \in N^-(u) \cap A_{k-1}} b_{wu} = \sum_{w \in N^-(u) \cap (A_k \setminus A_{k-1})} b_{wu}.$$

Therefore,

$$\Pr[u \in A_{k+1} \mid u \notin A_k] = \frac{\sum\limits_{w \in N^-(u) \cap (A_k \setminus A_{k-1})} b_{wu}}{1 - \sum\limits_{w \in N^-(u) \cap A_{k-1}} b_{wu}}.$$

Now, we consider the MC process. Since all in-neighbors influence u in mutually-exclusive way, we have

$$\Pr[u \notin A_k] = 1 - \sum_{w \in N^-(u) \cap A_{k-1}} p_{wu}$$

and

$$\Pr[(u \in A_{k+1}) \wedge (u \notin A_k)] = \sum_{w \in N^-(u) \cap (A_k \setminus A_{k-1})} p_{wu}.$$

Therefore,

$$\Pr[u \in A_{k+1} \mid u \notin A_k] = \frac{\sum_{w \in N^-(u) \cap (A_k \setminus A_{k-1})} p_{wu}}{1 - \sum_{w \in N^-(u) \cap A_{k-1}} p_{wu}}.$$

Since $b_{wu} = p_{wu}$, this completes the proof. $\square$

At the end of this section, we point out an important property of the MC model. Please look at Fig. 2.1. For the two networks in this figure, the probability that nodes 1 and 2 influence node 4 is the same, i.e., $(p_1 + p_2)p_3 = p_1 p_3 + p_2 p_3$. This means that for computing influence probability, the network can be split at node 3. This property is due to the mutually-exclusive property of influence from in-neighbors at each node. In general, we may describe this property as follows.

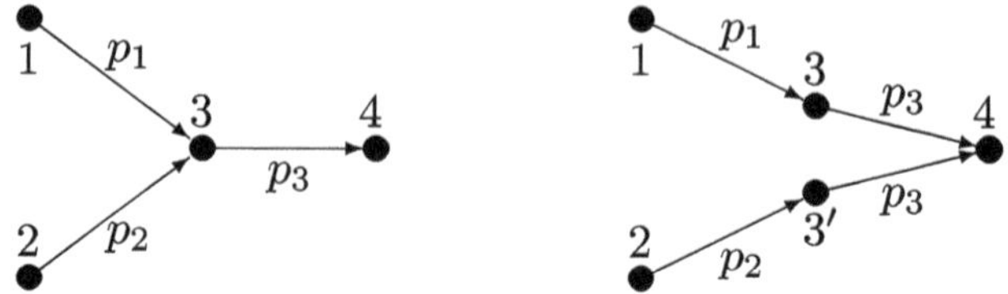

Fig. 2.1 A property of the MC model

Proposition 2.1.10 *Let P be the set of paths from a node u to another node v in network G. Then in the MC model or the LT model,*

$$Pr[v \text{ is influenced by } u \text{ through network } G]$$
$$= \sum_{p \in P} Pr[v \text{ is influenced by } u \text{ through path } p].$$

In the proof of Lemma 1.5.5, this property has been used once.

2.2 Bharathi-Kempe-Salek Conjecture

In 2007, Bharathi et al. [14] conjectured that the influence maximization is NP-hard on an *in-arborescence* (which is a directed tree such that every arc is directed to the root) under the IC model. What is the significance of this conjecture? To explain it, let us first study the influence maximization problem on an in-arborescence under the deterministic model an the LT model.

First, consider the deterministic model.

Lemma 2.2.1 *In the in-arborescence with deterministic diffusion model, the maximum influence spread can always be achieved by placing all seeds at leaves.*

Proof Let u be a non-leaf node with a seed. Suppose w is a leaf such that u lies on the path from w to the root r. If we move the seed at u to w, then u can still be activated in the deterministic process. Therefore, the influence spread would not be reduced. $\square$

Let $f(v, k)$ denote the maximum influence spread restricted to the subtree rooted at node v when at most k seeds are located on some leaves of this subtree. By Lemma 2.2.1, we can obtain the following recursive formula:

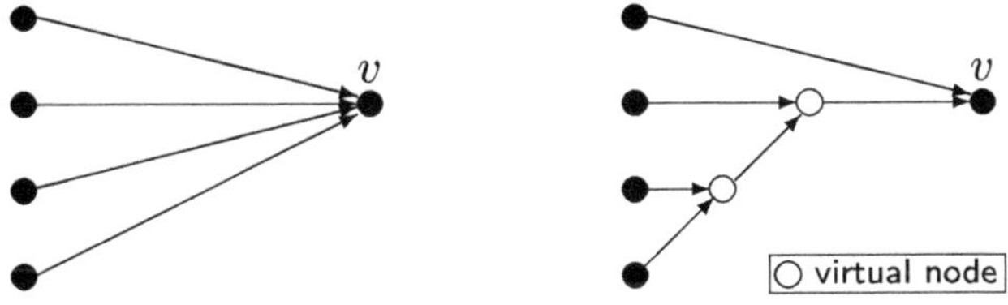

Fig. 2.2 Transform an in-arborescence into a binary in-arborescence

$$f(v, 0) = 0, \quad \text{and for } k \geq 1$$

$$f(v, k) = \begin{cases} 1, & \text{if } v \text{ is a leaf,} \\ 1 + \max\{ \sum\limits_{u \in N^-(v)} f(u, k_u) \mid \\ \qquad \sum\limits_{u \in N^-(v)} k_u = k, k_u \geq 0\}, & \text{otherwise.} \end{cases}$$

A dynamic program can be designed based on this recursive formula. However, this algorithm may not be able to run in polynomial-time since the maximum is taken over all possible compositions of integer k into $|N^-(u)|$ nonnegative integers, which has

$$\binom{k + |N^-(v)| - 1}{k}$$

items. How to overcome this trouble? An approach is to transform the in-arborescence into a binary in-arborescence by introducing virtual nodes as shown in Fig. 2.2, and introducing a node weight function as follows:

$$w(v) = \begin{cases} 0 & \text{if } v \text{ is a virtual node,} \\ 1 & \text{otherwise.} \end{cases}$$

Now, let $f(v, k)$ denote the maximum total weight of activated nodes in the subtree rooted at node v when at most k seeds are located on some leaves of this subtree. Then, we have the following recursive formula.

$$f(v, 0) = 0, \quad \text{and for } k \geq 1$$

$$f(v, k) = \begin{cases} w(v), & \text{if } v \text{ is a leaf,} \\ w(v) + \max\{ \sum\limits_{u \in N^-(v)} f(u, k_u) \mid \\ \qquad \sum\limits_{u \in N^-(v)} k_u = k, k_u \geq 0\}, & \text{otherwise.} \end{cases}$$

Since $|N^-(v)| \leq 2$, the minimization is taken over at most $k + 1$ items. Based on this formula, a dynamic program can be designed to compute $f(r, k)$ with running time $O(k^2(n + n'))$ where n' is the number of virtual nodes and r is the root. It is not hard to estimate that $n' \leq n - 1$. Therefore, we obtain the following result.

Theorem 2.2.2 *The influence maximization problem on an in-arborescence with deterministic model can be solved in time $O(k^2 n)$ where n is the number of nodes.*

Next, consider the MC (or LT) process. First, notice that Lemma 2.2.1 does not hold. To see so, look at Fig. 2.3. In the first case, a seed is placed at leaf 1 so that the influence spread is $1 + x + xy$. In the second case, a seed is placed at non-leaf node 2 so that the influence spread is $1 + y$. When x is enough small, we have $1 + x + xy < 1 + y$.

Actually, this fact does not bring too much trouble in establishing a recursive relation. The main trouble is about the root. We do not know what is the probability that the root gets activated. This main trouble could be solved by using the special property about splitting network described at the end of previous section.

As before, we introduce virtual node and node weight to transform the input in-arborescence to a binary in-arborescence. Then, we define labels for arcs incident with virtual nodes as follows. For an arc (u, v), if u is a virtual node, then set $p_{uv} = 1$; if u is not a virtual node, then along the path starting from v find the first non-virtual node w and set $p_{uv} = p_{uw}$ where p_{uw} is the label of arc (u, w) in the original in-arborescence.

Denote by T_v the sub-in-arborescence of T rooted at node v. Let $T_{v,i}$ be the union of sub-in-arborescence T_v and a coring of length i, where the coring is the directed path of i nodes, connected to v and directed away from v (not including node v). An illustration is given in Fig. 2.4.

Let $f(v, k, i)$ denote the maximum total weight of activated nodes of $T_{v,i}$ when at most k seeds are located in T_v (no seed is allowed to lie at the coring). Denote by $f(v, k, i \mid v$ is a seed$)$ the above value under the condition that v is a seed, and $f(v, k, i \mid v$ is not a seed$)$ the above value under the condition that v is not a seed. Then,

$$f(v, k, i) = \max \{ f(v, k, i \mid v \text{ is a seed}), f(v, k, i \mid v \text{ is not a seed}).\}$$

Fig. 2.3 Lemma 2.2.1 does not hold in the MC process

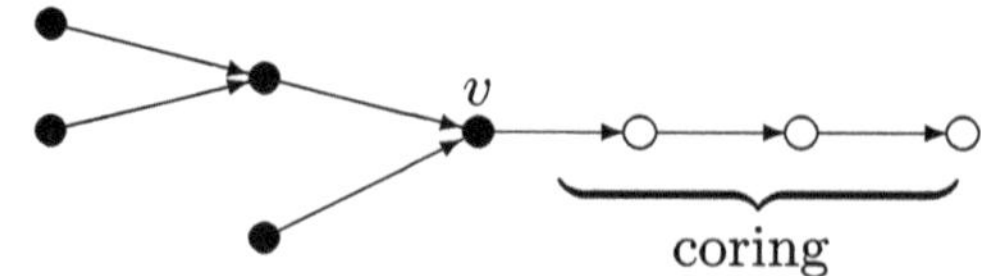

Fig. 2.4 Blackened nodes form T_v. White nodes form a coring of length 3. Attaching the coring to v form $T_{v,3}$

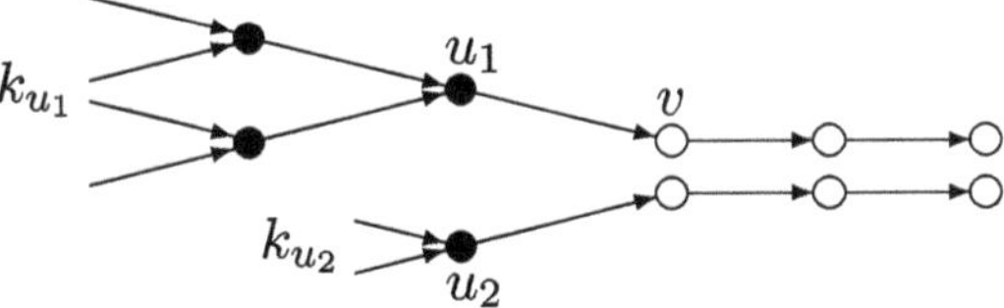

Fig. 2.5 v is not a seed

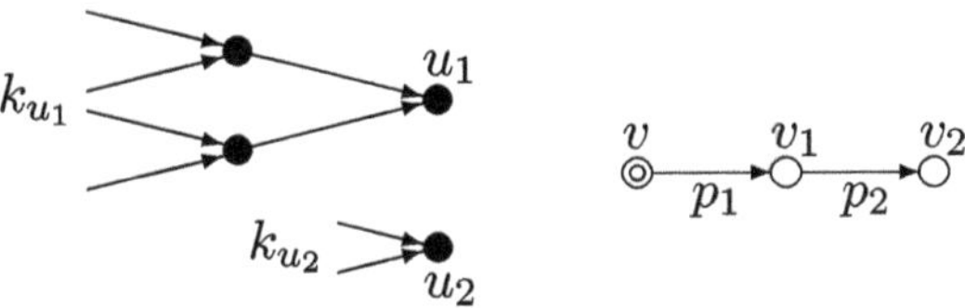

Fig. 2.6 A seed lies at v

Moreover, for $k \geq 1$ and non-leaf node v, we have (see Figs. 2.5 and 2.6 for an illustration)

$$f(v, k, i \mid v \text{ is not a seed})$$

$$= \min \left\{ \sum_{u \in N^-(v)} f(u, k_u, i+1) \mid \sum_{u \in N^-(v)} k_u = k, k_u \geq 0 \right\},$$

$$f(v, k, i \mid v \text{ is a seed})$$

$$= w(v, i) + \min \left\{ \sum_{u \in N^-(v)} f(u, k_u, 0) \mid \sum_{u \in N^-(v)} k_u = k - 1, k_u \geq 0 \right\},$$

where $w(v, i) = w(v) + p_{vv_1} w(v_1) + \cdots + p_{vv_1} p_{v_1 v_2} \cdots p_{v_{i-1} v_i} w(v_i)$ for coring $(v, v_1, \ldots, v_i)$.

For $k = 0$, we have $f(v, 0, i) = 0$, and for a leaf v and $k \geq 1$, we have $f(v, k, i) = w(v, i)$.

Based on the above recursive formulas, a dynamic program can be designed.

Theorem 2.2.3 *The influence maximization problem on an in-arborescence with the MC (or LT) model can be solved in time $O(k^2 n^2)$ where n is the number of nodes.*

We now come to discuss the significance of Bharathi-Kempe-Salek Conjecture. It is the first example that a problem has different computational complexities under the LT model and the IC model. Actually, this conjecture was proved by Lu et al. [274] in 2017.

Theorem 2.2.4 (Lu et al. [274]) *The influence maximization problem on an in-arborescence is NP-hard under the IC process.*

To prove this result, it suffices to show the NP-completeness of the following decision version: Given an in-arborescence $G = (V, E)$ with the IC model, an integer

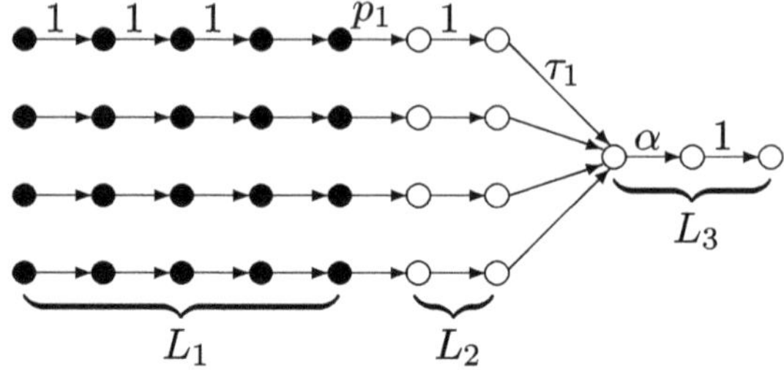

Fig. 2.7 The proof of Theorem 2.2.4

$k > 0$, and an integer $h > 0$, determine whether there exist k locations for seeds to make the influence spread bigger than or equal to h.

Lu et al. [274] employ the sub-sum problem, which is known to be NP-complete.

Problem 2.2.5 (*Sub-sum*) Given $n + 2$ nonnegative integers $a_1, a_2, \ldots, a_n, t$, and k, determine whether there exist k integers in $\{a_1, a_2, \ldots, a_n\}$ such that their sum equals t.

Without loss of generality, assume $a_i \le t$ for any $i = 1, 2, \ldots, n$. In fact, if $a_i > t$, then a_i can be ignored without affecting the answer.

The reduction from the sub-sum problem is constructed as shown in Fig. 2.7. The following is an explanation for this construction.

- There are n leaves $1, 2, \ldots, n$.
- The path from leaf i to the root r can be divided into three disjoint pieces with lengths L_1, L_2, L_3.
- The arc from the first piece to the second piece is labeled with p_i. The arc from the second piece to the third piece is labeled with τ_i. The first arc in the third piece is labeled with α.
- All remaining edges are labeled with 1.

Those parameters appearing in the above construction are selected as follows.

$$
\begin{aligned}
x_i &= a_i / t, \\
L_2 &> (1 - x_i)/x_i, \\
L_3 &= 3k, \\
L_1 &> L_2 + L_3, \\
p_i &= (L_2 - 1 + x_i)/L_2, \\
\tau_i &= (1 - e^{(x_i - 1)/k})/p_i, \\
\alpha &= (k \cdot e^{(k-1)/k} - 1)/(L_3 - 1).
\end{aligned}
$$

This selection implies the following lemma, which certifies that p_i, τ_i and α can indeed serve as probabilities.

Lemma 2.2.6 *For any* $i = 1, 2, \ldots, n$, $0 \le p_i \le 1$, $0 \le \tau_i \le 1$, *and* $0 \le \alpha \le 1$.

Proof Since $0 \le x_i \le 1$, we have $0 \le p_i \le 1$. Since $L_3 > ke$, we have $0 \le \alpha \le 1$. To see $0 \le \tau_i \le 1$, notice that

$$1 - e^{(x_i-1)/k} < 1 - (1 + (x_i - 1)/k) = (1 - x_i)/k \le 1 - x_i.$$

Thus, it suffices to show

$$1 - x_i \le p_i.$$

This is guaranteed by the choice of p_i and L_2. $\qquad\square$

Lemma 2.2.7 *In the in-arborescence constructed as the above, the maximum influence spread with the IC model can always be achieved by placing k seeds at leaves.*

Proof If a seed is placed at a node in the second piece or the third piece of the path P_i from leaf i to the root, then this seed can achieve an influence spread value (expected number of nodes activated by the seed) at most $L_2 + L_3$. Since $L_1 > L_3 + L_3$, moving this seed to leaf i may get more influence spread value because all nodes on the first piece of P_i can be definitely activated. If a seed is placed on the first piece but not at the leaf, then it is clear that moving it to the leaf is better. $\quad\square$

Now, we place all k seeds at leaves. Let A be the set of k locations of those seeds. Then

$$\sigma_{IC}(S) = \sum_{i \in A}(L_1 + p_i \cdot L_2) + \left(1 - \prod_{i \in A}(1 - p_i \cdot \tau_i)\right)(1 + \alpha \cdot (L_3 - 1))$$

$$= \sum_{i \in A}(L_1 + L_2 - 1 + x_i) + \left(1 - \prod_{i \in A}e^{(x_i-1)/k}\right) \cdot k \cdot e^{(k-1)/k}$$

$$= k(L_1 + L_2 - 1) + x - k \cdot e^{(x-1)/k} + k \cdot e^{(k-1)/k},$$

where $x = \sum_{i \in A} x_i$. The following lemma completes the proof of Theorem 2.2.4, by noticing that $x = 1$ if and only if the answer to the sub-sum instance is "yes."

Lemma 2.2.8 *Let $h = k(L_1 + L_2 - 2 + e^{(k-1)/k}) + 1$. Then $\sigma_{IC}(S) \le h$ holds for any seed set S. Furthermore, $\sigma_{IC}(S) = h$ if and only if $x = 1$.*

Proof Define $g(x) = x - k \cdot e^{(x-1)/k}$. Then $g(x)$ has a unique maximum point at $x = 1$ by observing

$$g'(x) = 1 - e^{(x-1)/k} \begin{cases} > 0 \text{ for } x < 1, \\ = 0 \text{ for } x = 1, \\ < 0 \text{ for } x > 1. \end{cases}$$

The lemma follows immediately from this property of $g(x)$. $\qquad\square$

2.3 General Threshold Model and General Cascade Model

The LT model can be generalized as follows.

General Threshold (GT) Model: This model is a probabilistic diffusion process consisting of discrete steps. Consider a directed graph $G = (V, E)$ and each node has two states, *active* and *inactive*. Each node v is provided with a monotone non-decreasing function $f_v : 2^{N^-(v)} \to [0, 1]$ where $N^-(v)$ is the set of all in-neighbors of v and $2^{N^-(v)}$ is the collection of all subsets of $N^-(v)$. The process with the GT model is as follows:

- Initially, every node v is in inactive state and randomly chooses a threshold θ_v from $[0, 1]$ with uniform distribution.
- To start the process, choose a subset of nodes as seeds and activate them.
- At each subsequent step, every inactive node evaluates the function value of f_v on the set of all active in-neighbors. If this value is at least its threshold, then its state becomes active; else, it keeps in inactive state.
- The process ends if no more node can be activated.

  The IC model can also be generalized.

General Cascade (GC) Model: This is a probabilistic diffusion process consisting of discrete steps. Consider a directed social network $G = (V, E)$. Each node has two states, *active* and *inactive* and is provided with a function $p_v : N^-(v) \times 2^{N^-(v)} \to [0, 1]$. The process of the GC model is as follows.

- Initially, every node is in inactive state.
- To start, choose a seed set and activate all seeds.
- At each subsequent step, each freshly-active node u tries to activate its inactive out-neighbor v, and v accepts influence from u with probability $p_v(u, A)$, where A is the set of active in-neighbors of v who have tried to activate v before u, and a node is said to be *freshly-active* if its state changes from inactive to active at the last step.
- An important condition is as follows

    (*) When v is activated by more than one in-neighbors $u_1, u_2, \ldots, u_k$, the probability that v becomes active is independent of ordering of trying.

- The process ends if no more node can be activated.

In the definition of the GC model, condition (*) is equivalent to saying that for any permutation τ of $(1, 2, \ldots, k)$, the probability that v becomes active when its active in-neighbors try to activate v in ordering $(u_1, u_2, \ldots, u_k)$ is equal to the probability that v becomes active when its active in-neighbors try to activate v in ordering $(u_{\tau(1)}, u_{\tau(2)}, \ldots, u_{\tau(k)})$. That is,

$$1 - (1 - p_v(u_1, \emptyset))(1 - p_v(u_2, \{u_1\})) \cdots (1 - p_v(u_k, \{u_1, u_2, \ldots, u_{k-1}\}))$$
$$= 1 - (1 - p_v(u_{\tau(1)}, \emptyset))(1 - p_v(u_{\tau(2)}, \{u_{\tau(1)}\}))$$
$$\cdots (1 - p_v(u_{\tau(k)}, \{u_{\tau(1)}, u_{\tau(2)}, \ldots, u_{\tau(k-1)}\})).$$

Notice that every permutation can be represented as a product of several consecutive swaps. Therefore, we have the following.

Proposition 2.3.1 *The condition (*) is equivalent to asking the function $p_v(u, A)$ to satisfy the property that for any node v, any subset A of in-neighbors of v, and any two in-neighbors u_1 and u_2 of v not in A,*

$$(1 - p_v(u_1, A))(1 - p_v(u_2, A \cup \{u_1\})) = (1 - p_v(u_2, A))(1 - p_v(u_1, A \cup \{u_2\})).$$

Proposition 2.3.1 is quite useful for verifying an information diffusion model to be a GC model. In the following, we give two examples.

Corollary 2.3.2 *The MC model is a special case of the GC model.*

Proof For any $A \subset N^-(v)$ and $u_1, u_2 \in N^-(v) \setminus A$,

$$(1 - p_v(u_1, A))(1 - p_v(u_2, A \cup \{u_1\}))$$
$$= \left(1 - \frac{p_{u_1 v}}{1 - \sum_{u \in A} p_{uv}}\right)\left(1 - \frac{p_{u_2 v}}{1 - (p_{u_1 v} + \sum_{u \in A} p_{uv})}\right)$$
$$= \frac{1 - (p_{u_1 v} + p_{u_2 v} + \sum_{u \in A} p_{uv})}{1 - \sum_{u \in A} p_{uv}}$$
$$= (1 - p_v(u_2, A))(1 - p_v(u_1, A \cup \{u_2\}))$$

Then the corollary follows from Proposition 2.3.1. $\qquad\square$

Corollary 2.3.3 *Consider the GT model. For any node v, any in-neighbor u of v, and any subset $A \subseteq N^-(v) \setminus \{u\}$, define*

$$p_v(u, A) = \frac{f_v(A \cup \{u\}) - f_v(A)}{1 - f_v(A)}.$$

Then $p_v(u, A)$ induces a GC model.

Proof For any $A \subset N^-(v)$ and $u_1, u_2 \in N^-(v) \setminus A$,

$$
(1 - p_v(u_1, A))(1 - p_v(u_2, A \cup \{u_1\}))
$$
$$
= \left(1 - \frac{f_v(A \cup \{u_1\}) - f_v(A)}{1 - f_v(A)}\right)\left(1 - \frac{f_v(A \cup \{u_1, u_2\}) - f_v(A \cup \{u_1\})}{1 - f_v(A \cup \{u_1\})}\right)
$$
$$
= \frac{1 - f_v(A \cup \{u_1, u_2\})}{1 - f_v(A)}
$$
$$
= (1 - p_v(u_2, A))(1 - p_v(u_1, A \cup \{u_2\})).
$$

Then the corollary follows from Proposition 2.3.1. $\qquad\square$

Corollary 2.3.3 explores a relationship between the GT model and the GC model. Actually, they are much closer to each other.

Theorem 2.3.4 *Consider a directed social network $G = (V, E)$. The following holds.*

(a) For any GT model m, there exists a GC model m' such that for any seed set S, $\sigma_m(S) = \sigma_{m'}(S)$.

(b) For any GC model m, there exists a GT model m' such that for any seed set S, $\sigma_m(S) = \sigma_{m'}(S)$.

Proof (a) In the GT model m, suppose that each node v has a threshold function f_v. For any in-neighbor u of v, and any subset $A \subseteq N^-(v) \setminus \{u\}$, define

$$
p_v(u, A) = \frac{f_v(A \cup \{u\}) - f_v(A)}{1 - f_v(A)}.
$$

By Corollary 2.3.3, $p_v(u, A)$ induces a GC model m'. We are going to prove that for any seed set S, $\sigma_m(S) = \sigma_{m'}(S)$. To do so, let $A_0 = S$, and for $k \geq 1$ let A_k be the set of active nodes at the end of step k. We will show, by induction on k, that A_k has the same distribution in the process of m and the process of m'. For $k = 0$, this is clearly true. For $k \geq 1$, it is sufficient to prove that

$$
\Pr[u \in A_{k+1} \mid u \notin A_k]
$$

has the same value in the process of models m and m'.

First, consider the process of model m. Since $u \notin A_k$, θ_u must lie in interval $(f_u(A_{k-1} \cap N^-(u)), 1]$. In order that $u \notin A_k$ and $u \in A_{k+1}$, θ_u must lie in interval $(f_u(A_{k-1} \cap N^-(u)), f_u(A_k \cap N^-(u))]$. Therefore,

$$
\Pr[u \in A_{k+1} \mid u \notin A_k] = \frac{f_u(A_k \cap N^-(u)) - f_u(A_{k-1} \cap N^-(u))}{1 - f_u(A_{k-1} \cap N^-(u))}.
$$

Next, consider the process of the GC model m'. For simplicity of notation, denote $A_k^u = A_k \cap N^-(u)$. Suppose $A_k^u \setminus A_{k-1}^u = \{w_1, w_2, \ldots, w_h\}$. Then

$$\Pr[u \in A_{k+1} \mid u \notin A_k]$$
$$= 1 - (1 - p_u(w_1, A_{k-1}^u))(1 - p_u(w_2, A_{k-1}^u \cup \{w_1\}))$$
$$\cdots (1 - p_u(w_h, A_{k-1}^u \cup \{w_1, \ldots, w_{h-1}\}))$$
$$= 1 - \left(1 - \frac{f_u(A_{k-1}^u \cup \{w_1\}) - f(A_{k-1}^u)}{1 - f_u(A_{k-1}^u)}\right)$$
$$\left(1 - \frac{f_u(A_{k-1}^u \cup \{w_1, w_2\}) - f(A_{k-1}^u \cup \{w_1\})}{1 - f_u(A_{k-1}^u \cup \{w_1\})}\right)$$
$$\cdots \left(1 - \frac{f_u(A_{k-1}^u \cup \{w_1, \ldots, w_h\}) - f(A_{k-1}^u \cup \{w_1, \ldots, w_{h-1}\})}{1 - f_u(A_{k-1}^u \cup \{w_1, \ldots, w_{h-1}\})}\right)$$
$$= \frac{f_u(A_{k-1}^u \cup \{w_1, \ldots, w_h\}) - f(A_{k-1}^u)}{1 - f_u(A_{k-1}^u)}$$
$$= \frac{f_u(A_k \cap N^-(u)) - f_u(A_{k-1} \cap N^-(u))}{1 - f_u(A_{k-1} \cap N^-(u))}.$$

(b) Given a GC model m with function $p_v(u, A)$ at each node v, define a threshold function

$$f_v(A) = 1 - \prod_{i=1}^{k}(1 - p_v(u_i, A_{i-1})),$$

where we assume $A = \{u_1, \ldots, u_k\}$, denote $A_i = \{u_1, \ldots, u_i\}$ for $i = 1, \ldots, k$, and let $A_0 = \emptyset$. This threshold is well-defined because the condition (*) holds for the GC model. Let m' be the GT mode with threshold function f_v for node v. We next prove that for any seed set S, $\sigma_m(S) = \sigma_{m'}(S)$. To do so, let $A_0 = S$, and for $k \geq 1$ let A_k be the set of active nodes at the end of step k. We will show, by induction on k, that A_k has the same distribution in the process of m and the process of m'. This is true for $k = 0$. For $k \geq 1$, it is sufficient to prove that

$$\Pr[u \in A_{k+1} \mid u \notin A_k]$$

has the same value in the process of models m and m'.

First, consider the process of the GC model m. Denote $A_k^u = A_k \cap N^-(u)$. Assume $A_k^u \setminus A_{k-1}^u = \{w_1, \ldots, w_h\}$. Then we have

$$\Pr[u \in A_{k+1} \mid u \notin A_k]$$
$$= 1 - (1 - p_u(w_1, A_{k-1}^u))(1 - p_u(w_2, A_{k-1}^u \cup \{w_1\}))$$
$$\cdots (1 - p_u(w_h, A_{k-1}^u \cup \{w_1, \ldots, w_{h-1}\})).$$

Now, consider the process of the GT model m'. Assume

$$A_k^u = \{w_1', \ldots, w_{g+h}'\}$$

where $w_{g+i}' = w_i$ for $i = 1, 2, \ldots, h$. Then

$$\Pr[u \in A_{k+1} \mid u \notin A_k]$$
$$= \frac{f_u(A_k^u) - f_u(A_{k-1}^u)}{1 - f_u(A_{k-1}^u)}$$
$$= [(1 - \prod_{i=1}^{g+h}(1 - p_u(w_i', \{w_1', \ldots, w_{i-1}'\})))) -$$
$$(1 - \prod_{i=1}^{g}(1 - p_u(w_i', \{w_1', \ldots, w_{i-1}'\}))))]/[\prod_{i=1}^{g}(1 - p_u(w_i', \{w_1', \ldots, w_{i-1}'\}))]$$
$$= 1 - \prod_{i=g+1}^{g+h}(1 - p_u(w_i', \{w_1', \ldots, w_{i-1}'\}))$$
$$= 1 - \prod_{i=1}^{h}(1 - p_u(w_i, A_{k-1}^u \cup \{w_1, \ldots, w_{i-1}\})).$$

The theorem is proved. $\qquad\square$

2.4　Kempe-Kleinberg-Tardös Conjecture

Let us now look at a negative result.

Theorem 2.4.1 *For and $0 < \varepsilon < 1$, there exists a class of social networks with a GT model such that there does not exist a polynomial-time approximation algorithm for the influence maximization problem with performance ratio $n^{\varepsilon-1}$ unless $NP = P$, where n is the number of nodes in the input social network and the computational time for the influence spread is not counted.*

To show this result, let us prove a lemma.

Lemma 2.4.2 *The set cover problem (Problem 2.1.3) is still NP-complete in the special case that $1/2 \cdot |X| \le |\mathcal{C}| \le 2|X|$ and $1/3 \cdot |X| \le k \le |X|$.*

Proof Notice that the vertex-cover problem (Research Problem 2.1.8) can be considered as a special case of the set-cover problem when we treat all edges as elements and each vertex as a subset of edges incident to the vertex. In a popular proof of the NP-completeness of the vertex-cover problem [75] (see Fig.2.8), a 3CNF F with n variables and m clauses is transformed to a graph with $2n + 3m$ vertices and $n + 6m$

Fig. 2.8 An NP-complete special case of the vertex-cover problem

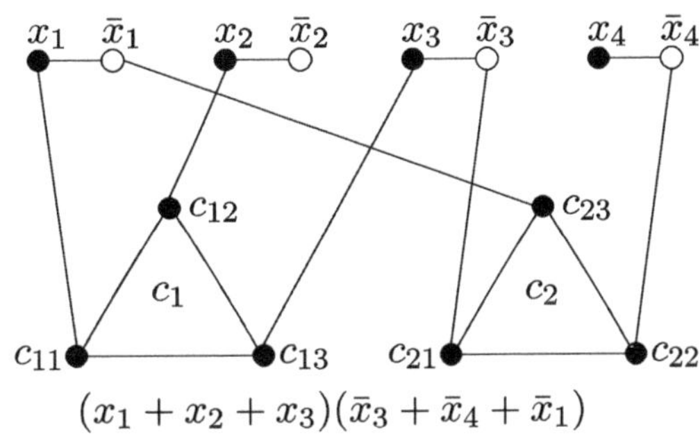

Fig. 2.9 A tripartite graph G

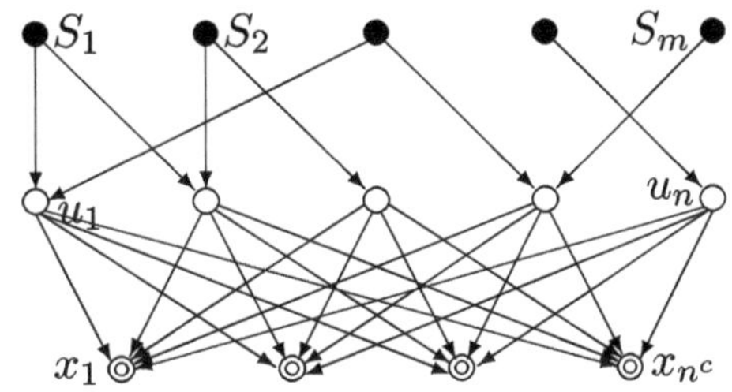

edges, and moreover $k = n + 2m$. This means that the set cover problem is NP-complete in the special case that $1/2 \cdot |X| = 1/2 \cdot (n + 6m) \le |\mathcal{C}| = 2n + 3m \le 2|X|$ and $1/3 \cdot |X| \le k \le |X|$. $\square$

Proof of Theorem 2.4.1. Consider an instance of the set cover problem, consisting of a finite set $X = \{u_1, u_2, \ldots, u_n\}$, a collection of subsets of X, $\mathcal{C} = \{S_1, S_2, \ldots, S_m\}$, and a positive integer k with $n/2 \le m \le 2n$ and $n/3 \le k \le n$. Construct a tripartite graph G as shown in Fig. 2.9. G contains three disjoint node sets $\mathcal{C}$, X, and $N = \{x_1, x_2, \ldots, x_{n^c}\}$, where c is a constant which will be chosen later. Edges in G are constructed as follows.

- For any $S_i \in \mathcal{C}$ and $u_j \in X$, there is an arc (S_i, u_j) if and only if $u_j \in S_i$.
- For any $u_j \in X$ and $x_h \in N$, there is an arc (u_j, x_h).
- No other edge exists.

Define threshold functions

$$f_{u_j}(S) = \begin{cases} 0, & \text{if } S = \emptyset, \\ 1, & \text{if } S \ne \emptyset, \end{cases}$$

and

$$f_{x_h}(S) = \begin{cases} 0, & \text{if } S \ne X, \\ 1, & \text{if } S = X. \end{cases}$$

Those threshold functions form a GT model. Then

(1) If $\mathcal{C}$ contains k subsets covering X, then the influence maximization problem has maximum value $k + n + n^c$.
(2) If $\mathcal{C}$ does not contain k subsets covering X, then the influence maximization problem has maximum value at most $k + n - 1$.

Notice that G has totally $m + n + n^c = n'$ nodes. Suppose a polynomial-time $(n')^{\varepsilon-1}$-approximation exists. Then in case (1), the algorithm will produce a solution with objective function value at least

$$\frac{k + n + n^c}{(n')^{1-\varepsilon}} \geq \frac{n^c}{(4n^c)^{1-\varepsilon}} \geq \frac{n^{c\varepsilon}}{4}.$$

Choosing $c > 2/\varepsilon$ and $n > 8$, we have

$$\frac{n^{c\varepsilon}}{4} > n + k - 1.$$

While in case (2), the algorithm will produce a solution with objective function value at most $n + k - 1$. Thus, this algorithm can solve the set cover problem for $n \geq 8$, which implies $NP = P$. $\qquad\square$

Notice that in the proof of Theorem 2.4.1, f_{u_j} is submodular, but f_{x_h} is not. What would happen if every threshold function is submodular? Kempe, Kleinburg, and Tadös [211] made the following conjecture, which was proved by Mossel and Roch in [297].

Theorem 2.4.3 *If in a GT model m, f_v is monotone nondecreasing and submodular for every node v, then the influence spread, $\sigma_m(A)$, is also monotone nondecreasing and submodular with respect to the seed set A.*

The proof in [297] got a little simplification [439] by using the following result [80].

Lemma 2.4.4 *Consider a real function f defined on 2^X. Then f is submodular if and only if for any two subsets $A \subset B$ and any element $x \in X - B$,*

$$\Delta_x f(A) \geq \Delta_x f(B),$$

where $\Delta_x f(A) = f(A \cup \{x\}) - f(A)$. Moreover, f is monotone nondecreasing if and only if for any two subsets $A \subset B$ and any element $x \in B - A$,

$$\Delta_x f(A) \geq \Delta_x f(B).$$

To start the proof of the conjecture, we first extend the domain of each threshold function f_v from $2^{N^-(v)}$ to 2^V by setting $f_v(A) = f_v(A \cap N^-(v))$ for any $A \subseteq V$. Let $Q(S)$ denote the process of GT with seed set S. Let

$$(T_t)_{t=0}^{2n-1} \sim Q(R_1, R_2)$$

denote the process as follows:

1. For each $v \in V$, pick up a threshold θ_v uniformly and independently from $[0, 1]$. Set $T_0 = R_1$.
2. $(T_t)_{t=0}^{n-1} \sim Q(R_1 \mid \theta)$,
   $(T_t)_{t=n}^{2n-1} \sim Q(T_{n-1} \cup R_2 \mid \theta)$.

where $Q(S \mid \theta)$ denotes the process of $Q(S)$ after threshold $\theta = (\theta_v)_{v \in V}$ is selected. Let

$$(S_t)_{t=0}^{2n-1} \sim Q_-(R_1, R_2)$$

denote the following process, called *antisense process*:

1. For each $v \in V$, pick up a threshold θ_v uniformly and independently from $[0, 1]$. Set $S_0 = R_1$.
2. $(S_t)_{t=0}^{n-1} \sim Q(R_1 \mid \theta)$.
3. Set $S_n = S_{n-1} \cup R_2$ and for $t = n+1, \ldots, 2n-1$, set

$$S_t = S_{t-1} \cup \{v \in V \setminus S_{t-1} \mid f_v(S_{t-1}) - f_v(S_{n-1}) \geq 1 - \theta_v\}.$$

The following three lemmas were proved by Mossel and Roch in [272].

Lemma 2.4.5 *(Piecemeal Growth) Let* (R_1, R_2) *be a partition of S. Let*

$$(S_t)_{t=0}^{n-1} \sim Q(S \mid \theta) \text{ and } (T_t)_{t=0}^{2n-1} \sim Q(R_1, R_2 \mid \theta).$$

Then $S_{n-1} = T_{2n-1}$.

Proof Let $(T_t')_{t=0}^{2n-1} \sim Q(S, \emptyset \mid \theta)$ and $(T_t'')_{t=0}^{2n-1} \sim Q(\emptyset, S \mid \theta)$. Then $T_{2n-1}'' \subseteq T_{2n-1} \subseteq T_{2n-1}'$ and $T_{2n-1}' = S_{n-1} = T_{2n-1}''$. It follows that $S_{n-1} = T_{2n-1}$. $\qquad\square$

Lemma 2.4.6 *(Need-to-Know Representation) Let*

$$(S_t)_{t=0}^{2n-1} \sim Q_-(R_1, R_2) \quad and \quad (T_t)_{t=0}^{2n-1} \sim Q(R_1, R_2).$$

Then S_{2n-1} *and* T_{2n-1} *have the same distribution.*

Proof Antisense process is equivalent to choose θ as follows:

$$\theta_v' = \begin{cases} \theta_v & \text{if } v \in S_{n-1} \\ f_v(S_{n-1}) + 1 - \theta_v & \text{if } v \notin S_{n-1}. \end{cases}$$

For $v \notin S_{n-1}$, as θ_v varies from $f_v(S_{n-1})$ to 1, θ_v' goes from 1 to $f_v(S_{n-1})$. Therefore, both θ_v and θ_v' have the uniform distribution in $[f_v(S_{n-1}), 1]$. $\qquad\square$

Lemma 2.4.7 *Let f be a monotone decreasing and submodular function. Assume $X \subseteq X'$ and $Y \subseteq Y'$. Then*

$$f(X \cup Y') - f(X) \geq f(X' \cup Y) - f(X').$$

Proof Since f is monotone nondecreasing and submodular, we have

$$f(X \cup Y') - f(X) \geq f(X' \cup Y') - f(X') \geq f(X' \cup Y) - f(X').$$

The lemma is proved. $\qquad\square$

Now, we are ready to give a natural and simple proof of Theorem 2.4.3.

Proof of Theorem 2.4.3. First, consider

$$\begin{aligned}
(A_t)_{t=0}^{n-1} &\sim Q(A \mid \theta), \\
(S_t)_{t=0}^{n-1} &\sim Q(A \cup \{x\} \mid \theta), \\
(T_t)_{t=0}^{2n-1} &\sim Q(A, \{x\} \mid \theta), \\
(A_t)_{t=0}^{2n-1} &\sim Q_-(A, \{x\} \mid \theta).
\end{aligned}$$

Clearly,

$$\sigma_m(A) = \int_0^1 \cdots \int_0^1 |A_{n-1}| \prod_{v \in V} d\theta_v \text{ and } \sigma_m(A \cup \{x\})$$

$$= \int_0^1 \cdots \int_0^1 |S_{n-1}| \prod_{v \in V} d\theta_v.$$

By Lemma 2.4.5, $|S_{n-1}| = |T_{2n-1}|$. By Lemma 2.4.6,

$$\int_0^1 \cdots \int_0^1 |T_{2n-1}| \prod_{v \in V} d\theta_v = \int_0^1 \cdots \int_0^1 |A_{2n-1}| \prod_{v \in V} d\theta_v,$$

i.e.,

$$\sigma_m(A \cup \{x\}) = \int_0^1 \cdots \int_0^1 |A_{2n-1}| \prod_{v \in V} d\theta_v.$$

Therefore,

$$\sigma_m(A \cup \{x\}) - \sigma_m(A) = \int_0^1 \cdots \int_0^1 (|A_{2n-1}| - |A_{n-1}|) \prod_{v \in V} d\theta_v.$$

Now, let $A \subset B$ and $x \in V \setminus B$. Consider

$$(A_t)_{t=0}^{2n-1} \sim Q_-(A, \{x\} \mid \theta)$$
$$(B_t)_{t=0}^{2n-1} \sim Q_-(B, \{x\} \mid \theta).$$

Then $A_{n-1} \subseteq B_{n-1}$ and

$$\sigma_m(A \cup \{x\}) - \sigma_m(A) = \int_0^1 \cdots \int_0^1 (|A_{2n-1}| - |A_{n-1}|) \prod_{v \in V} d\theta_v,$$
$$\sigma_m(B \cup \{x\}) - \sigma_m(B) = \int_0^1 \cdots \int_0^1 (|B_{2n-1}| - |B_{n-1}|) \prod_{v \in V} d\theta_v.$$

To show Theorem 2.4.3, by Lemma 2.4.4, it suffices to prove

$$|A_{2n-1}| - |A_{n-1}| \geq |B_{2n-1}| - |B_{n-1}|. \tag{2.1}$$

We claim that for $t = n, \ldots, 2n - 1$, the following properties hold.

$(1, t)$ $B_t \setminus B_{n-1} \subseteq A_t \setminus A_{n-1}$.

$(2, t)$ $f_v(B_t) - f_v(B_{n-1}) \leq f_v(A_t) - f_v(A_{n-1})$ for $v \in V$.

This claim can be proved by induction on t. For $t = n$, $B_n \setminus B_{n-1} = \{x\} = A_n \setminus A_{n-1}$. So, $(1, n)$ holds. Combining $A_{n-1} \subseteq B_{n-1}$ with Lemma 2.4.7, property $(1, t)$ implies property $(2, t)$ for any $t = n, \ldots, 2n - 1$. Moreover, by the definition of $Q_-(A, \{x\} \mid \theta)$ and $Q_-(B, \{x\} \mid \theta)$, property $(2, t)$ implies property $(1, t + 1)$. Therefore, our claim holds.

Then the theorem follows by observing that the claim for property $(1, 2n - 1)$ implies inequality (2.1). $\qquad\square$

Next, we use Theorem 2.4.3 to study influence spread in the deterministic model, the IC model, and the LT model, again.

The deterministic model can be considered as a special of the GT model, in which the threshold function at each node v is

$$f_v(A) = \begin{cases} 0 & \text{if } A = \emptyset, \\ 1 & \text{if } A \neq \emptyset. \end{cases}$$

Clearly, f_v is monotone nondecreasing and submodular and so is the influence spread.

In the LT model, the threshold function f_v at each node v is linear with nonnegative coefficients and hence is monotone nondecreasing and submodular, so is the influence spread.

The IC model is equivalent to a GT model with threshold function at each node v,

$$f_v(A) = 1 - \prod_{u \in A}(1 - p_{uv}).$$

For any two subsets A and B with $A \subset B$, and any node $x \in N^-(v) \setminus B$, we have

$$\Delta_x f_v(A) = p_{xv} \cdot \prod_{u \in A}(1 - p_{uv}) \geq p_{xv} \prod_{u \in B}(1 - p_{uv}) = \Delta_x f_v(B),$$

and for any $x \in B \setminus A$, we have

$$\Delta_x f_v(A) = p_{xv} \cdot \prod_{u \in A}(1 - p_{uv}) \geq 0 = \Delta_x f_v(B).$$

Hence, f_v is monotone nondecreasing and submodular, so is the influence spread. We next study the Triggering model.

Triggering Model: This model is a probabilistic diffusion process consisting of discrete steps. Consider a directed graph $G = (V, E)$. Each node has two states, *active* and *inactive*. Initially, each node v independently selects a set T_v from $2^{N^-(v)}$ with a probabilistic distribution. Then block all arcs which are not from T_v to v and keep all arcs from T_v to v. Choose a seed set S and proceed in the same way as in the deterministic model.

There are several interesting facts about the triggering model.

(1) Based on its definition, it is easy to see that the influence spread in the triggering model can be represented as

$$\sigma_T(S) = \sum_{H \subseteq G} \Pr[H]\sigma_D^H(S),$$

where $\sigma_D^H(S)$ denotes the influence spread in subgraph H under deterministic model and $\Pr[H]$ denotes the probability that all arcs in H are alive and all arcs not in H are blocked. Hence, its influence spread is monotone nondecreasing and submodular.

(2) The IC model is a special case of the triggering model. To see it, choose triggering set T_v with probability $\prod_{u \in T_v} p_{uv}$.

(3) The LT model is a special case of the triggering model. To see it, choose triggering set T_v with probability

$$\Pr[T_v] = \begin{cases} 0 & \text{if } |T_v| \neq 1, \\ b_{uv} & \text{if } T_v = \{u\}. \end{cases}$$

In the definition of the triggering model, triggering set T_v is selected independently at each node v. In general, we may remove "independently."

General Triggering (GTrig) Model: This model is a probabilistic diffusion process consisting of discrete steps. Consider a directed graph $G = (V, E)$. Each node has two states, *active* and *inactive*. Initially, at every node v, select a set T_v from $2^{N^-(v)}$ with a probabilistic distribution. Then block all arcs which are not from T_v to v and keep all arcs from T_v to v. Choose a seed set S and proceed as in the deterministic model.

For the GTrig model, we can obtain an additional property.

Theorem 2.4.8 *A probabilistic diffusion model m is a GTrig model if and only if its influence spread can be represented as*

$$\sigma_T(S) = \sum_{H \subseteq G} \Pr[H]\sigma_D^H(S),$$

where $\sigma_D^H(S)$ denotes the influence spread in subgraph H under deterministic model and $\Pr[H]$ denotes the probability that all arcs in H are alive and all arcs not in H are blocked.

Proof It is sufficient to note that a subgraph H is given if and only if all triggering sets at all nodes are selected. Therefore, in the definition of the GTig model, a subgraph is randomly selected in a probabilistic distribution which is equivalent to that all triggering sets randomly selected. $\square$

The Triggering model or the GTrig model may not be a special case of the GT model. However, it can be represented as a linear combination of GT models. In fact, the deterministic model is a special case of the GT model and we already see that the Triggering model and the GTrig model can be written as linear combination of the deterministic model.

2.5 Complexity of Influence Spread

In this section, we discuss the computational complexity of influence spread. Given a social network $G = (V, E)$ with an information diffusion model m, the influence spread is a function with respect to seed set S, i.e., $\sigma_m(S)$. How hard is it to compute $\sigma_m(S)$? When m is the deterministic model, $\sigma_m(S)$ is easy to compute. However, when m is the IC model or the LT model, computing $\sigma_m(S)$ is #P-hard. What is #P? What is #P-hard? The reader may refer to [79] for details. Here, let us make a brief explanation.

#P is a class of functions, for each function $f(x)$ in this class, there exists a polynomial-time nondeterministic Turing machine M such that on input x, M has exactly $f(x)$ paths ending at accepting state.

For two functions $f(x)$ and $g(y)$, if there is a polynomial-time algorithm for computing $f(x)$ which uses an algorithm for computing $g(y)$ as subroutine, then we say that f is polynomial-time Turing reducible to g, denoted by $f \leq_T^p g$. Clearly, this reduction has the property that: if g is polynomial-time computable, then f is also polynomial-time computable.

A function g is said to be #P-hard if for every $f \in$ #P, $f \leq_T^p g$. If a #P-hard function g is polynomial-time computable, then every function f in #P is polynomial-time computable. This means that for any #P function $f(x)$, there exists a polynomial-time nondeterministic Turing machine M, such that there is a polynomial-time algorithm to compute how many accepting paths that M has on input x. If the number of accepting paths is at least one, then it means "accept"; otherwise, it means "reject." Thus, giving a polynomial-time algorithm to a #P-hard problem means that every problem in NP is polynomial-time solvable. This is unlikely to happen. Therefore, we usually say that a #P-hard problem is unlikely to be polynomial-time computable.

Actually, a #P-hard problem seems harder to be solved than an NP-hard problem. For example, given a graph G, it is NP-hard to find out whether G contains a Hamiltonian cycle, and it is #P-hard to find out how many cycles that G contains.

To prove the #P-hardness of a function g, it is sufficient to select a #P-hard function f and prove $f \leq_T^p g$. This is because the reduction $\leq_T^p$ is transitive, that is, if $h \leq_T^p f$ and $f \leq_T^p g$, then $h \leq_T^p g$. For a #P-hard function f, any $h \in$ #P has $h \leq_T^p f$. Thus, $f \leq_T^p$ implies $h \leq_T^p g$ for every $h \in$ #P.

Now, let us show the #P-hardness of the influence spread in the IC and LT models.

Theorem 2.5.1 *(Chen et al. [50]) The influence spread is #P-hard in the IC model.*

Proof It is pointed out in [406] that the following problem is #P-hard (actually #P-complete since in addition, the problem belong to #P class.)

Problem 2.5.2 *(s-t Connectedness)* Given a directed graph $G = (V, E)$ and two nodes s and t, count the number of subgraphs which contains a path from s to t.

Assign every arc $(u, v) \in E$ with label $p_{uv} = 1/2$. Add a new node t' and a new arc (t, t') with label $p_{t,t'} = 1$. Denote the resulting graph by G' (Fig. 2.10). Let $N(s, t, G)$ be the number of subgraphs of G which contains a path from s to t. Notice that every subgraph of G appears with probability $1/2^{|E|}$. Then t' is activated by s with probability $N(s, t, G) \cdot 2^{-|E|}$. Therefore,

$$\sigma_{IC}^{G'}(\{s\}) = \sigma_{IC}^{G}(\{s\}) + N(s, t, G) \cdot 2^{-|E|}$$

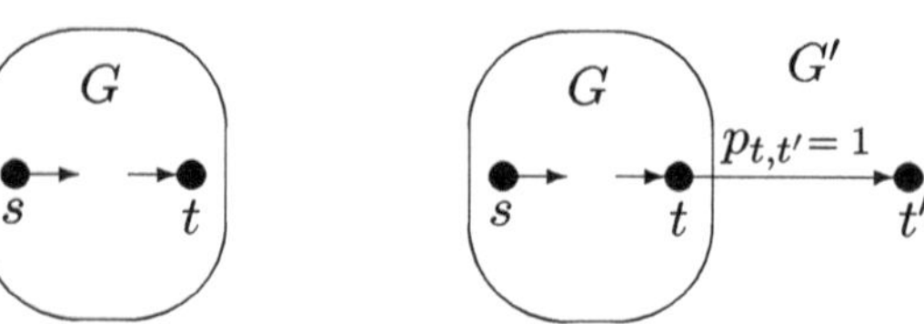

Fig. 2.10 The construction from graph G to graph G'

Fig. 2.11 A simple example

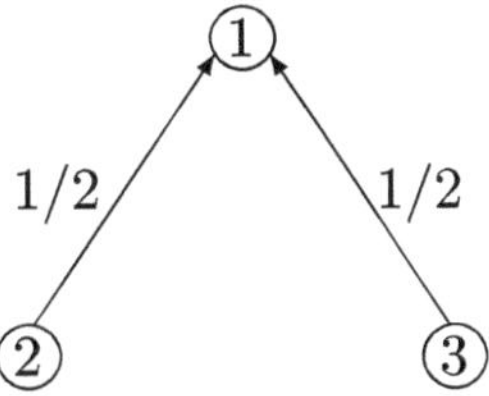

where $\sigma_{IC}^{G}(\{s\})$ denotes the influence spread in network G with the IC model when $\{s\}$ is the seed set. This means that the s-t connectedness problem is polynomial-time Turing reducible to the influence spread computation in the IC model. Hence, the influence spread is $\#P$-hard in the IC model. $\square$

Can the above argument be used for the LT model? The answer is NO. This is because in the LT or MC model, the probability for the appearance of a subgraph should be calculated in a different way. To see it, consider a simple example in Fig. 2.11. The probability of two edges appearing is 0 (since MC is a mutually-exclusive model) and the probability of exactly one edge appearing is $1/2$. So, the following proof employs a different argument.

Theorem 2.5.3 (Chen et al. [53]) *The influence spread is $\#P$-hard in the LT model.*

Proof It is pointed out in [406] that the following problem is $\#P$-hard.

Problem 2.5.4 (s-t Paths) Given a directed graph $G = (V, E)$ and two nodes s and t, count the number of simple paths connecting s to t.

Let d be the maximum in-degree of G. For a number $w \in (0, 1/d]$, assign every arc $(u, v) \in E$ with label $p_{uv} = w$. Add a new node t' and a new arc (t, t') with label $p_{t,t'} = 1$. Denote the resulting graph as G'_w. Let α_i be the number of simple paths from s to t in G with length i. Then by the special property of the LT (or MC) model (see Proposition 2.1.10), the probability that node t' is activated by s is $\sum_{i=1}^{n-1} \alpha_i w^i$ where $n = |V|$. Denote $f(w) = \sum_{i=1}^{n-1} \alpha_i w^i$. Then

$$\sigma_{LT}^{G'_w}(\{s\}) = \sigma_{LT}^{G_w}(\{s\}) + f(w). \tag{2.2}$$

This means that $f(w)$ can be computed by computing influence spread in the LT model.

For the s-t paths problem, the goal is to compute $f(1)$. However, w is not allowed to take value 1 unless the maximum in-degree of G is 1 (this is due to the requirement of the LT model). We use Lagrange interpolation to circumvent this trouble. Notice that $f(w)$ is a polynomial of degree at most $n - 1$, which can be uniquely determined by n distinct pairs $(w_i, f(w_i))$. So, we may select n distinct numbers $w_1, \ldots, w_n$ from $(0, 1/d)$ and compute $f(w_k)$ for $1 \le k \le n$ by using (2.2). Then determine polynomial $f(w)$ by Lagrange interpolation.

$$f(w) = \sum_{k=1}^{n} f(w_k) \cdot \frac{(w - w_1) \cdots (w - w_{k-1})(w - w_{k+1}) \cdots (w - w_n)}{(w_k - w_1) \cdots (w_k - w_{k-1})(w_k - w_{k+1}) \cdots (w_k - w_n)}.$$

Finally, compute $f(1)$. Therefore, the number of s-t simple paths in G can be computed by calculating the influence spread in the LT model. $\qquad\Box$

2.6 Continuous-Time IC Model

Social influence is an important topic in the study of social science. In fact, everyone tends to affect others' attitudes, beliefs, or behavior to meet the demands of a social environment. There are several types of social influence, such as conformity, compliance, and obedience.

- Conformity: Change behavior in response to a group.
- Compliance: Change behavior in response to a request.
- Obedience: Change behavior in response to an authority figure.

There are also several ways to study social influence based on its various aspects.

The information diffusion model plays a fundamental role in the study of the computational aspect of social influence. In fact, knowing the information diffusion model is the initial step in most of the research work on computational social networks, whereas there are only very few exceptions [390].

In previous sections, we studied a large class of discrete information diffusion models. However, we may encounter new models in the study of new problems. For example, when studying time delay in information diffusion, we may need to consider the continuous-time IC model.

Definition 2.6.1 (*Continuous-Time IC (CTIC) Model* [350]) Consider the (directed) social network $G = (V, E)$. In this model, each link $(u, v) \in E$ is associated with a time-delay parameter $r_{uv} > 0$ and a diffusion parameter $\kappa_{uv} \in (0, 1)$. Before the diffusion process, all nodes are set in inactive state. The diffusion process unfolds in continuous-time t as follows.

- Initially, at time 0, activate a subset of nodes, S, called a seed set.
- Suppose that u is a node that becomes active at time t. Then u has a single chance to activate each inactive out-neighbor v.
- First, select a delay-time δ from the exponential distribution with parameter r_{uv}. Then attempt to activate the inactive node v with the succeeding probability κ_{uv}. If success, then v becomes active at time $t + \delta$.

- If inactive node v is activated at time $t + \delta$ by more than one active in-neighbors, then they attempt to do activation in sequence independently in an arbitrary order.
- The process terminates if no more activation is possible,

Let $\phi(S)$ denote the number of active nodes at the end of the process and $\sigma(S)$ the expectation of $\phi(S)$, i.e., the influence spread.

The outcome of the random process for the CTIC model is described by the following data.

$$D =< D(t) \mid t \in \mathcal{T} >$$

where $D(t)$ is the set of nodes, each of which is activated at time t, so that

$$D(t) \neq \emptyset \forall t \in \mathcal{T},$$

$$\mathcal{T} =< t \mid D(t) \neq \emptyset >,$$

and

$$\phi(S) = |\cup_{t \in \mathcal{T}} D(t)|.$$

Define

$$r = (r_{uv})_{(u,v) \in E} \text{ and } \kappa = (\kappa_{uv})_{(u,v) \in E}.$$

Saito et al. [350] study a problem of learning r and κ from an observed data set of independent information diffusion results,

$$\mathcal{D}_M = \{D_1, D_2, \ldots, D_M\}.$$

Definition 2.6.2 (*Learning CTIC* [350]) Given a sequence of data, $\mathcal{D}_M$ generated from the independent information diffusion process for the CTIC model, estimate r and κ.

Saito et al. [350] present an approach consisting of two steps.

- Step 1. Establish the likelihood function $\mathcal{L}(r, \kappa, \mathcal{D}_M)$.
- Step 2. They intend to compute r and κ by maximizing the likelihood function. However, it is not easy to solve this maximization problem. Therefore, an iterated algorithm is designed to produce a sequence of approximate values for r and κ. This sequence makes the value of the likelihood function monotone nondecreasing and hence convergent to a certain level.

Now, let us first establish the likelihood function. For every node v, denote

$$B(v) = \{u \mid (u, v) \in E\}$$

and by $t_{m,v}$ the time at which v becomes active in the data D_m, that is,

$$v \in D_m(t_{m,v}).$$

For any time t and data D_m, denote by $C_m(t)$ the set of nodes activated before time t in D_m, i.e.,

$$C_m(t) = \cup_{t' < t, t' \in \mathcal{T}_m} D_m(t').$$

Then, the probability density that a node $u \in B(v) \cap C_m(t_{m,v})$ activates v at time $t_{m,v}$ can be expressed as

$$\mathcal{A}_{m,uv} = \kappa_{uv} r_{uv} \exp(-r_{uv}(t_{m,v} - t_{m,u})).$$

Thus, the probability that node v is not activated by any node $u \in B(v) \cap C_m(t_{m,v})$ within the time period $[t_{m,u} - t_{m,v}]$ can be expressed as

$$
\begin{aligned}
\mathcal{B}_{m,uv} &= 1 - \kappa_{uv} \int_{t_{m,u}}^{t_{m,v}} r_{uv} \exp(-r_{uv}(t - t_{m,u}))dt \\
&= \kappa_{uv} r_{uv} \exp(-r_{uv}(t_{m,v} - t_{m,u})) + (1 - \kappa_{uv}).
\end{aligned}
\tag{2.3}
$$

Due to continuous property, simultaneous activations on a node can be considered as a sequence of individual activations. Let $h_{m,v}$ be the probability density that node v is activated at time $t_{m,v}$. Then, we have

$$
\begin{aligned}
h_{m,v} &= \sum_{u \in B(v) \cap C(t_{m,v})} \mathcal{A}_{m,uv} \prod_{x \in (B(v) \cap C(t_{m,v})) \backslash \{u\}} \mathcal{B}_{m,xv} \\
&= \prod_{x \in B(v) \cap C(t_{m,v})} \mathcal{B}_{m,xv} \sum_{u \in B(v) \cap C(t_{m,v})} \mathcal{A}_{m,uv} \mathcal{B}_{m,uv}^{-1}
\end{aligned}
\tag{2.4}
$$

We now consider a node w such that $(v, w) \in E$ for $v \in C(\mathcal{T}_m)$, but $w \notin D_m$. Let $g_{m,vw}$ denote the probability that node w is not activated by node v within the observed time period $[t_m, T_m]$. Then, we have

$$g_{m,vw} = \kappa_{vw} \exp(-r_{vw}(T_m - t_{m,v})) + (1 - \kappa_{vw}).$$

We may assume that each information diffusion process finished much earlier than the final observed time, i.e., $T_m \gg \max\{t \mid D_m(t) \neq \emptyset\}$. In other words, we assume $T_m \to \infty$ and hence

$$g_{m,vw} = 1 - \kappa_{vw}.$$

Finally, we can establish the likelihood function.

$$\mathcal{L}(r, \kappa, \mathcal{D}_M) = \prod_{m=1}^{M} \left(\prod_{t \in \mathcal{T}_m} \prod_{v \in D_m(t)} h_{m,v} \prod_{v \in D_m} \prod_{w \in F(v) \setminus D_m} g_{m,vw} \right), \tag{2.5}$$

where $F(v) = \{w \mid (v, w) \in E\}$. $\mathcal{L}(r, \kappa, \mathcal{D}_M)$.

Next, we describe the second step, that is, derive an iterated method for finding r and κ to maximize the likelihood function. To do so, for $v \in D_m$ and $u \in B(v) \cap C_m(t_{m,v})$, define

$$\alpha_{m,uv} = \frac{\mathcal{A}_{m,uv} \mathcal{B}_{m,uv}^{-1}}{\sum_{x \in B(v) \cap C_m(t_{m,v})} \mathcal{A}_{m,uv} \mathcal{B}_{m,xv}^{-1}}.$$

Denote by $\bar{\mathcal{A}}_{m,uv}$, $\bar{\mathcal{B}}_{m,uv}$, $\bar{h}_{m,v}$, and $\bar{\alpha}_{m,uv}$ the values of $\mathcal{A}_{m,uv}$, $\mathcal{B}_{m,uv}$, $h_{m,v}$, and $\alpha_{m,uv}$ computed by using $\bar{r}$ and $\bar{\kappa}$, respectively. Then

$$\log \mathcal{L}(r, \kappa, \mathcal{D}_M) = Q(r, \kappa, \bar{r}, \bar{\kappa}) - H(r, \kappa, \bar{r}, \bar{\kappa})$$

where

$$Q(r, \kappa, \bar{r}, \bar{\kappa}) = \sum_{m=1}^{M} \left(\sum_{t \in \mathcal{T}_m} \sum_{v \in D_M(t)} Q_{m,v} + \sum_{v \in D_m} \sum_{w \in F(v) \setminus D_m} \log(1 - \kappa_{vw}) \right),$$

$$Q_{m,v} = \sum_{u \in B(v) \cap C_m(t_{m,v})} (\log(\mathcal{B}_{m,uv}) + \bar{\alpha}_{m,uv} \log(\mathcal{A}_{m,uv} \mathcal{B}_{m,uv}^{-1})),$$

$$F(v) = \{w \mid (v, w) \in E\},$$

and

$$H(r, \kappa, \bar{r}, \bar{\kappa}) = \sum_{m=1}^{M} \sum_{t \in \mathcal{T}_m} \sum_{v \in D_m(t)} \sum_{u \in B(v) \cap C_m(t_{m,v})} \bar{\alpha}_{m,uv} \log \alpha_{m,uv}.$$

Define

$$\bar{\beta}_{m,uv} = \bar{kappa}_{uv} \exp(-r_{uv}(t_{m,v} - t_{m,u})) / \bar{\mathcal{B}}_{m,uv}.$$

By maximizing $Q(r, \kappa, \bar{r}, \bar{\kappa})$, the following can be obtained.

$$r_{uv} = \frac{\sum_{m \in \mathcal{M}_{uv}^+} \bar{\alpha}_{m,uv}}{\sum_{m \in \mathcal{M}_{uv}^+} (\bar{\alpha}_{m,uv} + (1 - \bar{\alpha}_{m,uv})\bar{\beta}_{m,uv})(t_{m,v} - t_{m,u})},$$

$$\kappa_{uv} = \frac{1}{|\mathcal{M}_{uv}^+| + |\mathcal{M}_{uv}^-|} \cdot \sum_{m \in \mathcal{M}_{uv}^+} (\bar{\alpha}_{m,uv} + (1 - \bar{\alpha}_{m,uv})\bar{\beta}_{m,uv})$$

where

$$\mathcal{M}_{uv}^{+} = \{m \in \{1, \ldots, M\} \mid u, v \in D_m, v \in F(u), t_{m,u} < t_{m,v}\},$$

$$\mathcal{M}_{uv}^{-} = \{m \in \{1, \ldots, M\} \mid u \notin D_m, v \in F(u)\}.$$

Note that by using above formulas to update the values of r and κ, the value of the likelihood function in each iteration is not decreasing. Therefore, it is convergent. However, it is unknown whether this value would converge to the maximum value. This is an open problem.

2.7 Time-Delayed IC Model

Suppose that the diffusion of information from node u to node v can occur only if u and v meet in some way, such as face-to-face, phone call, facetime, and message. Then, although u has the ability to influence v, it has to wait for a meeting to come. Taking into account this scenery, Chen et al. [46] propose a variation of the IC model and a problem of maximization of the time-critical influence as follows.

Definition 2.7.1 (*Time-Delayed IC (TDIC) Model* [46]) Consider the (directed) social network $G = (V, E)$. In this model, each link $(u, v) \in E$ is associated with a meeting probability $m_{uv} \in [0, 1]$ and an influence probability $p_{uv} \in [0, 1]$. Before the diffusion process, all nodes are set in inactive state. The diffusion process consists of discrete times t as follows.

- Initially, at time 0, activate a subset of nodes, S, called a seed set.
- At any subsequent time $t > 0$, suppose that node u is activated. Then u will intend to influence each inactive neighbor in a meeting at time $t + 1$ or later. At any time, the meeting between u and v is held with probability m_{uv}. If the meeting is held for the first time, then u has a single chance to influence v with probability p_{uv}.
- The process terminates if no more activation is possible.

Definition 2.7.2 (*Time-Critical Influence Maximization* [46]) For a deadline $\tau \in \mathbb{Z}$ and seed set S, let $\sigma_\tau(S)$ denote the expected number of active nodes at the end of time step τ under the time-delayed IC model. Given $\tau \in \mathbb{Z}_+$ and $k \in \mathbb{Z}_+$, the problem is to find S with $|S| \leq k$ to maximize $\sigma_\tau(S)$.

Theorem 2.7.3 (Chen et al. [46]) *The problem of time-critical influence maximization is NP-hard.*

Proof Let τ be sufficiently large. Then the process would last until no further influence is possible, that is, at the end of the process under the IC model. In this special case, the problem is NP-hard. Therefore, in the general case, it is also NP-hard. $\square$

Theorem 2.7.4 (Chen et al. [46]) *For any $\tau \in \mathbb{Z}_+$, $\sigma_\tau(S)$ is monotone nondecreasing and submodular with respect to S.*

Proof An edge (u, v) is said to be alive if the influence occurs successfully through this edge. Let H be a subgraph of G, consisting of all alive edges. Then H appears with certain probability $\Pr[H]$ and $\sigma_\tau(S)$ can be expressed as

$$\sigma_{H \subseteq G} \Pr[H] \sigma^H(S)$$

where $\sigma^H(S)$ is the number of active nodes when starting from S, the influence process on H for deterministic information diffusion model. Since every $\sigma^H(S)$ is already known to be monotone nondecreasing and submodular with respect to S, so is $\sigma_\tau(S)$. $\qquad\square$

Corollary 2.7.5 *There exists a polynomial-time $(1 - 1/e)$-approximation algorithm for the time-critical influence maximization.*

Exercises

1. Consider the positive influence diffusion model as follows: Let $G = (V, E)$ be a directed graph. Each node has two states, active and inactive. Initially, every node is inactive. To start, activate a subset of nodes, called seeds. At each iteration, every inactive node v evaluates how many active in-neighbors exist. If the number of active in-neighbors is not less than the number of inactive in-neighbors, then the inactive node v would be activated, i.e., becomes active. This process repeats until no inactive node becomes a new active one. Please show, by giving the definition of threshold function f_v at each node v, that the positive influence is an instance of the general threshold model.
2. Show that the positive influence model is equivalent to an instance of the triggering model if and only if every node has in-degree at most two.
3. Show the following properties of the triggering model.

   (a) If the triggering set T_v at every node v is fixed, then it is equivalent to an instance of the general threshold model with monotone nondecreasing and submodular threshold function f_v at every node.
   (b) The triggering model is a linear combination of instances of the general threshold model.
   (c) The influence spread under the triggering model is monotone nondecreasing and submodular.

4. In the LT model, instead of selecting threshold θ_v randomly from $[0, 1]$ with normal distribution at each node v, we select threshold θ_v randomly from $[0, 1/2]$ with normal distribution. Define a general threshold model equivalent to this diffusion model such that in the general threshold model, the threshold function f_v at each node v is monotone nondecreasing and submodular.
5. Show that there exists a triggering model which cannot be equivalent to an instance of the general threshold model.

6. Describe a general cascade model which is equivalent to the positive influence model and prove the correctness of you definition.

7. Show that there exists a social network with the positive influence model such that its influence spread is neither submodular nor supermodular.

8. In the LT model, instead of selecting threshold θ_v randomly from $[0, 1]$ with normal distribution at each node v, we select threshold θ_v randomly from $[1/2, 1]$ with normal distribution. Define a general threshold model equivalent to this diffusion model such that in the general threshold model, the threshold function f_v at each node v is monotone nondecreasing.

9. Construct a counterexample to show that there exists a social network with the general threshold model such that (a) at every node v, the threshold function f_v is monotone nondecreasing and supermodular, however (b) its influence spread is not supermodular.

10. In the IC model, suppose that there are several active parent nodes (in-neighbors) $u_1, u_2, \ldots, u_k$ who intend to activate an inactive node v. Then, the outcome results from k independent events; each is that node u_i intends to activate v. Show that this is equivalent to say "their activation attempts are sequenced in an arbitrary order."

11. Consider a real function f over all subsets of a finite set X. f is *strictly submodular* if for any two subsets $A \subset B$ any element $x \notin B$,

$$\Delta_x f(A) > \Delta_x f(B)$$

where $A \subset B$ means that $A \subseteq B$ and $A \neq B$. Show that f is strictly submodular if and only if for any two subsets A and B of X with $A \nsubseteq B$ and $B \nsubseteq A$,

$$f(A) + f(B) > f(A \cup B) + f(A \cap B).$$

12. Consider a real function f over all subsets of a finite set X. f is *strictly increasing* if for any two subsets A and B of X

$$A \subset B \Rightarrow f(A) < f(B).$$

Show that for a strictly increasing and strictly submodular function f, if the optimal solution for unconstrained minimization exists, then it is unique.

13. Consider a social network $G = (V, E)$ with a general threshold diffusion model. Suppose at every node v, the threshold function $f_v : 2^{N^-(v)} \to [0, 1)$ is strictly increasing and strictly submodular. Show that if G is k-connected, then there exists a value, $\theta = (\theta_v)_{v \in V}$, such that on a neighborhood of θ,

$$|T_{2n-1}| - |T_{n-1}| > |T'_{2n-1}| - |T'_{n-1}|$$

where T_t and T'_t denote sets of active nodes at step t in $\mathcal{Q}_-(A, \{x\})$ and and $\mathcal{Q}_-(B, \{x\})$, respectively for $A \subset B$ with $|B| \leq k$, and element $x \notin B$.

14. A set function f is said to be *strictly k-submodular* if for any two subsets $A \subset B$ with $|B| \leq k$, and any element $x \notin B$, $\Delta_x f(A) > \Delta_x f(B)$. Let f be a submodular and strictly k-submodular function over all subsets of X. Show that for any two subsets $A \subset B$ with $|A| \leq k - 1$ and any element $x \notin B$,

$$\Delta_x f(A) > \Delta_x f(B).$$

15. Consider a social network $G = (V, E)$ with a general threshold diffusion model. Suppose at every node v, the threshold function $f_v : 2^{N^-(v)} \to [0, 1)$ is strictly increasing and strictly submodular. Show that if G is k-connected, then the influence spread $\sigma(A)$ is strictly k-submodular.

16. Let f be a monotone nondecreasing submodular and strictly k-submodular function over all sunsets of finite set X. Then the greedy algorithm generates a solution A satisfying

$$f(A) > (1 - e^{-1})opt$$

where opt is the objective function value of the optimal solution for the maximization problem $\max\{f(A) \mid |A| \leq k\}$.

17. Let f be a strictly increasing and strictly submodular function. Assume $X \subset X'$ and $Y \subseteq Y'$. Show that if $Y' \nsubseteq X$, then

$$f(X \cup Y') - f(X) > f(X' \cup Y) - f(X').$$

18. Using the s-t paths problem, show the $\#P$-hardness of following s simple paths problem: Given a directed graph $G = (V, E)$ and a node $s \in V$, count how many simple paths starting from s.

19. Using the s simple paths problem, show that the influence spread is $\#P$-hard in the LT model.

20. Consider a social network $G = (V, E)$ with a diffusion model defined as follows.

- Each node has two states, inactive and active. Before diffusion process, every node is in inactive state. Initially, a subset of nodes, called seeds, are activated. Meanwhile, An update schedule, i.e., a node sequence $\{v_1, \ldots, v_\ell\}$ of length ℓ, is selected. Each node may appear in the sequence multiple times.
- At step t, consider node v_t. Let S_{t-1} be the set of active nodes at the end of previous step. If v_t is active, then keep v_t in active state. In fact, no active node is allowed to change its state, i.e., the influence process is *progressive*. If v_t is inactive, then the state of v_t is changed to active with probability $h_v(S_{t-1})$ where $h_v(\cdot)$ satisfying the following conditions.

$$0 \leq h_v(S) \leq 1$$

  where S is a subset of nodes.
- The process ends after ℓ update steps.

Show the following

(a) The IC model can be transformed to this model such that the influence spread is the same.

(b) A social network G' can be constructed from G such that the influence spread in G with this model can be calculated from the influence spread in G' with a general threshold model.

21. (Strictly Submodularity [313]) Consider a real set function $f : 2^X \to \mathbb{R}$. f is *strictly submodular* if for any two subsets A and B of X with $A \not\subseteq B$ and $B \not\subseteq A$,

$$f(A) + f(B) > f(A \cap B) + f(A \cup B).$$

f is *strictly increasing* if for any two subsets A and B of X,

$$A \subset B \Longleftarrow f(A) < f(B).$$

Show the following.

(a) f is strictly submodular if and only if for any two subsets A and B with $A \subset B$, and $x \in X \setminus B$,

$$\Delta_x f(A) > \Delta_x f(B).$$

(b) Consider a social network $G = (V, E)$ with a general threshold diffusion model. Suppose at every node v, the threshold function $f_v : 2^{N^-(v)} \to [0, 1)$ is strictly increasing and strictly submodular. If G is k-connected, then the influence spread $\sigma(\cdot)$ is strictly submodular for any pair of subsets $A \subset B$ with $|B| \leq k$, i.e., for any subsets $A \subset B$ with $|B| \leq k$, and $x \in X \setminus B$,

$$\Delta_x f(A) > \Delta_x f(B).$$

22. (Topic-Aware Influence [255]) Consider a social network with the IC model. Each node $v \in V$ is affiliated with a topic distribution $< p_v^1, p_v^2, \ldots, p_v^Z >$ where $p_v^i \in [0, 1)$ denote the weight on topic i. A query is a pair $(\vec{\gamma}, k)$ where $\vec{\gamma} =< \gamma^1, \gamma^2, \ldots, \gamma^Z >$ is the topic distribution such that $\sum_{i=1}^{Z} \gamma^i = 1$. For any edge (u, v), the probability that inactive node v is activated by an active node u is

$$\mathrm{ap}(u, v \mid \vec{\gamma}) = \sum_{i=1}^{Z} \gamma^i \cdot p_v^i.$$

Each node has three states, such as inactive, informed, and active. If an inactive node v is not activated by an active in-neighbor u, then v is turned to be informed. Given a query Q, the problem is to find a seed set S of size k to maximize

$$\mathbb{E}[|A|] + \mathbb{E}[|M|]$$

where A is the set of active nodes and M is the set of informed nodes at the end of information process, i.e., no node is newly turned to be active. Please determine whether this problem is a monotone submodular maximization.

23. (IM with a Limited Time [185]) Consider a social network $G = (V, E)$ with the LT model. Given a time limit τ and a positive integer k, find k seeds to maximize the influence spread within time limit τ. Please show that this is a monotone submodular maximization problem.

24. (Target Users' Influence Maximization [448]) Given a social network $G = (V, E)$ with the IC model (or the LT model), a target node set, and a positive integer k, find k seeds to maximize the influence spread in the target set. Please show that this problem is a monotone submodular maximization.

25. (Minimum-Cost Information Dissemination [73]) Consider a directed social network $G = (V, E)$. Each edge (u, v) is associated with a forwarding probability $p_{(u,v)}$. Given a number $p \in (0, 1]$, find a minimum set of source nodes S to spread a message such that at least a percentage p of users will be likely to forward the message. Here, a user is likely to forward a message if the probability that the user the message from S is not smaller than a given threshold $\theta > 0$. Please determine whether this problem is a minimum submodular cover problem, or not.

26. (Time-Delayed LT Model [46]) Compared with the time-delayed IC model, please give a proper definition for the time-delayed LT model and show that the time-critical influence spread for the time-delayed LT model is monotone nondecreasing and submodular

Historical Notes

Social influence has attracted one's attention for a long time [30, 77, 138, 139, 347]. The IC model was first studied in [139] while the LT model was studied much early in [152, 356]. The influence maximization problem was first studied by Kempe, Kleinburg and Tardös [211]. They showed that in the IC and LT models, the problem is NP-hard and influence spread is monotone nondecreasing and submodular, and hence has greedy $(1 - 1/e)$-approximation solution. They also proposed the general cascade model and the general threshold model, and discussed the submodularity of influence spread in those general models. Especially, the general threshold model contains various threshold models similar to the LT model [13, 282, 283, 296, 326, 405, 432, 471]. In the discussion, they proposed the well-known Kempe-Kleinburg-Tardos conjecture which was proved by Mossel and Roch [297].

The #P-hardness of the influence spread was established in [50] for the IC model and in [53] for the LT model.

There exist many variations of the influence maximization [46, 134, 163, 179, 243, 289, 342, 343, 384, 486] in the literature. We will introduce some of them in later chapters.

Chapter 3
Monte Carlo Method

Creativity is the ability to introduce order into the randomness of nature.
—Eric Hoffer

Since the influence spread is $\#P$-hard in the IC and the LT models, that is, it is unlike to have a polynomial-time approach to compute the influence spread, which provides a platform for Monte Carlo Methods to play. In this chapter, we study fundamental knowledge related to Monte Carlo methods and the social influence.

3.1 Probabilistic Inequalities

To analyze Monte Carlo methods, probabilistic inequalities are important tools. In this section, they are introduced for preparation.

Theorem 3.1.1 (Markov's Inequality) *Let Z be a nonnegative random variable. Then, for any $a \in \mathbb{R}^+$,*

$$\Pr[Z \geq a] \leq \frac{\mathbb{E}[Z]}{a}.$$

Here, $\mathbb{R}^+$ is the set of positive real number.

Proof Let $f(z)$ be the probabilistic distribution of random variable Z. Then, we have

$$\mathbb{E}[Z] = \int_0^\infty z f(z)dz \geq \int_a^\infty z f(z)dz \geq a \int_a^\infty f(z)dz = a \Pr[Z \geq a].$$

$\square$

W. Wu et al., *Computational Aspects of Social Networks*, Springer Optimization and Its Applications 234, https://doi.org/10.1007/978-3-032-14833-9_3

Corollary 3.1.2 *Let Z be a nonnegative random variable. Then, for any $\delta > 0$,*

$$\Pr[Z \geq (1 + \delta)\mathbb{E}[Z]] \leq \frac{1}{1+\delta}.$$

Theorem 3.1.3 (Chebyshev's Inequality) *Consider a sequence of 0-1 random variables $X_1, X_2, \ldots, X_n$, i.e., $X_1, X_2, \ldots, X_n \in \{0, 1\}$. Denote $p_i = \mathbb{E}[X_i] = \Pr[X_i = 1]$. Let $X = \sum_{i=1}^{n} X_i$ and $\mu = \mathbb{E}[X] = \sum_{i=1}^{n} p_i$. If $X_1, X_2, \ldots, X_n$ are pairwise-independent, then for $z \in \mathbb{R}^+$,*

$$\Pr[(X - \mu)^2 \geq z] \leq \frac{\mu}{z}.$$

Moreover, if $p_i > 0$ for at least one i, then

$$\Pr[(X - \mu)^2 \geq z] < \frac{\mu}{z}.$$

Proof Set $Y_i = X_i - p_i$ and $Y = X - \mu$. Then, $\mathbb{E}[Y_i] = 0$. Hence,

$$
\begin{aligned}
\mathbb{E}[Y^2] &= \mathbb{E}[\sum_{i,j \in [1,n]} Y_i \cdot Y_j] \\
&= \sum_{i=1}^{n} \mathbb{E}[Y_i^2] + 2 \sum_{\leq i < j \leq n} \mathbb{E}[Y_i \cdot Y_j] \\
&= \sum_{i=1}^{n} \mathbb{E}[Y_i^2] + 2 \sum_{\leq i < j \leq n} \mathbb{E}[Y_i] \cdot \mathbb{E}[Y_j] \\
&= \sum_{i=1}^{n} \mathbb{E}[Y_i^2] \\
&= \sum_{i=1}^{n} (1 - p_i)^2 p_i + (-p_i)^2 (1 - p_i) \\
&= \sum_{i=1}^{n} p_i (1 - p_i) \\
&\leq \mu.
\end{aligned}
$$

Substituting it into Markov's Inequality, we obtain

$$\Pr[Y^2 \geq z] \leq \frac{\mu}{z}.$$

Moreover, if $p_i > 0$ for at least one i, then $\mathbb{E}[Y^2] < \mu$. Hence,

$$\Pr[Y^2 \geq z] < \frac{\mu}{z}.$$

Corollary 3.1.4 *Suppose $\mu > 0$. Then, for $\delta > 0$,*

$$\Pr[X \geq (1 + \delta)\mu] < \frac{1}{\delta^2 \mu},$$
$$\Pr[X \leq (1 - \delta)\mu] < \frac{1}{\delta^2 \mu}.$$

Proof By Chebyshev's Inequality,

$$\Pr[X \geq (1 + \delta)\mu] \leq \Pr[(X - \mu)^2 \geq \delta^2 \mu^2] \leq \frac{1}{\delta^2 \mu}$$

and

$$\Pr[X \leq (1 - \delta)\mu] \leq \Pr[(X - \mu)^2 \geq \delta^2 \mu^2] \leq \frac{1}{\delta^2 \mu}.$$

$\square$

Lemma 3.1.5 *Let $X_1, X_2, \ldots, X_n$ be independent 0-1 random variables. Denote $p_i = \mathbb{E}[X_i]$. Let $X = \sum_{i=1}^{n} X_i$ and $\mu = \sum_{i=1}^{n} p_i$. Then, for any $\alpha \geq 1$,*

$$\mathbb{E}[\alpha^X] \leq e^{(\alpha-1)\mu}.$$

Proof Note that $\mathbb{E}[\alpha^{X_i}] = p_i \alpha + (1 - p_i) = (\alpha - 1)p_i + 1 \leq e^{(\alpha-1)p_i}$ since $1 + t \leq e^t$ for any $t \in \mathbb{R}$. Therefore, since $X_1, X_2, \ldots, X_n$ are independent, we have

$$\mathbb{E}[\alpha^X] = \prod_{i=1}^{n} \mathbb{E}[\alpha^{X_i}] \leq e^{(\alpha-1)\mu}.$$

$\square$

Theorem 3.1.6 (Generic Chernoff Bounds) *Let $X_1, X_2, \ldots, X_n$ be independent 0-1 random variables. Denote $p_i = \mathbb{E}[X_i]$. Let $X = \sum_{i=1}^{n} X_i$ and $\mu = \sum_{i=1}^{n} p_i$. Then, for any $x \geq \mu$,*

$$\Pr[X \geq x] \leq e^{x-\mu} \left(\frac{\mu}{x}\right)^x,$$

and for any $x \leq \mu$,

$$\Pr[X \leq x] \leq e^{x-\mu} \left(\frac{\mu}{x}\right)^x.$$

Proof By Markov's Inequality, for $x \geq \mu$,

$$\Pr[X \geq x] = \Pr\left[\left(\frac{x}{\mu}\right)^X \geq \left(\frac{x}{\mu}\right)^x\right] \leq \frac{\mathbb{E}[(\frac{x}{\mu})^X]}{(\frac{x}{\mu})^x}.$$

By Lemma 3.1.5, $\mathbb{E}[(\frac{x}{\mu})^X] \leq e^{x-\mu}$. Therefore,

$$\Pr[X \geq x] \leq e^{x-\mu}\left(\frac{\mu}{x}\right)^x.$$

Similarly, for $x \leq \mu$,

$$\Pr[X \leq x] = \Pr\left[\left(\frac{x}{\mu}\right)^X \geq \left(\frac{x}{\mu}\right)^x\right].$$

By Markov's Inequality and Lemma 3.1.5, we obtain

$$\Pr[X \leq x] \leq e^{x-\mu}\left(\frac{\mu}{x}\right)^x.$$

$\square$

Lemma 3.1.7 *For $0 < x < 1$,*

$$\ln(1+x) > \frac{2x}{2+x}$$

and

$$(1-x)\ln(1-x) > -x + \frac{x^2}{2}.$$

Proof First, note that

$$\frac{2+x}{2x} = \frac{1}{x} + \frac{1}{2}.$$

Set $y = 1/x$. The first inequality becomes

$$\ln(1+y) - \ln y > \frac{1}{y + \frac{1}{2}}.$$

Define

$$f(y) = \ln(1+y) - \ln y - \frac{1}{y + \frac{1}{2}}.$$

Then

$$f'(y) = \frac{1}{1+y} - \frac{1}{y} + \frac{1}{(y+\frac{1}{2})^2} < 0.$$

Thus, $f(y)$ is monotone decreasing in $(0, 1]$. Hence, for $y \in (0, 1)$,

$$f(y) > f(1) = \ln 2 - \frac{2}{3} > 0.$$

To prove the second inequality, define

$$g(x) = (1 - x)\ln(1 - x) + x - \frac{x^2}{2}.$$

Then

$$g'(x) = -\ln(1 - x) - 1 + 1 - x = -\ln(1 - x) - x > 0.$$

Therefore, $g(x)$ is monotone increasing for $x \in [0, 1)$. Hence,

$$g(x) > g(0) = 0.$$

$\square$

Corollary 3.1.8 (Multiplicative Chernoff Bound) *Suppose $X_1, \ldots, X_r$ are independent random variables taking values in $[0,1]$. Let $X = \sum_{i=1}^{n} X_i$ and $\mu = E[X]$. Then, for $0 < \delta < 1$,*

$$\Pr[X - \mu \geq \delta\mu] \leq \exp\left(-\frac{\delta^2 \mu}{2 + \delta}\right),$$

$$\Pr[X - \mu \leq -\delta\mu] \leq \exp\left(-\frac{\delta^2 \mu}{2}\right).$$

Proof Let $x = (1 + \delta)\mu$ where $0 < \delta < 1$. By Generic Chernoff Upper Bound,

$$\Pr[X - \mu \geq \delta\mu] \leq \exp(\delta\mu - (1 + \delta)\mu \ln(1 + \delta)).$$

Note that $\ln(1 + \delta) \geq \frac{2\delta}{2+\delta}$. Thus,

$$\delta\mu - (1 + \delta)\mu \ln(1 + \delta) \leq \delta\mu\left(1 - (1 + \delta) \cdot \frac{2}{2 + \delta}\right)$$

$$= -\delta^2\mu \cdot \frac{1}{2 + \delta}.$$

Next, let $x = (1 - \delta)\mu$ where $0 < \delta < 1$. By Generic Chernoff Lower Bound,

$$\Pr[X - \mu \leq -\delta\mu] \leq \exp(-\delta\mu - (1 - \delta)\mu \ln(1 - \delta)).$$

Note that $(1 - \delta)\ln(1 - \delta) \geq -\delta + \frac{\delta^2}{2}$. Thus,

$$-\delta\mu - (1 - \delta)\mu\ln(1 - \delta) \leq \delta\mu\left(-1 - (1 - \delta)\left(-1 - \frac{\delta}{2}\right)\right)$$

$$\leq -\frac{\delta^2\mu}{2}.$$

$\square$

Lemma 3.1.9 (Hoeffding's Lemma) *Let X be a random variable taking value in $[a, b]$ with probability one. Then, for all $\lambda \in \mathbb{R}$,*

$$\mathbb{E}\left[e^{\lambda(X - \mathbb{E}[x])}\right] \leq \exp\left(\frac{\lambda^2(b - a)^2}{8}\right).$$

Proof First, assume $\mathbb{E}[X] = 0$. Thus, $a \leq 0 \leq b$. Since $e^{\lambda x}$ is convex with respect to x, we have

$$e^{\lambda x} \leq \frac{b - x}{b - a}e^{\lambda a} + \frac{x - a}{b - a}e^{\lambda b}$$

for $x \in [a, b]$. Thus,

$$\begin{aligned}
\mathbb{E}\left[e^{\lambda X}\right] &\leq \frac{b - \mathbb{E}[X]}{b - a}e^{\lambda a} + \frac{\mathbb{E}[x] - a}{b - a}e^{\lambda b} \\
&= \frac{b}{b - a}e^{\lambda a} + \frac{-a}{b - a}e^{\lambda b} \\
&= \exp(L(\lambda(b - a))),
\end{aligned}$$

where $L(y) = \frac{ay}{b-a} + \ln(1 + \frac{a - e^y a}{b - a})$. Note that $L(0) = L'(0) = 0$ and $L''(y) \leq 1/4$ for all y. By Taylor's Theorem, for $0 \leq \theta \leq 1$,

$$L(y) = L(0) + L'(0) + \frac{1}{2}y^2 L''(y\theta) \leq \frac{1}{8}y^2.$$

Therefore,

$$\mathbb{E}\left[e^{\lambda X}\right] \leq \exp\left(\frac{1}{8}\lambda^2(b - a)^2\right).$$

Now, consider the general case. Set $Y = X - \mathbb{E}[X]$. Then $\mathbb{E}[Y] = 0$. Applying above argument to Y, we can complete the proof. $\square$

Theorem 3.1.10 (Hoeffding's Inequality) *Let $X_1, \ldots, X_n$ be independent random variables such that $a_i \leq X_i \leq b_i$ holds with probability one. Let $S_n = X_1 + \cdots + X_n$. Then, for all $t > 0$,*

$$\Pr[S_n - \mathbb{E}[S_n] \geq t] \leq \exp\left(-\frac{2t^2}{\sum_{i=1}^{n}(b_i - a_i)^2}\right).$$

Proof Denote $\alpha = \frac{4t}{\sum_{i=1}^{n}(b_i - a_i)^2}$. By Markov's Inequality, independence of X_i, and Hoeffding's Lemma, we have

$$
\begin{aligned}
\Pr[S_n - \mathbb{E}[S_n] \geq t] &= \Pr[\exp(\alpha(S_n - \mathbb{E}[S_n])) \geq \exp(\alpha t)] \\
&\leq \exp(-\alpha t)\mathbb{E}[\exp(\alpha(S_n - \mathbb{E}[S_n]))] \\
&= \exp(\alpha t) \prod_{i=1}^{n} \mathbb{E}[\exp(\alpha(X_i - \mathbb{E}[X_i]))] \\
&\leq \exp(-\alpha t) \prod_{i=1}^{n} \exp(\tfrac{1}{8}\alpha^2(b_i - a_i)^2) \\
&= \exp(-\alpha t + \tfrac{1}{8}\alpha^2 \sum_{i=1}^{n}(b_i - a_i)^2) \\
&= \exp\left(-\frac{2t^2}{\sum_{i=1}^{n}(b_i - a_i)^2}\right).
\end{aligned}
$$

$\square$

Theorem 3.1.11 (Union Bound) *Let $A_1, A_2, \ldots, A_n$ be n random events. Then,*

$$\Pr[A_1 + A_2 + \cdots + A_n] \leq \Pr[A_1] + \Pr[A_2] + \cdots + \Pr[A_n].$$

Proof Note that

$$\Pr[A_1 + A_2] = \Pr[A_1] + \Pr[A_2] - \Pr[A_1 A_2] \leq \Pr[A_1] + \Pr[A_2].$$

Thus,

$$
\begin{aligned}
\Pr[A_1 + A_2 + \cdots + A_n] &\leq \Pr[A_1 + A_2 + \cdots + A_{n-1}] + \Pr[A_n] \\
&\leq \cdots \\
&\leq \Pr[A_1] + \Pr[A_2] + \cdots + \Pr[A_n].
\end{aligned}
$$

$\square$

3.2   Buffon's Needle

Since the influence spread is $\#P$-hard in the IC and the LT models, that is, it is unlike to have a polynomial-time approach to compute the influence spread, we get two questions:

1. How to compute a good approximation value for the influence spread?
2. If we use approximation value for the influence spread in the greedy algorithm, what is the performance of the greedy solution for the influence maximization problem?

For the first question, answer was suggested in [211] that use Monte Carlo method. What is the Monte Carlo method? Let us see a simple example: Use Buffon's needle to compute the real number π.

As shown in Fig. 3.1, take a needle. Suppose the length of the needle is ℓ. Draw a sequence of parallel lines with distance ℓ on a paper. Randomly drop the needle on the paper and count how many times the needle crosses a line. Suppose that this number is c. Then π can be approximated by $\frac{2c}{n}$, where n is the total number of drops. Of course, $\frac{2c}{n}$ cannot be the exact value of π since π is not a rational number. However, as n is large enough, $\frac{2c}{n}$ is close to π within certain accuracy with very high probability. The following is a detailed analysis.

Let x be the distance from the center of the needle to the closest parallel line. Then the uniform probability density function of x between 0 and ℓ is constant $2/\ell$.

Let ϕ be the acute angle between the needle and one of the parallel lines. Then uniform probability density function of ϕ between 0 and $\pi/2$ is constant $2/\pi$.

Two random variables x and ϕ are independent. Thus, their joint probability density function is

$$\frac{4}{\ell\pi} \quad \text{for } x \in [0, \ell/2], \phi \in [0, \pi/2].$$

The needle crosses a line if and only if $x \leq \frac{\ell}{2} \sin\phi$. Therefore, its probability is

$$\int_{\phi=0}^{\pi/2} \int_{x=0}^{(\ell/2)\sin\phi} \frac{4}{\ell\pi} dx d\phi = \frac{2}{\pi}.$$

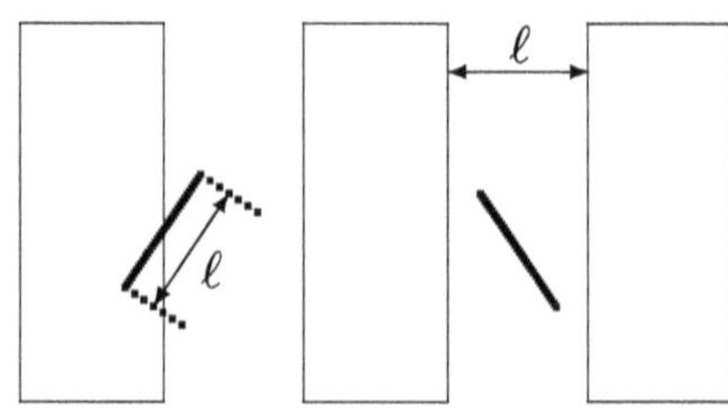

Fig. 3.1 Buffon's needle

By the multiplicative Chernoff bound,

$$\Pr\left[\left|\frac{c}{n} - \frac{2}{\pi}\right| \geq \delta \cdot \frac{2}{\pi}\right] \leq \exp\left(-\frac{\delta^2 n}{3} \cdot \frac{2}{\pi}\right) + \exp\left(-\frac{\delta^2 n}{2} \cdot \frac{2}{\pi}\right).$$

3.3 Influence Spread

To compute the influence spread in the IC or LT model, we may sample subgraphs based on the distribution determined by the probabilistic model. A very early suggestion is that to compute one value of the influence spread (actually, the increment of the influence spread in the greedy algorithm), 10,000 trials of the sampling may be required for a satisfactory precision [211].

How should we analyze the sampling? An important tool is the multiplicative Chernoff bound.

Now, let us look at what is the probability that 10,000 trials reach accuracy 1%. Set $r = 10000$ and $\delta = 1/100$. Let $\Delta_1, \ldots, \Delta_r$ be the value of $\Delta_u \sigma_m(S)$ in r trials, respectively. Consider random variables $X_1 = \Delta_1, \ldots, X_r = \Delta_r$ and $\mu = r \Delta_u \sigma_m(S)$. By the multiplicative Chernoff bound, we obtain

$$\Pr\left[\left|\frac{1}{r}\sum_{i=1}^{r}\Delta_i - \Delta_u\sigma_m(S)\right| \geq 0.01 \cdot \Delta_u\sigma_m(S)\right] \leq e^{-\frac{\Delta_u\sigma_m(S)}{2}} + e^{-\frac{\Delta_u\sigma_m(S)}{3}}.$$

Note that the right side becomes larger as S gets bigger.

However, it is interesting to note that computing $\sigma_m(S)$ has a different situation. Again, set $r = 10,000$ and $\delta = 1/100$. Let $\sigma_1, \ldots, \sigma_r$ be the number of influenced nodes in r trials, respectively. Consider random variables $X_1 = \sigma_1, \ldots, X_r = \sigma_r$ and $\mu = r\sigma_m(S)$. By the multiplicative Chernoff bound, we obtain

$$\Pr\left[\left|\frac{1}{r}\sum_{i=1}^{r}\sigma_i - \sigma_m(S)\right| \geq 0.01 \cdot \sigma_m(S)\right] \leq e^{-\frac{\sigma_m(S)}{2}} + e^{-\frac{\sigma_m(S)}{3}}.$$

Note that the right side becomes smaller as $\sigma_m(S)$ gets bigger. In this case, the suggestion of 10,000 trials does make sense. Although we have this note, we will, in later discussion, still directly compute $\Delta_u\sigma_m(S)$ with samplings in order to be consistent with the literature.

Next, let us consider the second question. Since the influence spread is unlikely to be computed exactly, we have to modify the greedy algorithm and its analysis for the submodular maximization problem, that is, an error of ε is allowed in each iteration of greedy choice.

Algorithm 9 Modified Greedy Algorithm for Submodular Maximization

---

Input: A polymatroid function f over 2^X, a positive integer k, and a small positive number ε.
Output: A set A_k with $|A_k| \leq k$.
1: $A_0 \leftarrow \emptyset$;
2: **for** $i = 1$ to k **do**
3:    choose $v \in X \setminus A_{i-1}$ such that
4:    $\Delta_v f(A_{i-1}) \geq \max_{w \in X \setminus A_{i-1}} \Delta_w f(A_{i-1}) - \varepsilon$
5:    and $A_i \leftarrow A_{i-1} \cup \{v\}$;
6: **end for**
7: **return** A_k.

Theorem 3.3.1 (Kempe et al. [210]) *Suppose A_k is produced by Algorithm 9. If $1 \leq k \leq opt$, then $f(A_k) \geq (1 - e^{-1} - \varepsilon) \cdot opt$, where opt is the optimal value.*

Proof Suppose $A_k = \{x_1, x_2, \ldots, x_k\}$ is the output of Algorithm 9, where $x_1, \ldots, x_k$ are in the ordering of the selection. Denote by $A_i = \{x_1, x_2, \ldots, x_i\}$ for $i = 1, \ldots, k$, and let $A_0 = \emptyset$. Suppose $A^* = \{u_1, u_2, \ldots, u_k\}$ is an optimal solution. Then for $i = 0, \ldots, k - 1$, we have

$$
\begin{aligned}
f(A^*) &\leq f(A^* \cup A_i) \\
&= f(A_i) + \Delta_{u_1} f(A_i) + \cdots + \Delta_{u_k} f(A_i \cup \{u_1, u_2, \ldots, u_{k-1}\}) \\
&\leq f(A_i) + \Delta_{u_1} f(A_i) + \cdots + \Delta_{u_k} f(A_i) \\
&\leq f(A_i) + k \cdot \Delta_{x_{i+1}} f(A_i) + k\varepsilon,
\end{aligned}
$$

where the first inequality is due to the monotonicity of f, the second inequality is due to the submodularity of f, and the third inequality is due to the greedy rule for selecting x_i in the algorithm.

Denote $a_i = f(A^*) - f(A_i)$. Then we have $a_i \leq k(a_i - a_{i+1}) + k\varepsilon$. Hence

$$
a_{i+1} \leq a_i(1 - 1/k) + \varepsilon \leq a_i \cdot e^{-1/k} + \varepsilon
$$

where the second inequality makes use of $1 + x \leq e^x$. Therefore,

$$
a_k \leq a_{k-1} e^{-1/k} + \varepsilon \leq \cdots \leq a_0 e^{-1} + \varepsilon(e^{-(k-1)/k} + \cdots + e^{-1/k} + 1)
$$

$$
= a_0 e^{-1} + \varepsilon \cdot \frac{1 - e^{-1}}{1 - e^{-1/k}}.
$$

Since $a_0 = f(A^*) - f(\emptyset) = f(A^*) = opt$, we have

$$
f(A_k) \geq (1 - e^{-1})opt - \varepsilon \cdot \frac{1 - e^{-1}}{1 - e^{-1/k}}.
$$

Note that $1 \leq k \leq opt$ implies

$$\frac{1 - e^{-1}}{1 - e^{-1/k}} \leq \frac{1 - e^{-1}}{\frac{1}{k} - \frac{1}{2k^2}} = k \cdot \frac{1 - e^{-1}}{1 - 1/2k} \leq 2(1 - e^{-1})k \leq k \leq opt.$$

Hence $f(A_k) \geq (1 - e^{-1} - \varepsilon) \cdot opt$. □

Corollary 3.3.2 *In Algorithm 9, replace line 4 by*

$$\Delta_v f(A_{i-1}) \geq \max_{w \in X \setminus A_{i-1}} \Delta_w f(A_{i-1}) - \varepsilon \cdot \frac{opt}{k}.$$

Then, the algorithm returns A_k satisfying

$$f(A_k) \geq (1 - e^{-1} - \varepsilon) \cdot opt.$$

Proof Let $\varepsilon' = \varepsilon \cdot \frac{opt}{k}$. It is similar to the proof of Theorem 3.3.1 except the last two lines. We would obtain

$$f(A_k) \geq (1 - e^{-1})opt - \varepsilon' \cdot k = (1 - e^{-1})opt - \varepsilon opt.$$

□

Next, let us consider the greedy algorithm (Algorithm 10) for influence maximization.

Algorithm 10 Modified Greedy Algorithm for Influence Maximization

---

Input: A social network G with diffusion model m, a small number $\varepsilon > 0$ and an integer $k > 0$.
Output: A seed set S with $|S| \leq k$.
1: $S \leftarrow \emptyset$;
2: **for** $i = 1$ to k **do**
3:     select $u_i \in V \setminus S$ such that
4:     $\Delta_{u_i} \sigma(S) \geq \max_{u \in V \setminus S} \Delta_u \sigma_m(S) - \varepsilon$
5:     and $S \leftarrow S \cup \{u_i\}$;
6: **end for**
7: **return** S.

The running time for Algorithm 10 can be analyzed as follows.

- Let r be the number of samplings for computing $\Delta_u \sigma_m(S)$.
- The algorithm runs k iterations.
- Each iteration requires to estimate $O(n)$ marginal values $\Delta_u \sigma_m(S)$.
- Each sampling needs $O(|E|)$ time to determine a subgraph.
- Total running time $O(k|E|nr)$.

Using the multiplicative Chernoff bound and Corollary 3.3.2, the following theorem can be proved.

Theorem 3.3.3 (Tang et al. [385]) *The greedy algorithm for the influence maximization (Algorithm 10) produces a* $(1 - e^{-1} - \varepsilon)$*-approximate solution with probability at least* $1 - n^{\ell}$ *if*

$$r \geq (8k^2 + 2k\varepsilon) \cdot \frac{(\ell + 1)\ln n + \ln k + \ln 2}{\varepsilon^2 \cdot opt}, \tag{3.1}$$

where opt is the maximum influence spread, i.e., the optimal value, and r *is the number of samplings for computing* $\Delta_u \sigma_m(S)$.

Proof By Corollary 3.3.2, for the purpose of the proof, we may relax condition in line 4 of Algorithm 10 to the following:

$$\Delta_{u_i} \sigma_m(S) \geq \max_{u \in V \setminus S} \Delta_u \sigma_m(S) - \varepsilon \cdot \frac{opt}{k}. \tag{3.2}$$

Let $\Delta_1, \ldots, \Delta_r$ be the value of $\Delta_u \sigma_m(S)$ in r trials, respectively. In order to have (3.2), we should have that

$$\left| \frac{1}{r} \sum_{i=1}^{r} \Delta_i - \Delta_u \sigma_m(S) \right| \leq \frac{\varepsilon}{2} \cdot \frac{opt}{k} \tag{3.3}$$

when evaluate $\Delta_u \sigma_m(S)$. In the Multiplicative Chernoff Bound, set $X_i = \Delta_i$, $\delta = opt \cdot \varepsilon / (2k \Delta_u \sigma(S))$, $\mu = r \Delta_u \sigma(S)$. Then we obtain

$$\Pr \left[\left| \frac{1}{r} \sum_{i=1}^{r} \Delta_i - \Delta_u \sigma(S) \right| > \varepsilon \cdot \frac{opt}{2k} \right] \leq 2 \exp \left(-\frac{\delta^2 \mu}{2 + \delta} \right)$$

$$\leq 2 \exp \left(-\frac{(opt/k)^2 \varepsilon^2 r}{8 \Delta_u \sigma(S) + 2(opt/k)\varepsilon} \right)$$

$$\leq 2 \exp \left(-\frac{(opt/k)^2 \varepsilon^2 r}{8 \cdot opt + 2(opt/k)\varepsilon} \right)$$

$$= 2 \cdot \exp \left(-\frac{opt \cdot \varepsilon^2 r}{8k^2 + 2k\varepsilon} \right).$$

Note that the inequality (3.3) is required to hold in every computation of $\Delta_u \sigma(S)$. By the Union Bound, this inequality (3.3) does not hold for some computation of $\Delta_u \sigma_m(S)$ with probability at most

$$kn \cdot 2 \cdot \exp \left(-\frac{opt \cdot \varepsilon^2 r}{8k^2 + 2k\varepsilon} \right).$$

Let

$$2kn \cdot \exp\left(-\frac{opt \cdot \varepsilon^2 r}{8k^2 + 2k\varepsilon}\right) \le n^{-\ell}.$$

Thus,

$$r \ge \frac{8k^2 + 2k\varepsilon}{opt \cdot \varepsilon^2}((\ell + 1)\ln n + \ln k + \ln 2).$$

$\square$

3.4 Reverse Influence Sampling

In Algorithm 10, every randomly generated graph is used once for the estimation of $\Delta_u \sigma_m(S)$. This is a type of waste. A clever way was already found, which is called *reverse influence sampling (RIS)*. It consists of two steps.

Step 1 Randomly generates θ RR-sets.
Step 2 Find k nodes to hit the maximum number of RR-sets.

What is RR-set? To explain it clearly. Let us consider a specific information diffusion model, the IC model. Each RR-set is generated as follows:

- Choose a node v from the input social network G uniformly at random.
- Initially, set $A \leftarrow \{v\}$, $R \leftarrow \emptyset$, and $g \leftarrow \emptyset$.
- **while** $A \ne \emptyset$ **do**
  choose $u \in A$;
  for every edge (w, u) **do**
  (a) put (w, u) into g and w into A with probability p_{wu}, and
  (b) move u from A to R.
  end-while
- Return R. R is the reverse reachable (RR) set for v, i.e., the set of nodes that can reach v in subgraph g.

There is another way to generate RR-sets.

- Select a node v from input social network G uniformly at random.
- Select a subgraph g randomly by deleting each edge (u, v) with probability $1 - p_{uv}$.
- The reverse reachable (RR) set for v is the set of nodes that can reach v in g.

These two ways produce different subgraphs g. We will use the first way in order to be consistent with the literature. However, we may discuss and compare them in some proper place later.

Lemma 3.4.1 *The probability that a seed set S hits a random RR-set for node v is equal to the probability that v gets activated by S.*

Proof Clearly, two events are equivalent. $\square$

Corollary 3.4.2 *Let $\mathcal{R}$ be the set of θ RR-sets. Let $F_{\mathcal{R}}(S)$ be the fraction of RR-sets in $\mathcal{R}$ hit by S, i.e.,*

$$F_{\mathcal{R}}(S) = \frac{|\{R \in \mathcal{R} \mid R \cap S \neq \emptyset\}|}{\theta}.$$

Then $\mathbb{E}[n \cdot F_{\mathcal{R}}(S)] = \sigma_{IC}(S)$.

Lemma 3.4.3 *Suppose*

$$\theta \geq \lambda/opt \tag{3.4}$$

where

$$\lambda = (8 + 2\varepsilon)n \cdot \frac{\ell \ln n + \ln \binom{n}{k} + \ln 2}{\varepsilon^2}.$$

Then, with probability $1 - n^{-\ell}/\binom{n}{k}$,

$$|n \cdot F_{\mathcal{R}}(S) - \sigma_{IC}(S)| < opt \cdot \varepsilon/2 \tag{3.5}$$

for any fixed S of size k.

Proof Let ρ be the probability that S hits a random RR-set. Then, $\theta \cdot F_{\mathcal{R}}(S)$ can be considered as the sum of θ independent 0-1 random variables each with a mean ρ. Thus,

$$\rho = \mathbb{E}[F_{\mathcal{R}}(S)] = \sigma_{IS}(S)/n \leq opt/n,$$

where n is the total number of nodes. Therefore,

$$\Pr[|n \cdot F_{\mathcal{R}}(S) - \sigma_{IS}(S)| \geq opt \cdot \varepsilon/2]$$
$$= \Pr[|F_{\mathcal{R}}(S) - \rho| \geq opt \cdot \varepsilon/(2n)]$$
$$= \Pr\left[|\theta \cdot F_{\mathcal{R}}(S) - \theta\rho| \geq \frac{opt \cdot \varepsilon}{2n\rho} \cdot \rho\theta\right].$$

Set $\delta = opt \cdot \varepsilon/(2n\rho)$. By Multiplicative Chernoff Bounds, we obtain

$$\Pr\left[|\theta \cdot F_{\mathcal{R}}(S) - \theta\rho| \geq \frac{opt \cdot \varepsilon}{2n\rho} \cdot \rho\theta\right]$$

$$< 2 \cdot \exp\left(-\frac{\delta^2}{2+\delta} \cdot \rho\theta\right)$$

$$= 2 \cdot \exp\left(-\frac{\varepsilon^2 \cdot opt^2}{8n^2\rho + 2\varepsilon n \cdot opt} \cdot \theta\right)$$

$$\leq 2 \cdot \exp\left(-\frac{\varepsilon^2 \cdot opt^2}{8n \cdot opt + 2\varepsilon n \cdot opt} \cdot \theta\right)$$

$$= 2 \cdot \exp\left(-\frac{\varepsilon^2 \cdot opt}{8n + 2\varepsilon n} \cdot \theta\right)$$

$$\leq \frac{1}{\binom{n}{k} \cdot n^\ell}.$$

$\square$

Note that

$$f(S) = |\{A \in \mathcal{R} \mid A \cap S \neq \emptyset\}|$$

is a monotone nondecreasing and submodular set function. Therefore, the maximization of $f(S)$ has a greedy $(1 - e^{-1})$-approximation.

Theorem 3.4.4 (Tang et al. [385]) *Suppose θ satisfies condition (3.4). Then the reverse sampling algorithm produces a $(1 - e^{-1} - \varepsilon)$-approximate solution with probability $1 - n^{-\ell}$.*

Proof Let S_k be the node set obtained in Step 2. Let S^+ be the node set hitting the maximum number of RR-sets. Let S^* be the optimal solution for the influence maximization problem. Then $\sigma_{IC}(S^*) = opt$ and we have

$$F_{\mathcal{R}}(S_k) \geq (1 - e^{-1})F_{\mathcal{R}}(S^+) \geq (1 - e^{-1})F_{\mathcal{R}}(S^*).$$

By Lemma 3.4.3, with probability at least $1 - n^{-\ell}/\binom{n}{k}$, (3.5) holds for any fixed S of size k. Thus, by Union Bound, we can see that with probability at least $1 - n^{-\ell}$, (3.5) holds for all size-k set simultaneously. In particular, we have

$$n \cdot F_{\mathcal{R}}(S^*) \geq \sigma_{IC}(S^*) - opt \cdot \varepsilon/2 = (1 - \varepsilon/2) \cdot opt,$$

and

$$\sigma_{IC}(S_k) \geq n \cdot F_{\mathcal{R}}(S_k) - opt \cdot \varepsilon/2$$
$$\geq (1 - e^{-1})n F_{\mathcal{R}}(S^*) - opt \cdot \varepsilon/2$$
$$\geq (1 - e^{-1})(1 - \varepsilon/2) \cdot opt - opt \cdot \varepsilon/2$$
$$\geq (1 - e^{-1} - \varepsilon) \cdot opt.$$

The theorem is proved. $\square$

Note that the classical greedy algorithm for the influence maximization problem requires $O(rnk)$ samplings. (Each sampling takes $O(|E|)$ time.) Comparing the bound (3.1) for r with the bound (3.4) for θ, we can see that $O(rnk)$ is about $O(k^3)$ times of θ. Therefore, the reverse sampling algorithm is better.

What information diffusion model can reverse sampling be employed in? Note that to generate an RR-set, we need to choose a node and a subgraph randomly. Thus, each subgraph should have an appearance with a certain probability. By Theorem 2.4.8, this occurs if and only if it is in the general triggering model. In later sections of this chapter, we will not specialize what model is considered. Unless we specially mention, the model can be any of them in the general triggering. In particular, the influence spread will be written as $\sigma(S)$ instead of $\sigma_{IC}(S)$.

To end this section, let us make a remark on the analysis of reverse influence sampling. The outcome in Theorem 3.4.4 can be improved if Lemma 3.4.3 is decomposed into two parts, i.e., establish the following:

1. If $\theta \geq \theta_1$, then
$$\Pr[n \cdot F_{\mathcal{R}}(S) - \sigma_{IC}(S) > \varepsilon_1 \cdot opt] \leq \delta_1.$$

2. If $\theta \geq \theta_2$, then

$$\Pr[n \cdot F_{\mathcal{R}}(S) - \sigma_{IC}(S) < -\varepsilon_2 \cdot opt] \leq \delta_2.$$

This is because the Multiplicative Chernoff Bound has two different bounds in the two cases. Moreover, set $\theta_1 = \theta_2$. Using these two parts on S_k and S^*, respectively, we would obtain the result with probability $1 - \delta_1 - \delta_2$,

$$\sigma_{IC}(S_k) \geq n \cdot F_{\mathcal{R}}(S_k) - \varepsilon_2 \cdot opt$$
$$\geq (1 - e^{-1})n \cdot F_{\mathcal{R}}(S^*) - \varepsilon_2 \cdot opt$$
$$\geq (1 - e^{-1})\sigma_{IC}(S^*) - ((1 - e^{-1})\varepsilon_1 + \varepsilon_2) \cdot opt$$
$$= (1 - e^{-1} - \varepsilon) \cdot opt$$

where $\varepsilon = (1 - e^{-1})\varepsilon_1 + \varepsilon_2$. This idea about decomposition has been used in [310, 386].

3.5 Martingale

Instead of Multiplicative Chernoff Bound, there exists another tool, martingale, for analysis of the reverse influence sampling.

Definition 3.5.1 (*Martingale*) A sequence of random variables $Y_1, Y_2, \ldots$ is a martingale, if $\mathbb{E}[|Y_i|] < +\infty$ and $\mathbb{E}[Y_i \mid Y_1, Y_2, \ldots, Y_{i-1}] = Y_{i-1}$ for every i.

Let $R_1, R_2, \ldots, R_\theta$ be the sequence of random RR sets generated by the reverse sampling. Let

$$x_i = \begin{cases} 1 \text{ if } S \cap R_i \neq \emptyset, \\ 0 \text{ otherwise.} \end{cases}$$

Then, by Corollary 3.4.2,

$$\sigma(S) = \frac{n}{\theta} \cdot \mathbb{E}[\sum_{i=1}^{\theta} x_i],$$

and for $i \in [1, \theta]$,

$$\mathbb{E}[x_i \mid x_1, x_2, \ldots, x_{i-1}] = \mathbb{E}[x_i] = \sigma(S)/n.$$

Denote $p = \sigma(S)/p$. Set $M_i = \sum_{j=1}^{i}(x_j - p)$. Then, we have $\mathbb{E}[M_i] = 0$ and $M_i = M_{i-1} + (x_i - p)$. Therefore

$$\mathbb{E}[M_i \mid M_1, M_2, \ldots, M_{i-1}] = M_{i-1} + \mathbb{E}[x_i - p] = M_{i-1}.$$

This means that the sequence of random variables $M_1, M_2, \ldots$ is a martingale. The following is an inequality about martingale.

Lemma 3.5.2 ([59]) *Consider a martingale* $\{Y_1, Y_2, \ldots\}$. *Suppose that* $|Y_1| \leq a$, $|Y_j - Y_{j-1}| \leq a$ *for* $j \in [2, i]$, *and*

$$Var[Y - 1] + \sum_{j=2}^{i} Var[Y_j \mid Y_1, Y_2, \ldots, Y_{j-1}] \leq b.$$

Then, for any $\lambda > 0$,

$$\Pr[Y_i - \mathbb{E}[Y_i] \geq \gamma] \leq \exp\left(-\frac{\gamma^2}{\frac{2}{3}a\gamma + 2b}\right).$$

This inequality can imply the following.

Lemma 3.5.3 *For any $\varepsilon > 0$,*

$$\Pr\left[\sum_{i=1}^{\theta} x_i - \theta p \geq \varepsilon \cdot \theta p\right] \leq \exp\left(-\frac{\varepsilon^2}{2 + \frac{2}{3}\varepsilon} \cdot \theta p\right)$$

and

$$\Pr\left[\sum_{i=1}^{\theta} x_i - \theta p \leq -\varepsilon \cdot \theta p\right] \leq \exp\left(-\frac{\varepsilon^2}{2} \cdot \theta p\right).$$

Proof For martingale $\{M_1, M_2, \ldots, M_\theta\}$, we have $|M_1| \leq 1$ and for any $j \in [2, \theta]$, $|M_j - M_{j-1}| \leq 1$. Moreover, we also have

$$\mathrm{Var}[M_1] + \sum_{j=2}^{\theta} \mathrm{Var}[M_j \mid M_1, M_2, \ldots, M_{j-1}] = \sum_{j=1}^{\theta} \mathrm{Var}[M_j]$$

$$= \theta p(1 - p).$$

Note that $M_\theta = \sum_{i=1}^{\theta}(x_i - p)$. By Lemma 3.5.2, we obtain

$$\Pr\left[\sum_{i=1}^{\theta} x_i - \theta p \geq \varepsilon \cdot \theta p\right] \leq \exp\left(-\frac{(\varepsilon\theta p)^2}{\frac{2}{3}\varepsilon\theta p + 2\theta p(1 - p)}\right)$$

$$\leq \exp\left(-\frac{\varepsilon^2}{2 + \frac{2}{3}\varepsilon} \cdot \theta p\right).$$

Moreover, by applying Lemma 3.5.2 to martingale $\{-M_1, -M_2, \ldots, -M_\theta\}$, we obtain

$$\Pr\left[\sum_{i=1}^{\theta} x_i - \theta p \leq -\varepsilon \cdot \theta p\right] \leq \exp\left(-\frac{\varepsilon^2}{2} \cdot \theta p\right).$$

$$\square$$

This lemma gives a little improvement to Multiplicative Chernoff bounds. Therefore, using this lemma, we can also obtain a little improvement for Lemma 3.4.3 and Theorem 3.4.4.

3.6   Lower Bound of *opt*

Since *opt* is unknown, we have to find a lower bound for *opt* in order to give a lower bound or sufficient number for θ. In [385, 386], several approaches are provided for computing a lower bound of *opt*. Let us introduce one of them, provided in [386]. The idea is to provide a test $B(x)$ to tell whether $opt \geq x$, that is,

$$B(x) = \begin{cases} \text{true} & \text{if } opt \geq x \\ \text{false} & \text{otherwise.} \end{cases}$$

The following lemma gives a specific $B(x)$.

Lemma 3.6.1 (Tang et al. [386]) *Let $\varepsilon, \delta \in (0, 1)$ and $x \in [0, 1]$. Let $\mathcal{R}$ be the set of θ RR-sets produced by reverse sampling where*

$$\theta \geq \frac{(2 + \frac{2}{3}\varepsilon)(\ln \binom{n}{k} + \ln(1/\delta))}{\varepsilon^2} \cdot \frac{n}{x}.$$

Let S_k^+ be the $(1 - 1/e)$-approximation solution obtained by greedy algorithm applying to $\mathcal{R}$ for size-k node set with maximum hitting. Then, if $opt < x$, we have $n \cdot F_{\mathcal{R}}(S_k^+) < (1 + \varepsilon)x$ with probability $1 - \delta$.

Proof Denote $p = \mathbb{E}[F_{\mathcal{R}}(S_k^+)]$. Define

$$y_i = \begin{cases} 1 & \text{if } R_i \cap S_k^+ \neq \emptyset \\ 0 & \text{otherwise.} \end{cases}$$

Then, we have

$$p = \mathbb{E}[F_{\mathcal{R}}(S_k^+)] = \sigma(S_k^+)/n \leq opt/n < x/n.$$

Denote $\zeta = \frac{(1+\varepsilon)x}{np} - 1$. Then $\zeta \geq (1 + \varepsilon) - 1 = \varepsilon$. By Lemma 3.5.3,

$$\begin{aligned}
&\Pr[n \cdot F_{\mathcal{R}}(S_k^+) \geq (1 + \varepsilon)x] \\
&= \Pr[\theta \cdot F_{\mathcal{R}}(S_k^+) - \theta p \geq \zeta \cdot \theta p] \\
&= \Pr\left[\sum_{i=1}^{\theta} y_i - \theta p \geq \zeta \cdot \theta p\right] \\
&\leq \exp\left(-\frac{\zeta^2}{2 + 2\zeta/3} \cdot \theta p\right) \\
&\leq \exp\left(-\frac{\varepsilon \cdot x/(np)}{2/\zeta + 2/3} \cdot \frac{(2 + 2\varepsilon/3) \cdot \ln(\binom{n}{k}/\delta)}{\varepsilon^2} \cdot \frac{np}{x}\right) \\
&\leq \exp\left(-\frac{1}{2/\varepsilon + 2/3} \cdot \frac{(2 + 2\varepsilon/3) \cdot \ln(\binom{n}{k}/\delta)}{\varepsilon}\right) \\
&= \delta / \binom{n}{k}.
\end{aligned}$$

Note that S_k^+ may not be unique. However, there are at most $\binom{n}{k}$ candidates for S_k^+. Above proof indicates that for each possible S_k^+, $n \cdot F_{\mathcal{R}}(S_k^+) \geq (1 + \varepsilon)x$ with probability at most $\delta/\binom{n}{k}$. Therefore, everyone satisfies $n \cdot F_{\mathcal{R}}(S_k^+) \leq (1 + \varepsilon)x$ with probability $1 - \delta$. $\qquad\qquad\square$

With test $B(x)$ stated in Lemma 3.6.1, an algorithm can be designed to compute a lower bound of *opt* as shown in Algorithm 11.

Algorithm 11 Lower Bound of *opt* [386]

Input: G, k, ε, ℓ.
Output: LB.
1: $\mathcal{R} \leftarrow \emptyset, LB \leftarrow 1$;
2: $\varepsilon' \leftarrow \varepsilon\sqrt{2}$;
3: **for** $i = 1$ to $\log_2 n - 1$ **do**
4: $\quad x \leftarrow n/2^i$;
5: $\quad \theta_i \leftarrow \lambda/x$ where

$$\lambda = \frac{(2 + 2\varepsilon'/3)(\ln\binom{n}{k} + \ell\ln n + \ln\log_2 n) \cdot n}{(\varepsilon')^2}$$

6: $\quad$ **while** $|\mathcal{R}| \leq \theta_i$ **do**
7: $\quad\quad$ Select a node v from G uniformly at random, generate an RR-set for v, and insert in $\mathcal{R}$;
8: $\quad$ **end while**
9: $\quad S_i \leftarrow (1 - 1/e)$-approximation solution for hitting $\mathcal{R}$ maximally;
10: $\quad$ **if** $n \cdot F_{\mathcal{R}}(S_i) \geq (1 + \varepsilon')x$ **then**
11: $\quad\quad$ **return** $LB \leftarrow n \cdot F_{\mathcal{R}}(S_i)/(1 + \varepsilon')$;
12: $\quad\quad$ Stop.
13: $\quad$ **end if**
14: **end for**

Usually, a further modification may be considered when employing a lower bound of *opt* in the reverse sampling Algorithm. For example, Algorithm 12 is designed by modifying the combination of the algorithm in Sect. 3.4 and lower bound of *opt* in this section.

Algorithm 12 Reverse Sampling Algorithm

Input: A social network G, a knapsack constraint, a partition matroid constraint, two parameters ε and ℓ.
Output: S^+.
1: $\mathcal{R} \leftarrow \emptyset, S \leftarrow \emptyset$;
2: Generate λ/n RR-sets and put them in $\mathcal{R}$ where

$$\lambda = \frac{(8 + 2\varepsilon)(\ln\binom{n}{k} + \ell\ln n + \ln 4) \cdot n}{\varepsilon^2}$$

3: Let $Cov_{\mathcal{R}}(S)$ denote the number of RR-sets, in $\mathcal{R}$, hit by S.
4: **while** $Cov_{\mathcal{R}}(S) < (1 + \varepsilon)\lambda$ **do**
5: $\quad$ Generate $|\mathcal{R}|$ more RR-sets and put in $\mathcal{R}$.
6: $\quad$ Compute $(1/3)$-approximation solution S for hitting $\mathcal{R}$ maximally with the knapsack constraint and the matroid constraint.
7: **end while**
8: **return** $S^+ = S$

The following theorem states the performance of Algorithm 12.

Theorem 3.6.2 *Algorithm 12 produces an approximation solution S^+ such that*

$$\sigma(S^+) \geq (1/3 - \varepsilon) \cdot opt$$

with probability $1 - n^{-\ell}$.

Proof Denote $\delta = n^{-\ell}/2$. At the end of Algorithm 12, we have

$$Cov_{\mathcal{R}}(S) \geq (1 + \varepsilon)\lambda.$$

Denote $\theta = |\mathcal{R}|$ and $x = \frac{\lambda}{\theta}$. Then,

$$n \cdot F_{\mathcal{R}}(S_i) \geq (1 + \varepsilon)x.$$

We first claim that $opt \geq x$ with probability at least $1 - \delta$.

Suppose $x > opt$. Denote $p = \mathbb{E}[F_{\mathcal{R}}(S)]$ for a fixed S with $w(S) \leq k$. Then, we have

$$p = \mathbb{E}[F_{\mathcal{R}}(S)] = \sigma_{ST}(S)/n \leq opt/n < x/n.$$

Define

$$y_i = \begin{cases} 1 & \text{if } R_i \cap S \neq \emptyset \\ 0 & \text{otherwise.} \end{cases}$$

Denote $\zeta = \frac{(1+\varepsilon)x}{np} - 1$. Then $\zeta \geq (1 + \varepsilon) - 1 = \varepsilon$. Therefore,

$$
\begin{aligned}
& \Pr[n \cdot F_{\mathcal{R}}(S) \geq (1 + \varepsilon)x] \\
\leq\ & \Pr[n \cdot F_{\mathcal{R}}(S) \geq (1 + \varepsilon)np] \\
=\ & \Pr[\theta \cdot F_{\mathcal{R}}(S) - \theta p \geq \zeta \cdot \theta p] \\
=\ & \Pr[\sum_{i=1}^{\theta} y_i - \theta p \geq \zeta \cdot \theta p] \\
\leq\ & \exp\left(-\frac{\zeta^2}{2 + \zeta} \cdot \theta p\right) \qquad \text{(by Multiplicative Chernoff Bound)} \\
\leq\ & \exp\left(-\frac{\varepsilon \cdot x/(np)}{2/\zeta + 1} \cdot \frac{(8 + 2\varepsilon) \cdot \ln(\binom{n}{k}/\delta)}{\varepsilon^2} \cdot \frac{np}{x}\right) \\
\leq\ & \exp\left(-\frac{1}{2/\varepsilon + 1} \cdot \frac{(8 + 2\varepsilon) \cdot \ln(\binom{n}{k}/\delta)}{\varepsilon}\right) \\
=\ & \delta/\binom{n}{k}.
\end{aligned}
$$

Since there are at most $\binom{n}{k}$ possible S, we have that for every possible S, $n \cdot F_{\mathcal{R}}(S) \geq (1 + \varepsilon)x$ with probability at most δ. In particular, $n \cdot F_{\mathcal{R}}(S_i) \geq (1 + \varepsilon)x$ with probability at most δ. Hence, $x > opt$ occurs with probability at most δ, that is, $x \leq opt$ with probability at least $1 - \delta$.

Therefore, with probability at least $1 - \delta$,

$$\theta \geq \frac{(8 + 2\varepsilon)(\ln \binom{n}{k} + \ln(1/\delta))}{\varepsilon^2} \cdot \frac{n}{opt} \tag{3.6}$$

Our second claim is that if (3.6) holds, then with probability $1 - \delta/\binom{n}{k}$,

$$|n \cdot F_{\mathcal{R}}(S) - \sigma_{ST}(S)| < opt \cdot \varepsilon/2 \tag{3.7}$$

for any fixed S with $w(S) \leq k$.

Let ρ be the probability that S hits a random RR-set. Then, $\theta \cdot F_{\mathcal{R}}(S)$ can be considered as the sum of θ independent random 0-1 variables each with a mean ρ. Thus,

$$\rho = \mathbb{E}[F_{\mathcal{R}}(S)] = \sigma_{ST}(S)/n \leq opt/n,$$

where $n = |V|$. Therefore,

$$\Pr[|n \cdot F_{\mathcal{R}}(S) - \sigma_{ST}(S)| \geq opt \cdot \varepsilon/2]$$
$$= \Pr[|F_{\mathcal{R}}(S) - \rho| \geq opt \cdot \varepsilon/(2n)]$$
$$= \Pr\left[|\theta \cdot F_{\mathcal{R}}(S) - \theta\rho| \geq \frac{opt \cdot \varepsilon}{2n\rho} \cdot \rho\theta\right].$$

Set $\tau = opt \cdot \varepsilon/(2n\rho)$. By Multiplicative Chernoff Bounds, we obtain

$$\Pr\left[|\theta \cdot F_{\mathcal{R}}(S) - \theta\rho| \geq \frac{opt \cdot \varepsilon}{2n\rho} \cdot \rho\theta\right]$$
$$< v2 \cdot \exp\left(-\frac{\tau^2}{2 + \tau} \cdot \rho\theta\right)$$
$$= 2 \cdot \exp\left(-\frac{\varepsilon^2 \cdot opt^2}{8n^2\rho + 2\varepsilon n \cdot opt} \cdot \theta\right)$$
$$\leq 2 \cdot \exp\left(-\frac{\varepsilon^2 \cdot opt^2}{8n \cdot opt + 2\varepsilon n \cdot opt} \cdot \theta\right)$$
$$= 2 \cdot \exp\left(-\frac{\varepsilon^2 \cdot opt}{8n + 2\varepsilon n} \cdot \theta\right)$$
$$\leq \frac{\delta}{\binom{n}{k}}.$$

Therefore, with probability at most $\frac{\delta}{\binom{n}{k}}$, (3.7) does not hold for a fixed S. Hence, with probability at most δ, (3.7) does not hold for some possible S. It follows that with probability at least $1 - \delta$, (3.7) holds for all possible S, simultaneously, under condition that (3.6) holds. After removing this condition, we know that with probability at least $1 - 2\delta$, (3.7) holds for all possible S, simultaneously.

Let S^* be the node set hitting the maximum number of RR-sets. Let OPT be the optimal solution for the influence maximization problem. Then $\sigma_{ST}(OPT) = opt$ and we have

$$F_{\mathcal{R}}(S^+) \geq (1 - e^{-1/2})F_{\mathcal{R}}(S^*) \geq (1 - e^{-1})F_{\mathcal{R}}(OPT).$$

Since with probability at least $1 - 2\delta$, (3.7) holds for all possible S with $w(S) \leq k$, simultaneously. By setting $S = OPT$ and $S = S^+$, we obtain

$$n \cdot F_{\mathcal{R}}(OPT) \geq \sigma_{ST}(OPT) - opt \cdot \varepsilon/2 = (1 - \varepsilon/2) \cdot opt,$$

and

$$\begin{aligned}
\sigma(S^+) &\geq n \cdot F_{\mathcal{R}}(S^+) - opt \cdot \varepsilon/2 \\
&\geq (1/3)n F_{\mathcal{R}}(OPT) - opt \cdot \varepsilon/2 \\
&\geq (1/3)(1 - \varepsilon/2) \cdot opt - opt \cdot \varepsilon/2 \\
&\geq (1/3 - \varepsilon) \cdot opt.
\end{aligned}$$

The theorem is proved. $\qquad\square$

3.7 Monte Carlo Estimation

Buffon's Needle is a simple example of using the Monte Carlo approach to estimate an unknown quantity μ. Typically, this approach consists of the following steps.

- Step 1. Design an experiment that produces a random variable Z distributed in $[0, 1]$ with $\mathbb{E}[Z] = \mu$.
- Step 2. Run this experiment independently a number of times and use the average of the outcomes as an estimate.

In this section, we introduce fundamental theoretical results about this approach and use those results in analysis of the reverse sampling algorithm.

In many applications, one looks for (ε, δ)-approximation as follows.

Definition 3.7.1 ((ε, δ)-*Approximation*) Let Z_1, Z_2, ... be independently and identically distributed samples according to 0-1 random variable with mean μ_Z. A Monte Carlo estimator of μ_Z,

$$\hat{\mu}_Z = \frac{1}{T} \sum_{i=1}^{T} Z_i$$

is said to be an (ε, δ)-Approximation of μ_Z if

$$\Pr[-\varepsilon\mu_Z \leq \hat{\mu}_Z - \mu_Z \leq \varepsilon\mu_Z] \geq 1 - \delta. \tag{3.8}$$

The following results can be found in [69] about (ε, δ)-approximation.

Theorem 3.7.2 (0-1 Estimator Theorem [69]) *Let θ^* be the optimal number of samples that guarantee an (ε, δ)-approximation of μ_Z. Define $\gamma(\varepsilon, \delta/2) = 2c \ln(2/\delta)/\varepsilon^2$ with $c = 2(e - 2)$ and $Cov(Z) = \sum_{i=1}^{T} Z_i$. If $T = \gamma(\varepsilon, \delta/2)/\mu_Z$, then $\hat{\mu}_Z = Cov(Z)/T$ is an (ε, δ)-approximation of μ_Z and*

$$T \leq c_1 \cdot \theta^*$$

where c_1 is a constant. Moreover, if only one side of event in (3.8) is required, then $\gamma(\varepsilon, \delta/2)$ should be replaced by

$$\gamma(\varepsilon, \delta) = \frac{2c}{\varepsilon^2} \cdot \ln \frac{1}{\delta}.$$

Theorem 3.7.3 (Dagum et al. [69]) *Let θ^* be the optimal number of samples that guarantee an (ε, δ)-approximation of μ_Z. Define $\gamma(\varepsilon, \delta/2) = 2c \ln(2/\delta)/\varepsilon^2$ with $c = 2(e - 2)$. Assume that T is the number of samples at which*

$$Cov(Z) \geq 1 + (1 + \varepsilon)\gamma(\varepsilon, \delta/2).$$

Then $\hat{\mu}_Z$ is an (ε, δ)-approximation of μ_Z and $T \leq (1 + \varepsilon)c_1 \cdot \theta^*$ where c_1 is the constant. Moreover, if only one side of event in (3.8) is required, then $\gamma(\varepsilon, \delta/2)$ should be replaced by

$$\gamma(\varepsilon, \delta) = \frac{2c}{\varepsilon^2} \cdot \ln \frac{1}{\delta}.$$

In previous section, we see that although the martingale provides an alternative tool to establish estimation for the number of samplings, it cannot give a significantly better result than Multiplicative Chernoff Bound. Now, we can make a shortcut passing over Multiplicative Chernoff Bound in analysis of the RIS (reverse influence sampling) algorithm, .

Let denote $\hat{\sigma}(S) = n F_{\mathcal{R}}(S)$. First, we prove the following.

Lemma 3.7.4 *Consider a social network G and $0 \leq \varepsilon, \delta \leq 1$. If $|\mathcal{R}| \geq \gamma(\varepsilon/2, \delta/4) \cdot n$, then*

$$\Pr[\sigma(S_k) \geq ((1 - 1/e - \varepsilon)opt] \geq 1 - \delta$$

where S_k is the set obtained by the RIS algorithm, i.e., a $(1 - e^{-1})$-approximation solution for hitting RR-sets.

Proof For any set S, we may define

$$Z_i = \begin{cases} 1 & \text{if } R_i \cap S \neq \emptyset \\ 0 & \text{otherwise.} \end{cases}$$

Then,

$$\hat{\mu}_Z = \frac{Cov(S)}{|\mathcal{R}|}, \quad \mu_Z = \frac{\sigma(S)}{n}.$$

Since $|\mathcal{R}| \geq \gamma(\varepsilon/2, \delta/4) \cdot n$, we have

$$|\mathcal{R}| \geq \gamma(\varepsilon/2, \delta/4) \cdot \frac{n}{\sigma(S)} = \gamma(\varepsilon/2, \delta/4) \cdot \frac{1}{\mu_Z}.$$

By Theorem 3.7.2, for any set S, $\hat{\mu}_Z$ is $(\varepsilon/2, \delta/2)$-approximation of of μ_Z, i.e., $\hat{\sigma}(S)$ is $(\varepsilon/2, \delta/2)$-approximation of $\sigma(S)$. Thus,

$$\Pr[\hat{\sigma}(S_k) \leq (1 + \varepsilon/2)\sigma(S_k)] \geq 1 - \delta/2. \tag{3.9}$$

$$\Pr[(1 - \varepsilon/2) \cdot opt \leq \hat{\sigma}(S_k^*)] \geq 1 - \delta/2, \tag{3.10}$$

where S_k^* is an optimal solution and $opt = \sigma(S_k^*)$.

Let S_{kmax} be the set, satisfying the size constraint $|S| \leq k$, hitting the maximum number of RR sets in $\mathcal{R}$. Then

$$\hat{\sigma}(S_k) \geq (1 - e^{-1}) \cdot \hat{\sigma}(S_{kmax}).$$

Note that $\hat{\sigma}(S_{kmax}) \geq \hat{\sigma}(S_k^*)$. Hence,

$$\hat{\sigma}(S_k) \geq (1 - e^{-1}) \cdot \hat{\sigma}(S_k^*).$$

Combined with (3.9), we obtain

$$\Pr[(1 - e^{-1}) \cdot \hat{\sigma}(S_k^*) \leq (1 + \varepsilon/2)\sigma(S_k)] \geq 1 - \delta/2.$$

Combined with (3.10), we obtain

$$\Pr[(1 - e^{-1}) \cdot (1 - \varepsilon/2)opt \leq (1 + \varepsilon/2)\sigma(S_k)] \geq 1 - \delta.$$

Since $\sigma(S_k) \leq opt$, we have

$$\Pr[\sigma(S_k) \geq (1 - e^{-1} - \varepsilon)opt] \geq 1 - \delta.$$

$\square$

Let us present a simpler version of algorithm in [310] as shown in Algorithm 13 and an analysis on it in Theorem 3.7.5

Algorithm 13 Simplified RIS Algorithm in Nguyen et al. [310]

Input: Social Network $G = (V, E)$, $0 \leq \varepsilon, \delta \leq 1$, and a budget $k > 0$.
Output: An $(1 - 1/e - \varepsilon)$-approximation solution S_k with probability at least $1 - \delta$.
1: Compute $\varepsilon_1 = \varepsilon/2$, $\varepsilon_2 = \varepsilon/(2(1 - 1/e))$
2: $\theta \leftarrow \max(1 + (1 + \varepsilon_1)\gamma(\varepsilon_1, \delta/3), (1 + \varepsilon_1)\gamma(\varepsilon_2, \delta/3))$
3: $\mathcal{R} \leftarrow$ generate θ random RR sets by RIS
4: **repeat**
5:    $S_k \leftarrow$ generate $(1 - 1/e)$-approximation for hitting maximum number of RR sets in $\mathcal{R}$
6:    **if** $Cov_{\mathcal{R}}(S_k) \geq \theta$ **then**
7:       **return** S_k
8:    **end if**
9:    Generate $|\mathcal{R}|$ more RR sets and add to $\mathcal{R}$
10: **until** $|\mathcal{R}| \geq \gamma(\varepsilon/2, \delta/4)n$
11: **return** S_k

Theorem 3.7.5 *Algorithm 13 returns a seed set S_k satisfying*

$$\Pr[\sigma(S_k) \geq (1 - 1/e - \varepsilon)opt] \geq 1 - \delta.$$

Proof First, consider the case that S_k is returned at line 7. Then we have

$$Cov_{\mathcal{R}}(S_k) \geq \theta.$$

Thus, $Cov_{\mathcal{R}}(S_k) \geq 1 + (1 + \varepsilon_1)\gamma(\varepsilon_1, \delta/3)$. By Lemma 3.7.3, we have

$$\Pr[\hat{\sigma}(S_k) \leq (1 + \varepsilon_1)\sigma(S_k)] \geq 1 - \delta/3. \tag{3.11}$$

Note that

$$\hat{\sigma}(S_k) = \frac{n \cdot Cov_{\mathcal{R}}(S_k)}{|\mathcal{R}|} \geq \frac{n(1 + \varepsilon_1)\gamma(\varepsilon_2, \delta/3)}{|\mathcal{R}|}.$$

Thus,

$$\Pr\left[\frac{n \cdot \gamma(\varepsilon_2, \delta/3)}{|\mathcal{R}|} \leq \sigma(S_k)\right] \geq 1 - \delta/3.$$

Since $\sigma(S_k) \leq opt$, we have

$$\Pr\left[\frac{n \cdot \gamma(\varepsilon_2, \delta/3)}{|\mathcal{R}|} \leq opt\right] \geq 1 - \delta/3,$$

i.e.,

$$\Pr\left[\frac{n \cdot \gamma(\varepsilon_2, \delta/3)}{opt} \leq |\mathcal{R}|\right] \geq 1 - \delta/3.$$

By Theorem 3.7.2, we have

$$\Pr[\hat{\sigma}(S_k^*) \geq (1 - \varepsilon_2)opt] \geq 1 - 2\delta/3.$$

Note that

$$\hat{\sigma}(S_k) \geq (1 - e^{-1})\hat{\sigma}(S_k^*).$$

Combined with (3.11), we obtain

$$\Pr[\sigma(S_k) \geq (1 - 1/e - \varepsilon)opt] \geq 1 - \delta.$$

If S_k is returned from line 10, then conclusion follows from Lemma 3.7.4.   □

In Fig. 3.2, let us summarize tools provided in this chapter for analysis of the RIS algorithm.

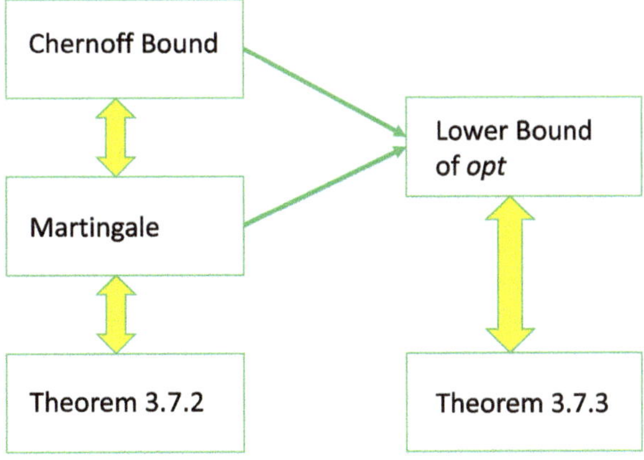

Fig. 3.2 Tools for analysis of RIS algorithms

Exercises

1. For two polynomials $f(x)$ and $g(x)$ of order r, we want to judge whether they are identical, without knowing their coefficients. Consider the following randomized algorithm: uniformly and randomly select an integer i from the set $\{1, 2, \ldots, n\}$; if $f(i) = g(i)$, then return "yes"; otherwise, return "no."

   (a) Estimate the probability of obtaining an erroneous answer.
   (b) Design a randomized algorithm to reduce the probability of an erroneous answer.

2. We consider the problem of distributing n balls into m bins. The key factors to consider are: whether the balls are identical, whether the bins are identical, and whether bins can be empty. There are 8 cases in total. Complete the following table by filling in the "Number of ways" column for each case.

Case	Identical balls?	Identical bins?	Bins can be empty?	Number of ways
1	Yes	No	Yes	
2	Yes	No	No	
3	No	Yes	No	
4	No	Yes	Yes	
5	No	No	Yes	
6	No	No	No	
7	Yes	Yes	Yes	
8	Yes	Yes	No	

3. We consider identical balls being placed independently into m distinct bins. What is the expected number of balls needed for every bin to contain at least one ball? Additionally, estimate the probability that all bins contain at least one ball after placing n balls.

4. Consider identical balls being placed independently into m distinct bins, where each ball is placed into a bin uniformly at random. Define a "collision" as an event where two balls are placed into the same bin. Answer the following:

   (a) What is the expected number of collisions when n balls are placed?
   (b) What is the expected number of empty bins when n balls are placed?
   (c) For a specified bin, what is the probability that it contains exactly k balls?

5. Let $X_1, X_2, \ldots, X_n$ be 0-1 pairwise independent random variables. Let $X = X_1 + X_2 + \cdots + X_n$, $\mu = \mathbb{E}[X]$, and $\sigma = \sqrt{\mathrm{Var}(X)}$. Show that for $k > 0$,

$$\Pr[|X - \mu| \geq k\sigma] \leq \frac{1}{k^2}.$$

6. (Bernstein Inequality) Let $X_1, X_2, \ldots, X_n$ be independent Bernoulli random variables taking values $+1$ and -1 with probability $1/2$. Show that for for every positive ε,

$$\Pr\left[\left|\frac{1}{n}\sum_{i=1}^{n} X_i\right| > \varepsilon\right] \leq 2 \cdot \exp\left(-\frac{n\varepsilon^2}{2 \cdot (1 + \frac{\varepsilon}{3})}\right).$$

7. Prove that a sum of martingales is a martingale.
8. Let $x_1, x_2, \ldots$ be a sequence of independent random variables with zero mean

$$S_n = \sum_{i=1}^{n} x_i \text{ and } T_n^2 = \sum_{i=1}^{n} \sigma_i^2.$$

Show that $\{S_n^2 - T_n^2\}_{n\in\mathbb{N}}$ is a martingale.
9. Show that each of the following is a martingale.

   (a) (Pólya's urn) Consider an urn which contains a number of marbles in different colors. At each iteration, a marble is randomly selected from the urn and replaced by several more in the same color. For any given color, the fraction of marbles in such color forms a martingale.
   (b) Suppose that every $M^{(i)}$ for $1 \leq i \leq d$ is a martingale. Then, the d-dimensional process $M = (M^{(1)}, \ldots, M^{(d)})$ is also a martingale.

10. (Azuma Inequality) Let $\{X_i \mid i = 0, 1, 2, \ldots\}$ be a martingale such that

$$|X_i - x_{i-1}| \leq c_i$$

almost surely. Show that for every positive integer n and any positive real ε,

$$\Pr[X_n - X_0 \geq \varepsilon] \leq \exp\left(\frac{-\varepsilon^2}{2\sum_{i=1}^{n} c_i^2}\right)$$

and

$$\Pr[X_n - X_0 \leq -\varepsilon] \leq \exp\left(\frac{-\varepsilon^2}{2\sum_{i=1}^{n} c_i^2}\right).$$

11. (Hassidim and Singer [182]) For an immediate application of probabilistic inequalities, let us give an analysis for a stochastic greedy algorithm 14. This algorithm produces approximation solution for maximizing $f(S)$ subject to $|S| \leq k$. Suppose that at each iteration, we can only find an element with approximately maximal marginal contribution in expectation. In other words, Let $\mathcal{D}$ be a probability distribution, at each iteration, a value ξ is sampled from $\mathcal{D}$. Then, an element can be selected to have marginal contribution which is a ξ-approximation to the maximal marginal contribution. Prove the following.

Algorithm 14 Stochastic Greedy Algorithm

Input: A monotone nondecreasing submodular function $f : 2^X \to \mathbb{R}_+$ and a positive integer k.
Output: A subset S of X, with $|S| \leq k$.
1: $S \leftarrow \emptyset$
2: **while** $|S| < k$ **do**
3:    $\xi \sim \mathcal{D}$
4:    choose x such that $\Delta_x f(S) \geq \xi \max_{x \in X} \Delta_x f(S)$
5:    $S \leftarrow S \cup \{x\}$
6: **end while**
7: **return** S

(a) Let S be the set of k elements generated by Algorithm 14. Let ξ_i be selected from $\mathcal{D}$ in the ith iteration. Denote $\hat{\mu} = \frac{1}{k} \sum_{i=1}^{k} \xi_i$. Then

$$f(S) \geq \left(1 - \frac{1}{e^{\hat{\mu}}}\right) \cdot opt.$$

(b) Consider a monotone nondecreasing submodular function f, which is evaluated with uncertainty coming from a distribution $\mathcal{D}$ with mean μ. For any $\varepsilon \in (0, 1)$, suppose that Algorithm 14 runs with $k \geq 2/(\varepsilon^2 \mu)$. Then, with probability at least $1 - \exp(-\varepsilon\mu/2)$, the algorithm returns S satisfying

$$f(S) \geq \left(1 - \frac{1}{e^{(1-\varepsilon)\mu}}\right) \cdot opt.$$

12. (Tang et al. [385]) Let $\mathcal{V}^*$ be a probability distribution over nodes in social network G, such that for each node v^*, its probability mass is the probability that v^* belongs to a RR set. Define EPT to be the expected number of coin tosses required for generating an RR set, i.e., the number of edges can be reversely reached from a randomly selected node. Let n be the number of nodes in G and m the number of edges in G. Please prove the following.

(a)
$$\frac{n}{m} EPT = \mathbb{E}_{v^* \in \mathcal{V}^*}[\sigma_{IC}(v^*)].$$

(b)
$$\frac{n}{m} EPT \leq opt.$$

(Therefore, we may select
$$\theta = \lambda \cdot \frac{m}{n \cdot EPT}$$

where EPT can be computed by sampling (Monte Carlo method). This lower bound of opt is independent from k and hence, is too far from opt. In fact, opt

is monotonically increasing as k goes to larger. This means that when k is large, opt goes away from $\frac{n}{m}EPT$ large and large. The second approach is motivated from this observation.)

13. (Tang et al. [385]) Let S^* be a subset of k nodes selected from $\mathcal{V}^*$. Let KPT be the mean of the expected spread of S^*. It is easy to see that

$$\frac{n}{m}EPT \le KPT \le opt.$$

Thus, using KPT, instead of $\frac{n}{m}EPT$, for a lower bound of opt is a better choice. Prove the following results about KPT.

(a) Let R be a random RR set. Then

$$KPT = n \cdot \mathbb{E}_R[\kappa(R)]$$

where

$$\kappa(R) = 1 - \left(1 - \frac{w(R)}{m}\right)^k$$

and

$$w(R) = \sum_{v \in R}(\text{the indegree of } v \text{ in } G).$$

(b) By (a), KPT can be estimated by measuring $n \cdot \kappa(R)$ for a set of random RR sets and then taking their average. By the Multiplicative Chernoff Bound, if we want to obtain an estimate with error not exceeding $\delta \in (0, 1)$ with probability at least $1 - n^{-\ell}$, then the number of RR sets should be at least $O(\ell n \log n \cdot \delta^{-2}/KPT)$, where KPT is exactly what we want to know.

Algorithm 15 KPT Estimation

Input: A social network G and an integer $k > 0$.
Output: A value of KPT^*.
1: **for** $i = 1$ to $\log_2 n - 1$ **do**
2:    $sum \leftarrow 0$;
3:    set $c_i = (6\ell \log_2 n + 6\log_2(\log_2 n)) \cdot 2^i$;
4:    **for** $j = 1$ to c_i **do**
5:      generate a random RR set R and compute $\kappa(R)$;
6:      $sum \leftarrow sum + \kappa(R)$;
7:    **end for**
8:    **if** $sum/c_i > 1/2^i$ **then**
9:      **return** $KPT^* = n \cdot sum/(2c_i)$;
10:   **end if**
11: **end for**
12: **return** $KPT^* = 1$.

To break this logic cycle, Algorithm 15 is designed to determine the number of RR sets adaptively. Please give proofs for analysis of this algorithm as follows:

- Let $\mathcal{K}$ be the distribution of $\kappa(R)$ over all random RR sets. Let $\mu = KPT/n$. Denote by s_i the sum of c_i samples independently selected from $\mathcal{K}$. Let $\mu \leq 2^{-j}$. Then, for any $i \in [1, j-1]$,

$$\Pr\left[\frac{s_i}{c_i} > \frac{1}{2^i}\right] < \frac{1}{n^\ell \cdot \log_2 n}.$$

  This fact indicates that if $KPT \leq 2^{-j}$, then it is unlikely that Algorithm 15 would terminate before the jth iteration, which keeps KPT^* not far from KPT.

- Suppose μ and c_i as defined as above. Assume $\mu \geq 2^{-j}$. Then, for any $i \geq j + 1$,

$$\Pr\left[\frac{s_i}{c_i} > \frac{1}{2^i}\right] > 1 - n^{-\ell \cdot 2^{i-j-1}} / \log_2 n.$$

(c) Let $n \geq 2$ and $\ell \geq 1/2$. Then, Algorithm 15 runs in expected time $O(\ell(m + n)\log n)$ and returns $KPT^* \in [KPT/4, opt]$ with probability at least $1 - n^{-\ell}$. Moreover, $\mathbb{E}\left[\frac{1}{KPT^*}\right] < \frac{12}{KPT}$.

14. (Tang et al. [385]) Algorithm 16 provides an approach to improve the lower bound KPT^* for opt.

Algorithm 16 Refine KPT

Input: A social network G with IC model, a positive integer k, KPT^* obtained from Algorithm 15, and a positive number $\varepsilon' > 0$.
Output: KPT^+.
1: Let $\mathcal{R}'$ be the set of all RR sets produced in the previous iteration of Algorithm 15.
2: $S'_k \leftarrow \emptyset$;
3: **for** $j = 1$ to k **do**
4:     find a node v_j to cover the maximum number of RR sets in $\mathcal{R}'$;
5:     $S'_k \leftarrow S'_k \cup \{v_j\}$;
6:     remove from $\mathcal{R}'$ all RR sets covered by v_j;
7: **end for**
8: set $\lambda' = (2 + \varepsilon')\ell n \log n \cdot (\varepsilon')^{-2}$
9: and $\theta' = \lambda'/KPT^*$;
10: generate $theta'$ random RR sets and put them in $\mathcal{R}'$;
11: let f be the fraction of RR sets in $\mathcal{R}'$ that are covered by S'_k;
12: set $KPT' = f \cdot n/(1 + \varepsilon')$;
13: **return** $KPT^+ = \max\{KPT', KPT^*\}$.

Please prove the following: Suppose

$$\mathbb{E}\left[\frac{1}{KPT^*}\right] < \frac{12}{EPT}$$

and

$$KPT^* \in [KPT/4, opt].$$

Then, within expected time

$$O(\ell(m+n)\log n/(\varepsilon')^2),$$

Algorithm 16 returns $KPT^+ \in [KPT^*, opt]$ with probability at least $1 - n^{-\ell}$.

15. In Sect. 3.4, replace IC model by triggering model, prove the following.

   (a) Consider a node subset S and a node v. Let $\mathcal{G}$ be the triggering graph distribution for G. Suppose that an RR set R is generated for v on a graph g sampled from $\mathcal{G}$. Then, the probability that S overlaps with R is equal to the probability that S, as a seed set, can activate v.
   (b) Under triggering model, Algorithm 10 runs in expected time $O((k+\ell)(m+n)\log n\varepsilon^{-2})$ and returns a $(1 - 1/e - \varepsilon)$-approximation solution with probability $1 - 2 \cdot n^{-\ell}$.

16. Show that Algorithm 10 produces $(1 - e^{-1} - \varepsilon)$-approximation solution with probability $1 - n^\ell$ if

$$r \geq (8k^2 + \frac{4}{3} \cdot k\varepsilon) \cdot \frac{(\ell+1)\ln n + \ln k + \ln 2}{\varepsilon^2 \cdot opt}$$

where opt is the maximum influence spread and r is the number of samplings for computing $\Delta_u \sigma_m(S)$.

17. (Tang et al. [386]) Prove the following.

   (a) Let $\delta_1 \in (0, 1)$, $\varepsilon_1 > 0$, and

$$\theta = |\mathcal{R}| \geq \theta_1 = \frac{2n \cdot \log(1/\delta_1)}{opt \cdot \varepsilon_1^2}.$$

Then, with probability at least $1 - \delta_1$, we have

$$n \cdot F_{\mathcal{R}}(S_k^*) \geq (1 - \varepsilon_1) \cdot opt$$

where S_k^* is the size-k node set with maximum influence spread, i.e., $opt = \sigma(S_k^*)$.

(b) Let $\delta_2 \in (0, 1)$, $\varepsilon_1 < \varepsilon$, and

$$\theta_2 = \frac{(2 - 2/e) \cdot n \cdot \log(\binom{n}{k}/\delta_2)}{opt \cdot (\varepsilon - (1 - 1/e) \cdot \varepsilon_1)^2}.$$

If $\theta = |\mathcal{R}| \geq \max\{\theta_1, \theta_2\}$, then, with probability at least $1 - \delta_1 - \delta_2$, we have

$$\sigma(S_k^+) \geq (1 - 1/e - \varepsilon) \cdot opt$$

where S_k^+ is the $(1 - 1/e)$-approximation solution obtained by greedy algorithm applying to $\mathcal{R}$ for size-k node set with maximum hitting.

(c) Let

$$\theta \geq \frac{2n((1 - 1/e)\alpha + \beta)^2}{\varepsilon^2 \cdot opt},$$

where

$$\alpha = \sqrt{\ell \ln n + \ln 2}$$

$$\beta = \sqrt{(1 - 1/e)(\ln \binom{n}{k} + \ell \ln n + \ln 2)}.$$

Then, with probability at least $1 - n^{-\ell}$, the reverse sampling algorithm returns a $(1 - 1/e - \varepsilon)$-approximation solution for the influence maximization problem.

(d) Given ℓ and ε, let θ^* be the smallest θ satisfying conditions stated in (a). Let θ^- be the lower bound of θ stated in (c). Prove that

$$\theta^* \leq \theta^- \leq \theta^* \cdot \frac{\ell \ln n + \ln 2}{\ell \ln n}.$$

18. (Tang et al. [386]) Let $x, \varepsilon, \delta. \mathcal{R}$ and S_k^+ be as defined in Lemma 3.6.1. Show that

$$opt \geq n \cdot \frac{F_{\mathcal{R}}(S_k^+)}{1 + \varepsilon}$$

with probability $1 - \delta$.

19. (Tang et al. [386]) Let $x, \varepsilon, \delta. \mathcal{R}$ and S_k^+ be as defined in Lemma 3.6.1. Let $c \geq 1$. Assume $opt \geq c \cdot \frac{(1+\varepsilon)^2}{1-1/e} \cdot x$. Show that $n \cdot F_{\mathcal{R}}(S_k^+) < (1 + \varepsilon)x$ with probability at most δ^c.

20. Show the correctness of Algorithm 11.

21. (RIS Threshold [310]) Consider a social network G and $0 \leq \varepsilon_a, \varepsilon_b, \delta_a, \delta_b \leq 1$. A number $N(\varepsilon_a, \varepsilon_b, \delta_a, \delta_b)$ is called an *RIS Threshold* in G with respect to $\varepsilon_a, \varepsilon_b, \delta_a, \delta_b$ if $|\mathcal{R}| \geq N(\varepsilon_a, \varepsilon_b, \delta_a, \delta_b)$ implies the following

$$\Pr[\hat{\sigma}(S_k) \leq (1 + \varepsilon_a)\sigma(S_k)] \geq 1 - \delta_a \tag{3.12}$$

and

$$\Pr[\hat{\sigma}(S_k^*) \geq (1 - \varepsilon_b)opt] \geq 1 - \delta_b \tag{3.13}$$

where S_k^* is an optimal solution for the influence maximization, i.e., $opt = \sigma(S_k^*)$, and S_k is returned from the RIS algorithm.

Show the following: Consider a social network G and $0 \leq \varepsilon_a, \varepsilon_b, \delta_a, \delta_b \leq 1$. Suppose that $N(\varepsilon_a, \varepsilon_b, \delta_a, \delta_b)$ is an RIS Threshold. Let $\varepsilon \geq \varepsilon_a + (1 - 1/e)\varepsilon_b$ and $\delta \geq \delta_a + \delta_b$. Then, the RIS returns a seed set S_k satisfying

$$\Pr[\sigma(S_k) \geq (1 - 1/e - \varepsilon)opt] \geq 1 - \delta.$$

Historical Notes

There are many publications on the influence maximization problems [7, 51, 63, 147–150, 205, 214, 225, 238, 250, 252, 273, 311, 321, 363, 416, 429, 445, 505]. Before reverse sampling appears, most of works are emphasis on computational experiments and after reverse sampling appears, theoretical analysis plays an important role.

The reverse sampling was first proposed by Borgs et al. [22], and later improved by Tang et al. [385, 386] and by Nguyen et al. [310]. In [386], the martingale is involved. To understand this approach better, it is worth reading an excellent article given by Chung and Lu [59].

There are also a lot of works extending the problem to other optimizations regarding social influence [14, 46, 90, 168, 253, 270, 312, 358, 402, 480].

Chapter 4
Viral Marketing

One viral post can catapulte your brand.
— Germany Kent

The Best Marketing Is Education!
— Nylus Stanton

An important application of social networks is the viral marketing, that is, advertise products through mouth to mouth in online social networks. In this chapter, we study several models about viral marketing.

4.1   Multiple Products

A company may have more than one products. In consideration of promotion, a budget may set up for total spending on advertisement of those products. Suppose there are n products $1, 2, \ldots, n$. A seed about a product means to distribute a free sample or a discount coupon to a customer. Suppose that for product i, any seed has cost b_i. Let B be the budget of total seed cost and S_i the set of seeds distributed for product i. Then the budget restriction can be written as a knapsack constraint (Fig. 4.1)

$$b_1|S_1| + b_2|S_2| + \cdots + b_n|S_n| \leq B \quad (Fig. 4.1).$$

Suppose all products are independently advertised in the online social network. For product i, if a customer gets influenced, so that this customer adopts product i, then we count to have value w_i. Thus, we obtain an objective function for maximization

$$f(S_1, S_2, \ldots, S_n) = w_1 \cdot \sigma_m(S_1) + w_2 \cdot \sigma_m(S_2) + \cdots + w_m \cdot \sigma_m(S_n).$$

W. Wu et al., *Computational Aspects of Social Networks*, Springer Optimization and Its Applications 234, https://doi.org/10.1007/978-3-032-14833-9_4

Fig. 4.1 Multiple products

$$b_1|S_1| + b_2|S_2| + \cdots + b_n|S_n| \le B.$$

Knapsack Constraint

When $\sigma_m(S)$ is monotone nondecreasing and submodular, so is f. Hence, we obtain a maximization of monotone nondecreasing submodular function with a knapsack constraint. Its standard form is as follows:

Problem 4.1.1 (*Submodular Maximization with a Knapsack Contraint*) Let f be a polymatroid function on 2^X, i.e., f is monotone nondecreasing, submodular and $f(\emptyset) = 0$. The maximization problem is as follows.

$$\begin{aligned} \max \quad & f(S) \\ \text{subject to} \quad & \sum_{x \in S} b_x x \le B \\ & S \in 2^X \end{aligned}$$

where X is the universe set, and $b_x > 0$ is the budget cost of item x.

There are two classical approximation algorithms for this problem. They both are motivated from approximation algorithms for the classical knapsack problem with linear objective function as follows.

$$\begin{aligned} \max \quad & a_1 x_1 + a_2 x_2 + \cdots + a_n x_n \\ \text{subject to} \quad & b_1 x_1 + b_2 x_2 + \cdots + b_n x_n \le B \\ & x_1, x_2, \ldots, x_n \in \{0, 1\}, \end{aligned} \qquad (4.1)$$

where $B > 0$ and $a_i > 0, b_i > 0$ for $i = 1, 2, \ldots, n$, i.e., $i \in [n]$.

First, let us look at the (1/2)-approximation for the knapsack problem.

Algorithm 17 (1/2)-Approximation for Knapsack

---

input: The volume B of a knapsack and a set of n items, $[n]$ with nonnegative volume b_i and nonnegative value a_i for each item $i \in [n]$, satisfying $b_i \leq B$ for every $i \in [n]$.
output: A subset S of items, satisfying knapsack constraint.
1: add a virtual item $n + 1$ with $b_{n+1} = +\infty$ and $a_{n+1} = 0$
2: sort $\frac{a_1}{b_1} \geq \frac{a_2}{b_2} \geq \cdots \geq \frac{a_n}{b_n} \geq \frac{a_{n+1}}{b_{n+1}} = 0$
3: choose k such that $\sum_{i=1}^{k} b_i \leq B < \sum_{i=1}^{k+1} b_i$
4: **if** $a_{k+1} > \sum_{i=1}^{k} a_i$ **then**
5:     $S \leftarrow \{k + 1\}$
6: **else**
7:     $S \leftarrow [k]$
8: **end if**
9: **return** S

Theorem 4.1.2 *Algorithm 17 is (1/2)-approximation for the knapsack problem (4.1).*

Proof Note that if $a_{k+1} > \sum_{i=1}^{k} a_i$, then we must have $a_{k+1} > 0$ and hence $k + 1 \leq n$. Therefore, S is always a feasible solution. Moreover, from the sorting ordering and the choice of k, we see

$$opt < \sum_{i=1}^{k+1} a_i \leq 2 \sum_{i \in S} a_i$$

where *opt* is the objective function value of an optimal solution. Hence, S is (1/2)-approximation.   □

Now, we extend the idea for design of above algorithm to the problem (4.1.1).

Algorithm 18 $((1 - e^{-1})/2)$-Approximation for Problem (4.1.1)

---

input: A polymatroid function f on 2^X and knapsack constraint coefficients $0 \leq b(x) \leq B$ for $x \in X$.
output: A subset S_g of X.
1: $S \leftarrow \emptyset, Y \leftarrow X$;
2: **while** $Y \neq \emptyset$ **do**
3:     choose $x \in Y \setminus S$ to maximize $\frac{\Delta_x f(S)}{b(x)}$ and set $S \leftarrow S \cup \{x\}$;
4:     $B \leftarrow B - b(x)$;
5:     Remove all element e from Y with $b(e) > B$;
6: **end while**
7: $x \leftarrow \mathrm{argmax}_{e \in X \setminus S} \frac{\Delta_e f(S)}{b(e)}$;
8: **if** $f(\{x\}) > f(S)$ **then**
9:     $S_g \leftarrow \{x\}$
10: **else**
11:     $S_g \leftarrow S$
12: **end if**
13: **return** S_g

Theorem 4.1.3 (Leskovec et al. [238]) *Algorithm 18 is $((1 - e^{-1})/2)$-approximation for problem (4.1.1).*

Proof At the end of the algorithm, let $S = \{x_1, x_2, \ldots, x_k\}$ and $x_{k+1} = x$. Denote $S_i = \{x_1, x_2, \ldots, x_i\}$. Then for $1 \leq i \leq k$

$$x_{i+1} = \operatorname{argmax}_{x \in X \setminus S_i} \frac{\Delta_x f(S_i)}{b(x)},$$

and $b(S_k) \leq B < b(S_{k+1})$. Let $S^* = \{y_1, y_2, \ldots, y_h\}$ be an optimal solution. Denote $S_j^* = \{y_1, y_2, \ldots, y_j\}$. Then

$$
\begin{aligned}
f(S^*) &\leq f(S_i \cup S^*) \\
&= f(S_i) + \Delta_{y_1} f(S_i) + \Delta_{y_2} f(S_i \cup S_1^*) + \cdots + \Delta_{y_h} f(S_i \cup S_{h-1}^*) \\
&\leq f(S_i) + \Delta_{y_1} f(S_i) + \Delta_{y_2} f(S_i) + \cdots + \Delta_{y_h} f(S_I) \\
&\leq f(S_i) + \sum_{j=1}^{h} \frac{b(y_j)}{b(x_{i+1})} \cdot \Delta_{x_{i+1}} f(S_i) \\
&= f(S_i) + \frac{b(S^*)}{b(x_{i+1})} \cdot (f(S_{i+1}) - f(S_i)).
\end{aligned}
$$

Denote $\alpha_i = opt - f(S_i)$. Then we have

$$\alpha_i \leq \frac{b(S^*)}{b(x_{i+1})} \cdot (\alpha_i - \alpha_{i+1}).$$

Hence,

$$\alpha_{i+1} \leq \left(1 - \frac{b(x_{i+1})}{b(S^*)}\right) \alpha_i \leq \alpha_i \cdot e^{-\frac{b(x_{i+1})}{b(S^*)}}.$$

Therefore,

$$opt - f(S_{k+1}) = \alpha_{k+1} \leq \alpha_0 \cdot e^{-\frac{b(S_{k+1})}{b(S^*)}} \leq opt \cdot e^{-1}.$$

Hence,

$$(1 - e^{-1}) \cdot opt \leq f(S_{k+1}) \leq f(S_k) + f(\{x_{k+1}\}).$$

Thus,

$$(1 - e^{-1})/2 \cdot opt \leq \max(f(S_k), f(\{x_{k+1}\})).$$

$\square$

Next, we study the second algorithm.

Algorithm 19 $(1 - e^{-1})$)-Approximation for Problem (4.1.1)

---

input: A polymatroid function f on 2^X and knapsack constraint coefficients $0 \le b(x) \le B$ for $x \in X$.
output: A subset S_g of X.
1: **for** each $I \subseteq X$ with $|I| = d$ **do**
2:    $S \leftarrow I, Y \leftarrow X$;
3:    **while** $Y \ne \emptyset$ **do**
4:       choose $x \in X \setminus S$ to maximize $\frac{\Delta_x f(S)}{b(x)}$ and set $S \leftarrow S \cup \{x\}$
5:       $B \leftarrow B - b(x)$;
6:       Remove every element e from Y with $b(e) > B$;
7:    **end while**
8:    $S(I) \leftarrow S$
9: **end for**
10: **return** $S_g = \mathrm{argmax}\, f(S(I))$

Theorem 4.1.4 (Sviridenko [374]) *For $d = 3$, Algorithm 19 is $(1 - e^{-1})$-approximation for problem (4.1.1).*

__Proof__ Let $S^* = \{u_1, u_2, \ldots, u_h\}$ be an optimal solution in ordering

$$u_i = \mathrm{argmax}_u (f(\{u_1, u_2, \ldots, u_{i-1}, u\}) - f(\{u_1, u_2, \ldots, u_{i-1}\})).$$

Consider $I = \{u_1, u_2, \ldots, u_d\}$. Define $g(A) = f(A \cup I) - f(I)$. Then g is a polymatriod function since f is monotone nondecreasing submodular. Note that the computation of $S(I)$ in Algorithm 19 can be seen as a greedy algorithm applying to maximization of g. Assume that this computation produces $S(I) \setminus I = \{v_1, v_2, \ldots, v_k\}$. Next, consider two cases.

Case 1. $S(I) \setminus I \supseteq S^* \setminus I$. In this case, $S(I) \setminus I = S^* \setminus I$ and $g(S(I) \setminus I) = g(S^* \setminus I)$.

Case 2. There exists $v_{k+1} \in (S^* \setminus I) \setminus (S(I) \setminus I)$. In this case,

$$a(I \cup \{v_1, v_2, \ldots, v_k\}) \le a_0 < a(I \cup \{v_1, v_2, \ldots, v_{k+1}\}).$$

By argument in the proof of Theorem 4.1.3, we can obtain

$$g(\{v_1, v_2, \ldots, v_{k+1}\}) \ge (1 - 1/e)g(S^* \setminus I).$$

Therefore,

$$f(I \cup \{v_1, v_2, \ldots, v_{k+1}\}) \ge (1 - 1/e)f(S^*) + (1/e)f(I).$$

Note that $S(I) = I \cup \{v_1, v_2, \ldots, v_k\}$. Hence,

$$f(S(I)) \ge (1 - 1/e)f(S^*) + (1/e)f(I) - \Delta_{v_{k+1}} f(S(I)).$$

Since f is monotone nondecreasing and submodular, we have

$$\Delta_{v_{k+1}} f(S(I)) \leq \Delta_{v_{k+1}} f(\{u_1, \ldots, u_{d-1}\}).$$

Moreover, since $v_{k+1} \in S^*$, we have

$$\Delta_{v_{k+1}} f(\{u_1, \ldots, u_{d-1}\}) \leq \Delta_{u_d} f(\{u_1, \ldots, u_{d-1}\})$$
$$\leq \cdots$$
$$\leq \Delta_{u_1} f(\emptyset) = f(\{u_1\}).$$

Therefore,

$$\Delta_{v_{k+1}} f(S(I)) \leq \frac{1}{d}(\Delta_{u_d} f(\{u_1, \ldots, u_{d-1}\}) + \cdots + \Delta_{u_1} f(\emptyset)) = \frac{1}{d} f(I).$$

Thus,

$$f(S(I)) \geq (1 - 1/e) f(S^*) + (1/e) f(I) - \frac{1}{d} f(I) \geq (1 - 1/e) f(S^*)$$

for $d \geq 3$. $\qquad\square$

Now, let us design a randomized algorithm by using reverse sampling. The algorithm consists of two components.

- Randomly generates a collection of RR-sets.
- Find a set of seeds to hit the maximum number of RR-sets under a knapsack constraint.

Before describe them in detail, let us refresh our problem.

Consider a social network $G = (V, E)$ with an IC model. There are k products with unit value $w_1, \ldots, w_k$, respectively. There is a budget upper bound B for the ad. A seed is initiated by distributing a free sample or a discount coupon, and therefore it has a cost b_i for the product i. Our problem is find seed sets $S_1, \ldots, S_k$, respectively for k products to maximize

$$w_1 \sigma_{IC}(S_1) + \cdots + w_k \sigma_{IC}(S_k)$$

subject to

$$b_1|S_1| + \cdots + b_k|S_k| \leq B.$$

First, let us describe how to generate RR-sets.

- Denote $p_i = w_i / \sum_{i=1}^{k} w_i$. Select i with probability p_i.
- Select a node u from V with probability $1/|V|$.

- Select a subgraph H of G with probability $p(H)$. In IC model, $p(G) = \prod_{(x,y)\in E(H)} p_{xy} \cdot \prod_{(x,y)\in E\setminus E(H)}(1 - p_{xy})$ where $E(H)$ is the arc set of subgraph H.
- Generate an RR-set $R = \{v^i \mid H$ contains a path from v to $u\}$.

By using techniques in Chap. 3, we can obtain the following.

Theorem 4.1.5 (Chen et al. [43]) *By generating θ RR-sets, we can obtain a $(1 - e^{-1} - \varepsilon)$-approximation solution with probability at least $1 - (nk)^{-l}$ where*

$$\theta \geq (8k + 2\varepsilon)nk^2 w_{\max} \cdot \frac{l \log(nk) + \log(2k\alpha) + \log \binom{nk}{\alpha}}{\varepsilon^2 \cdot opt},$$

$w_{\max} = \max_{i\in[k]} w_i$ *and* $\alpha = B/\min_{i\in[k]} b_i$.

In the remainder of this chapter, each algorithm for solving viral marketing problem is similar to the above one, consisting of two components. One is the reverse sampling to generate RR-sets and the other one is to solve the deterministic version of the problem. To simplify our content, we will describe only the deterministic part.

Recently, several approximation algorithms have been designed for submodular maximization with a knapsack constraint [86, 468]. To end this section, let us introduce one of them, with the performance ratio 1/2 and the running time $O(n^2)$. The algorithm is described in Algorithm 20.

Algorithm 20 Greedy + Max: $\frac{1}{2}$-Approximation for Problem (4.1.1)

---

input: A polymatroid function f on 2^X and knapsack constraint coefficients $0 \leq b(x) \leq B$ for $x \in X$.
output: A subset S of X.
1: $G \leftarrow \emptyset, S \leftarrow \emptyset$;
2: **while** $X \neq \emptyset$ **do**
3:     $X' \leftarrow X, G' \leftarrow G$;
4:     $s \leftarrow \mathrm{argmax}_{e\in X} \frac{\Delta_e f(G)}{b(e)}$;
5:     $G \leftarrow G \cup \{s\}$;
6:     $B \leftarrow B - b(s)$;
7:     Remove all elements $e \in X$ with $b(e) > B$;
8: **end while**
9: $a \leftarrow \mathrm{argmax}_{e\in X'} \Delta_e f(G')$;
10: $S \leftarrow G' \cup \{a\}$;
11: **return** S

Let G be obtained by Algorithm 20 and $G = \{s_1, s_2, \ldots, s_m\}$ where $s_1, s_2, \ldots, s_m$ are elements added in G, in the selection order of the algorithm. Note that S always contains one more element than G. Let s_{m+1} be this element. Denote $G_i = \{s_1, s_2, \ldots, s_i\}$. Next, we define three functions.

Definition 4.1.6 (*Greedy Performance Function (Fig. 4.2)*) For $x \in [0, b(G_m))$, let i be the smallest index such that $b(G_i) > x$. Define greedy performance function $g(x)$ by

$$g(x) = f(G_{i-1}) + (x - b(G_{i-1})) \cdot \frac{\Delta_{s_i} f(G_{i-1})}{b(s_i)}.$$

Clearly, $g(x)$ is a monotone increasing, piecewise linear function and its right derivative exists, denoted by $g'(x)$.

Definition 4.1.7 (*Greedy + Max Performance Function (Fig. 4.3)*) For $x \in [0, b(G_m))$, let i be the smallest index such that $b(G_i) > x$. Define greedy+max performance function $g_+(x)$ by

$$g_+(x) = g(x) + \Delta_v f(G_{i-1})$$

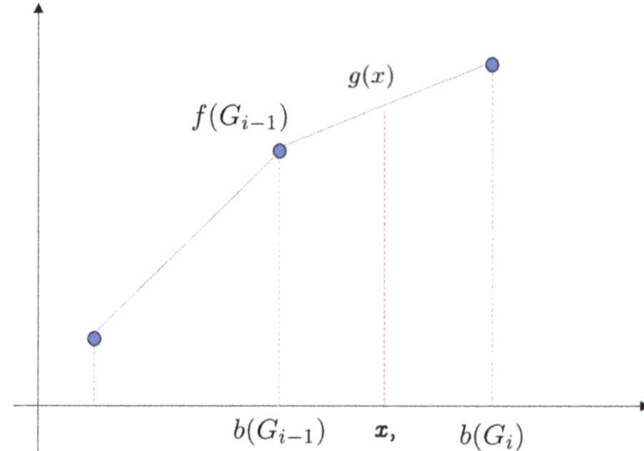

Fig. 4.2 Greedy performance function

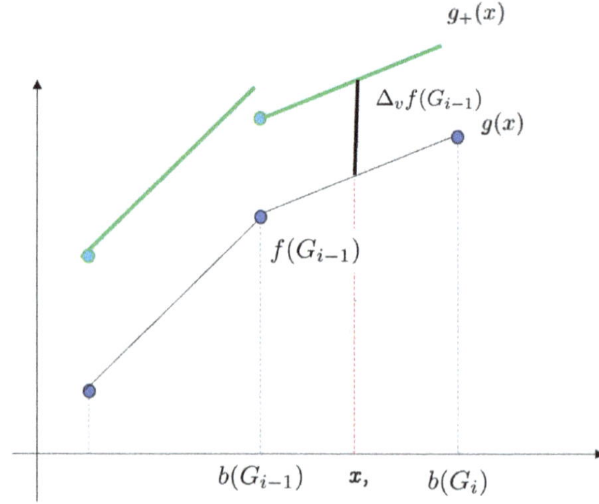

Fig. 4.3 Greedy + Max performance function

where

$$v = \operatorname*{argmax}_{e \in X \setminus G_{i-1} : b(\{e\} \cup G_{i-1}) \le B} \Delta_e f(G_{i-1}).$$

Let $o_1 = \operatorname*{argmax}_{o \in OPT} b(o)$. Let $b^* \ge 0$ such that there exists an index i^* satisfying

$$b(s_1) + \cdots + b(s_{i^*-1}) = B - b(o_1) - b^*$$

and

$$b(s_1) + \cdots + b(s_{i^*-1}) + b(s_{i^*}) > B - b(o_1).$$

Definition 4.1.8 (*Greedy + Max Performance Lower Bound (Fig. 4.4)*) For $x \in [0, B - b(o_1) - b^*]$, let i be the smallest index such that $b(G_i) > x$. Define

$$g_1(x) = g(x) + \Delta_{o_1} f(G_{i-1}).$$

Clearly, $g_1(x) \le g_+(x)$.

Lemma 4.1.9 (Yaroslavtsev et al. [468]) *For $x \in [0, B - b(o_1) - b^*]$,*

$$g_1(x) + (B - b(o_1))g'(x) \ge opt.$$

Proof Note that $g'(x)$ is a constant in $[b(G_{i-1}), b(G_i))$ and $g_1(x)$ is monotone increasing. It is sufficient to prove that the inequality holds for $x = b(G_{i-1})$.

Since $g_1(b(G_{i-1})) = g(b(G_{i-1})) + \Delta_{o_1} f(G_{i-1}) = f(G_{i-1} \cup \{o_1\})$, we have

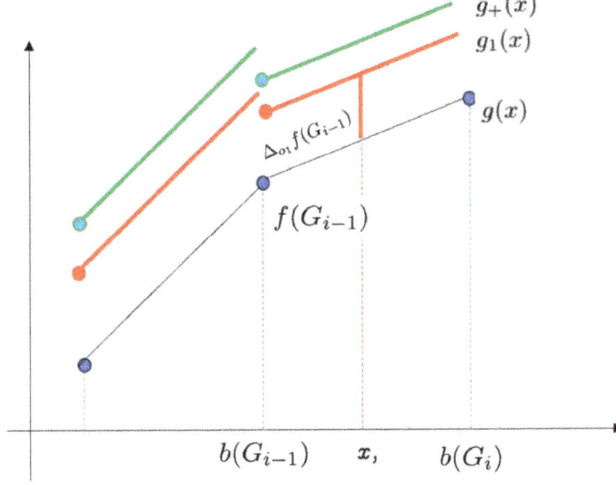

Fig. 4.4 Greedy + Max performance lower bound

$$
\begin{aligned}
opt &\leq f(G_{i-1} \cup OPT) \\
&= f(G_{i-1} \cup \{o_1\}) + \Delta_{OPT\setminus(\{o_1\}\cup G_{i-1})} f(G_{i-1} \cup \{o_1\}) \\
&\leq g_1(x) + \sum_{e\in OPT\setminus(\{o_1\}\cup G_{i-1})} \Delta_e f(G_{i-1} \cup \{o_1\}) \\
&= g_1(x) + \sum_{e\in OPT\setminus(\{o_1\}\cup G_{i-1})} b(e) \cdot \frac{\Delta_e f(G_{i-1} \cup \{o_1\})}{b(e)} \\
&\leq g_1(x) + \sum_{e\in OPT\setminus(\{o_1\}\cup G_{i-1})} b(e) \cdot \frac{\Delta_e f(G_{i-1})}{b(e)} \\
&\leq g_1(x) + \sum_{e\in OPT\setminus(\{o_1\}\cup G_{i-1})} b(e) \cdot g'(x)
\end{aligned}
$$

where the first three inequalities hold due to submodularity of f and for $x \in [0, 1 - b(o_1) - b^*]$, all items in $OPT \setminus (\{o_1\} \cup G_{i-1})$ are within knapsack constraint. The fourth inequality holds due to the greedy rule. Thus,

$$
\begin{aligned}
opt &\leq g_1(x) + g'(x) \cdot \sum_{e\in OPT\setminus(\{o_1\}\cup G_{i-1})} b(e) \\
&= g_1(x) + g'(x) \cdot b(OPT \setminus (\{o_1\} \cup G_{i-1})) \\
&\leq g_1(x) + (B - b(o_1))g'(x)
\end{aligned}
$$

where the inequality is due to $b(OPT) \leq B$ and $o_1 \in OPT$. $\square$

Theorem 4.1.10 (Yaroslavtsev et al. [468]) *Algorithm 20 gives a $\frac{1}{2}$-approximation for Problem 4.1.1 with running time $O(n^2)$.*

Proof By Lemma 4.1.9 at $x = B - b(o_1) - b^*$,

$$
g_1(B - b(o_1) - b^*) + (B - b(o_1))g'(B - b(o_1) - b^*) \geq opt.
$$

Consider two cases.

Case 1. $g_1(B - b(o_1) - b^*) \geq (1/2)opt$. In this case, we have

$$
\begin{aligned}
f(S) &\geq g_+(b(G_{i^*-1})) \\
&\geq g_1(b(G_{i^*-1})) \\
&= g_1(B - b(o_1) - b^*) \geq (1/2)opt.
\end{aligned}
$$

Case 2. $(B - b(o_1))g'(B - b(o_1) - b^*) \geq (1/2)opt$. By the submodularity, it can be seen that $g'(x)$ is monotone non-increasing. Since $g(0) = 0$, we have

$$
g(x) = \int_0^x g'(y)dy \geq \int_0^x g'(x)dy = g'(x) \cdot x.
$$

Set $x = B - b(o_1) - b^*$. We obtain

$$g(B - b(o_1) - b^*) \geq g'(B - b(o_1) - b^*) \cdot (B - b(o_1) - b^*). \qquad (4.2)$$

Note that $f(G_{i*-1}) = g(B - b(o_1) - b^*)$ and $b^* \leq b(s_{i*})$. Thus,

$$\begin{aligned}
f(S) &\geq f(G_{i*}) \\
&= f(G_{i*-1}) + b(s_{i*}) \cdot \frac{\Delta_{s_{i*}} f(G_{i*-1})}{b(s_{i*})} \\
&= g(B - b(o_1) - b^*) + b(s_{i*}) \cdot g'(B - b(o_1) - b^*).
\end{aligned}$$

By (4.2),

$$\begin{aligned}
f(S) &\geq g(B - b(o_1) - b^*) + b(s_{i*}) \cdot g'(B - b(o_1) - b^*) \\
&\geq g'(B - b(o_1) - b^*) \cdot (B - b(o_1) - b^*) + b^* \cdot g'(B - b(o_1) - b^*) \\
&= g'(B - b(o_1) - b^*)(B - b(o_1)) \\
&\geq (1/2)opt.
\end{aligned}$$

For estimation of running time, we note that there exist at most n iterations and in each iteration, the running time is $O(n)$. Therefore, the total time is $O(n^2)$. $\square$

4.2 Balanced Seed-Distribution

In social networks, each community consists of a group of people with similar interests. Therefore, when doing viral marketing, we may wish to distribute seeds (coupons or free samples) in a well-balanced way, i.e., not put too many seeds in some communities and too less seeds in some other communities. For this purpose, we may set an upper bound for the number of seeds distributed in each community. For different communities, upper bounds may be different since communities may have different sizes and different potentials for adopting the product. Motivated from this consideration, we can formulate the following problem.

Problem 4.2.1 (*Balanced Seed-Distribution*) Consider a social Network $G = (V, E)$ with community partition $(V_1, V_2, \ldots, V_p)$. Let $k_1, K_2, \ldots, k_p$ and k be $p + 1$ positive integers with $k_i \leq k$ for $i = 1, 2, \ldots, p$. The problem is as follows.

$$\begin{aligned}
\max \quad & \sigma_m(S) \\
\text{subject to} \quad & |S \cap V_i| \leq k_i \text{ for } 1 \leq i \leq p, \\
& |S| \leq k,
\end{aligned}$$

where m is a diffusion model.

The following lemma indicates that the constraint in Problem 4.2.1 is the intersection of two matroids.[1]

Lemma 4.2.2 *Let* $\mathcal{I}_1 = \{S \subseteq V \mid |S \cap V_i| \le k_i \forall i \in [p]\}$. *Then,* $(V, \mathcal{I}_1)$ *is a matroid.*

Proof Clearly, $\mathcal{I}_1$ satisfies the hereditary property. To show the augmentation property, consider $A, B \in \mathcal{I}_1$ with $|A| < |B|$. Then, there exists $i \in [p]$ such that $|A \cap V_i| < |B \cap V_i|$. Therefore, there exists a node $x \in B \cap V_i \subseteq A \cap V_i$. This implies that $|(A \cup \{x\}) \cap V_i| \le |B \cap V_i| \le k_i$. Thus, $A \cup \{x\} \in \mathcal{I}_1$. $\square$

Let $\mathcal{I}_2 = \{S \subseteq V \mid |S| \le k\}$. By Lemma 4.2.4, $(V, \mathcal{I}_2)$ is also a matroid. Hence, Problem 4.2.1 can be represented as follows.

$$\max \quad \sigma_m(S) \tag{4.3}$$
$$\text{subject to} \quad S \in \mathcal{I}_1 \cap \mathcal{I}_2.$$

Consider a diffusion model m such that the influence spread $\sigma_m(S)$ is a monotone nondecreasing submodular function. Then, Problem (4.3) is a special case of the following.

Problem 4.2.3 (*Submodular Maximization with Matroid Constraints*) Let f be a monotone nondecreasing submodular function over 2^X. Let $(X, \mathcal{I}_i)$ be a matroid for every $i \in [h]$. The problem is as follows.

$$\max \quad f(S)$$
$$\text{subject to} \quad S \in \mathcal{I}_i \quad \forall i \in [h].$$

Denote $\mathcal{I} = \cap_{i=1}^k \mathcal{I}_i$. Then $(X, \mathcal{I})$ may not be a matroid, however it is an independent system since $A \subset B \in \mathcal{I}$ implies $A \in \mathcal{I}$. In an independent system $(X, \mathcal{I})$, every set in $\mathcal{I}$ is still called an *independent set*. The following is a property of independent systems $(X, \mathcal{I})$ which is the intersection of k matroids $(X, \mathcal{I})$.

Lemma 4.2.4 *Consider* $A, B \in \mathcal{I}$. *Suppose that I is a maximal independent subset of $A \cup B$, containing B. Then* $|A \setminus I| \le k|B \setminus A|$.

Proof Consider matroid $(X, \mathcal{I}_i)$. Let J_A^i be a maximal independent set in $A \cup B$, containing A. Let J_I^i be a maximal independent set in $A \cup B$, containing I. Then, $|J_A^i| = |J_I^i|$. Therefore, $|A \setminus J_I^i| = |B \setminus J_A^i| \le |B \setminus A|$.

---

[1] For knowledge about matriod and independent system, please see [81].

Now, consider independent system $(X, \mathcal{I})$. Since I is a maximal independent set in $A \cup B$, containing B and $\cap_{i=1}^{k} J_I^i$ is an independent set, containing I, we have $I = \cap_{i=1}^{k} J_I^i$. Therefore,

$$|A \setminus I| = |A \setminus \cap_{i=1}^{k} J_I^i| \leq \sum_{i=1}^{k} |A \setminus J_I^i| \leq k|B \setminus A|.$$

$\square$

For Problem 4.2.3, there is a result as follows.

Theorem 4.2.5 (Fisher et al. [113]) *For Problem 4.2.3, there exists a greedy approximation with performance ratio $\frac{1}{h+1}$.*

Proof Consider Greedy Algorithm 21. We will prove that this algorithm generates an approximation solution with performance ratio $\frac{1}{h+1}$.

Algorithm 21 Greedy Approximation for Problem 4.2.3

Input: a monotone nondecreasing function f on 2^X with $f(\emptyset) = 0$, and k matroids $(X, \mathcal{I}_i)$ for $1 \leq i \leq k$..
Output: A subset $A \in \mathcal{I}$ and $f(A)$ where $\mathcal{I} = \cap_{i=1}^{k} \mathcal{I}_i$.
1: $A \leftarrow \emptyset$;
2: **while** A is not a maximal independent set in $(X, \mathcal{I})$ **do**
3:     choose $x \in X \setminus A$ to maximize $\Delta_x f(A)$ subject to $A \cup \{x\} \in \mathcal{I}$, and
4:     set $A \leftarrow A \cup \{x\}$
5: **end while**
6: **return** A and $f(A)$.

During computational process of the algorithm, suppose that set A is assigned with $A_0 = \emptyset, A_1, \ldots, A_g$. Then, $|A_i \setminus A_{i-1}| = 1$ for $i = 1, \ldots, g$. Let H_0 be an optimal solution. For $i = 1, 2, \ldots, g$, let H_i be a maximal independent set of $H_{i-1} \cup A_i$, containing A_i. By Lemma 4.2.4,

$$|H_{i-1} \setminus H_i| \leq k|A_i \setminus H_{i-1}| \leq |A_i \setminus A_{i-1}| \leq k.$$

Suppose $H_{i-1} \setminus H_i = \{v_1, \ldots, v_r\}$ for $1 \leq r \leq k$. Denote $H_{i-1}^j = H_{i-1} \setminus \{v_1, \ldots, v_j\}$. Then

$$f(A_i) - f(A_{i-1}) \geq f(\{v_j\} \cup A_{i-1}) - f(A_{i-1})$$
$$= \Delta_{v_j} f(A_{i-1})$$
$$\geq \Delta_{v_j} f(H_{i-1}^j \cup A_{i-1}).$$

Thus,

$$r(f(A_i) - f(A_{i-1})) \geq \sum_{j=1}^{r} \Delta_{v_j} f(H_{i-1}^{j} \cup A_{i-1})$$
$$= f(H_{i-1}^{0} \cup A_{i-1}) - f(H_{i-1}^{r} \cup A_{i-1})$$
$$= f(H_{i-1}) - f(H_i).$$

Thus,

$$k[f(A_g) - f(A_0)] \geq r \sum_{i=1}^{g} [f(A_i) - f(A_{i-1})]$$
$$\geq \sum_{i=1}^{g} [f(H_{i-1}) - f(H_i)]$$
$$= f(H_0) - f(H_g).$$

Note that $A_g = H_g$ and $f(A_0) = f(\emptyset) = 0$. Therefore,

$$(k + 1)f(A_g) \geq f(H_0) = opt.$$

$\square$

4.3   Continuous Greedy

Consider the balanced seed-distribution problem again. When $k_1 + k_2 + \cdots + k_p \leq k$, the constraint $|S| \leq k$ can be removed. In this case, the balanced seed-distribution problem can be represented as the maximization of a monotone nondecreasing submodular function with a matroid constraint. By Theorem 4.2.5, there exists a greedy $(1/2)$-approximation. However, we can do better by using another type of algorithm, continuous greedy. The continuous greedy is a relaxation algorithm consisting of three steps.

- Make multilinear extension of a set function and establish some properties of multilinear extension of submodular function. Those properties will play important roles in analysis of computational results in the 2nd and the 3rd steps.
- Compute a fractional solution for the relaxed problem.
- By the pipage rounding, find an integer solution from the fractional solution.

In this section, we study the first two steps and put the 3rd step in the next section.

Let $f(S)$ be a set function $f(S)$ over 2^X. Suppose $X = [n]$. For each element i, let x_i be an indicator indicating whether i belongs to subset S or not, i.e.,

$$
x_i = \begin{cases} 1 \text{ if } i \in S, \\ 0 \text{ otherwise.} \end{cases}
$$

Now, we consider $f(S)$ as a function $F(x) = f(S)$ defined on $x = (x_1, \ldots, x_n) \in \{0, 1\}^n$, and then extend $F(x)$ to $x \in [0, 1]^n$ through the multilinear extension as follows.

For each $x \in [0, 1]^n$, let S be a random set that element i belongs to S with probability x_i. Define $F(x)$ to be the expectation of function value of F, that is,

$$
F(x) = \mathbb{E}_{S \sim x}[f(S)] = \sum_{S \in 2^X} f(S) \left(\prod_{i \in S} x_i \right) \left(\prod_{i \notin S} (1 - x_i) \right).
$$

The multilinear extension F of f has the following properties.

Theorem 4.3.1 *Suppose F is the multilinear extension of f. Then,*

1. *If f is monotone nondecreasing, then F is monotone nondecreasing along any direction $d \geq 0$.*
2. *If f is submodular, then F is concave along any line $d \geq 0$.*
3. *If f is submodular, then F is convex along line $e_i - e_j$ $(i \neq j)$ where e_i has its ith component equal 1 and others equal 0.*

Proof Note the following representations of $\frac{\partial F(x)}{\partial x_i}$ and $\frac{\partial^2 F(x)}{\partial x_i \partial x_j}$.

Let R be the random subset of $X \setminus \{i\}$ such that R contains an element $j \in X \setminus \{i\}$ with probability x_j. Then,

$$
\frac{\partial F(x)}{\partial x_i} = F(x_1, \ldots, x_{i-1}, 1, x_{i+1}, \ldots, x_n) - F(x_1, \ldots, x_{i-1}, 0, x_{i+1}, \ldots, x_n)
$$

$$
= E[f(R \cup \{i\})] - E[f(R)].
$$

Let R' be a random subset of $X \setminus \{i, j\}$ such that R' contains an element $k \in X \setminus \{i, j\}$ with probability x_k. Then for $i \neq j$,

$$
\frac{\partial^2 F(x)}{\partial x_i \partial x_j} = E[f(R' \cup \{i, j\})] - E[f(R' \cup \{i\})] - E[f(R' \cup \{j\})] + E[f(R')]
$$

$$
= E[f(R' \cup \{i, j\}) - f(R' \cup \{i\}) - f(R' \cup \{j\}) + f(R')].
$$

1. If f is monotone nondecreasing, we have

$$
\frac{\partial F(x)}{\partial x_i} \geq 0.
$$

Let $x \in [0, 1]^n$ and $d \in [0, +\infty)^n$. Then

$$\frac{dF(x + \alpha d)}{d\alpha} = \langle d, \nabla F(x + \alpha d)\rangle \geq 0.$$

Therefore, along direction d, $F(x)$ is monotone nondecreasing.

2. If f is submodular, then for $i \neq j$,

$$\frac{\partial^2 F(x)}{\partial x_i \partial x_j} \geq 0$$

and

$$\frac{\partial^2 F(x)}{\partial x_i^2} = 0.$$

Therefore, for $x \in [0, 1]^n$ and $d \in [0, +\infty)^n$,

$$\frac{d^2 F(x + \alpha d)}{d\alpha^2} = d^T H_f(x + \alpha d)d \leq 0$$

where $H_f(x + \alpha d)$ is Hessian matrix of f at point $x + \alpha d$. It implies that $F(x)$ is concave along line d.

3. Note that for $d = e_i - e_j$, we have

$$\frac{d^2 F(x + \alpha d)}{d\alpha^2} = d^T H_f(x + \alpha d)d = -2\frac{\partial^2 F(x)}{\partial x_i \partial x_j} \geq 0.$$

Therefore, F is convex along $e_i - e_j$. $\square$

In the continuous greedy algorithm, we need to compute $F(x)$ and $\nabla F(x)$. How to compute them? Since in the definition of $F(x)$, the sum is over all subsets of X, i.e., it has $2^{|X|}$ terms, computing $F(x)$ for a single x may take exponential time.

This trouble can be overcome by Monte Carlo method. In fact, its representation through expectation $F(x) = E_{S \sim x}[f(S)]$ suggests a computation with random sampling. By Chernoff bounds, we can obtain

$$\left| \frac{1}{t} \sum_{i=1}^{t} f(S_i) - F(x) \right| \leq \varepsilon \cdot f(X)$$

with probability at least $1 - e^{t\varepsilon^2/4}$, where $S_1, \ldots, S_t$ are random subsets based on element selection probability x. Therefore, if $t = O(\frac{1}{\varepsilon^2})$, then with a constant probability, we can compute an approximation of $F(x)$ within $\varepsilon F(X)$ error.

Similarly, the sampling can be employed for computing $\nabla F(x)$ since

$$\frac{\partial F(x)}{\partial x_i} = E[f(R \cup \{i\})] - E[f(R)].$$

Next, we consider the following problem.

Problem 4.3.2 (*Monotone Submodular Maximization with a Matriod Constraint*)

$$\max \quad f(S)$$
$$\text{subject to} \quad S \in \mathcal{I}$$

where f is a monotone nondecreasing, submodular function on 2^X for a finite set X and $\mathcal{I}$ is the family of independent sets of a matroid $\mathcal{M} = (X, \mathcal{I})$.

When $f(S)$ is extended to $F(x)$, the constraint $S \in \mathcal{I}$ should be also extended to a polytope which is a convex hull of $\mathbf{1}_S$ for $S \in \mathcal{I}$ where

$$(\mathbf{1}_I)_i = \begin{cases} 1 & \text{if } i \in I, \\ 0 & \text{otherwise.} \end{cases}$$

This polytope is denoted by $P(\mathcal{M})$, called the polytope of matroid $\mathcal{M}$, which also be represented as

$$P(\mathcal{M}) = \{x \geq 0 \mid \forall S \in 2^X : \sum_{j \in S} \leq r_{\mathcal{M}}(S)\}$$

and

$$r_{\mathcal{M}}(S) = \max\{|I| \mid I \subseteq S, I \in \mathcal{I}\}.$$

Now, Problem 4.3.2 is relaxed into the following.

$$\max \quad F(x) \tag{4.4}$$
$$\text{subject to} \quad x \in P(\mathcal{M}).$$

This relaxation can be solved by a continuous greedy algorithm as shown in Algorithm 22.

Algorithm 22 Continuous Greedy

---

input $F(x)$, $P(\mathcal{M})$, and T.
output $x(1)$.
1: $\alpha \leftarrow \frac{1}{T}$.
2: $x(0) \leftarrow 0 \in R^n$.
3: **for** $t \leftarrow 1$ to T **do**
4:     $v \leftarrow v_{\max}(x((t-1)\alpha))$
5:     where $v_{\max}(x) = \text{argmax}_{v \in P(\mathcal{M})} \langle v, \nabla F(x) \rangle$.
6:     $x(t\alpha) \leftarrow x((t-1)\alpha) + \alpha v$
7: **end for**
8: **return** $x(1)$.

Let us explain how to compute $v_{\max}(x) = \text{argmax}_{v \in P(\mathcal{M})} \langle v, \nabla F(x) \rangle$. It looks like a trouble, but not a real trouble. From matriod theory, there exist several ways to represent $P(\mathcal{M})$. However, in general, everyone involves a large number of constraints. For example, there are exponential number of inequalities in following representation

$$P(\mathcal{M}) = \{x \geq 0 \mid \forall S \in 2^X : \sum_{j \in S} x_j \leq r_{\mathcal{M}}(S)\}.$$

Why it is not a real trouble? The reason is that $P(\mathcal{M})$ is the convex hull of $\mathbf{1}_S$ for $S \in \mathcal{I}$. Since optimal solution of linear programming can be found in vertices, $v_{\max}(x)$ can be a solution of $\max\{\langle \mathbf{1}_S, \nabla F(x) \rangle \mid S \in \mathcal{I}\}$. This can be solved by a greedy algorithm since $\nabla F(x) \geq 0$.

Next, we analyze Algorithm 22.

Lemma 4.3.3 *For any $x \in R^n$, $\langle v_{\max}(x), \nabla F(x) \rangle \geq opt - F(x)$, where opt is the objective function value of optimal solution for the relaxation (4.4), and $v_{\max}(x) = argmax_{v \in P(\mathcal{M})} \langle v, \nabla F(x) \rangle$.*

Proof Let v be an optimal solution of the relaxation (4.4), i.e., $F(v) = opt$. We show that $\langle v, \nabla F(x) \rangle \geq F(v) - F(x)$.

To do so, consider $d = (v - x) \vee 0$, where $x \vee y$ denotes the coordinate-wise max of x and y, i.e., $(x \vee y)_i = \max(x_i, y_i)$. Since $d \geq 0$, F is concave along direction d. Thus,

$$\langle d, \nabla F(x) \rangle \geq F(x + d) - F(x).$$

Since $d \leq v$ and $v \in P(\mathcal{M})$, we have $d \in P(\mathcal{M})$. Hence,

$$\langle d, \nabla F(x) \rangle \leq \langle v_{\max}(x), \nabla F(x) \rangle.$$

Therefore,

$$\langle v_{\max}(x), \nabla F(x) \rangle \geq \langle d, \nabla F(x) \rangle \geq F(x + d) - F(x) \geq F(v) - F(x).$$

The last inequality sign holds because $x + d \geq v$ and $F(\cdot)$ is monotone non-decreasing along d, which results from the monotone nondecreasing property of f. $\qquad\square$

Lemma 4.3.4 $x(1) \in P(\mathcal{M})$ *and*

$$F(x(1)) \geq \left(1 - \left(1 - \frac{1}{T}\right)^T\right) \cdot opt - \frac{c}{T}$$

for some constant c.

Proof Note that

$$x(1) = \frac{1}{T} \sum_{t=0}^{T-1} v_{\max}\left(x\left(\frac{t}{T}\right)\right),$$

that is, $x(1)$ is a convex combination of points in $P(\mathcal{M})$. Therefore, $x(1) \in P(\mathcal{M})$.

Next, we employ Taylor expansion on $F(x((t+1)\alpha) = F(x(t\alpha) + \alpha \cdot v_{\max}(x(t\alpha)))$. For some constant c, we have

$$\begin{aligned}
F(x((t+1)\alpha)) &\geq F(x(t\alpha)) + \alpha \cdot \langle v_{\max}(x(t\alpha)), \nabla F(x(t\alpha))\rangle - c\alpha^2 \\
&\geq F(x(t\alpha)) + \alpha[opt - F(x(t\alpha))] - c\alpha^2 \\
&= (1 - \alpha)F(x(t\alpha)) + \alpha \cdot opt - c\alpha^2.
\end{aligned}$$

The second inequality is due to Lemma 4.3.3. Therefore,

$$\begin{aligned}
opt - F(x((t+1)\alpha)) &\leq (1 - \alpha)(opt - F(x(t\alpha))) + c\alpha^2 \\
&\leq (1 - \alpha)^{t+1}(opt - F(x(0))) + c\alpha^2(t + 1) \\
&= (1 - \alpha)^{t+1} \cdot opt + c\alpha^2(t + 1)
\end{aligned}$$

since $F(x(0)) = 0$. Setting $t + 1 = T$, we have

$$F(x(1)) \geq (1 - (1 - \alpha)^T) \cdot opt - \frac{c}{T}.$$

$\qquad\square$

Theorem 4.3.5 (Calinescu et al. [34]) *Set $T = O(1/\varepsilon)$ in the continuous greedy algorithm. Then we can obtain*

$$F(x(1)) \geq (1 - e^{-1}) \cdot opt - \varepsilon.$$

Proof It follows immediately from Lemma 4.3.4. $\qquad\square$

How to round $x(1)$ into an integer solution? By Theorem 4.3.1, F is convex along line $e_i - e_j$. Therefore, we can apply pipage rounding to obtain an integer solution $\hat{x}$ such that $F(\hat{x}) \geq F(x(1))$.

Before studying the pipage rounding in next section, let us give the continuous greedy more explanation as follows.

Consider the following algorithm.

Algorithm 23 Non-discretized Continuous Greedy

---

input $F(x)$ and $P(\mathcal{M})$.
output $x(1)$.
1: Define $v_{\max}(x) = \mathrm{argmax}_{v \in P(\mathcal{M})} \langle v, \nabla F(x) \rangle$.
2: $x(0) \leftarrow 0 \in R^n$.
3: **for** $t \in [0, 1]$ **do**
4:    $x'(t) = v_{\max}(x(t))$
5: **end for**
6: **return** $x(1)$.

Let us analyze it.

Theorem 4.3.6 *Let F be the multilinear extension of a nondecreasing submodular function f. Then $x(1)$ obtained by Algorithm 23 satisfies the following.*

1. *$x(1) \in P(\mathcal{M})$.*
2. *$F(x(1)) \geq (1 - e^{-1}) \cdot opt$, where opt is the value of F at an optimal solution of the relaxation (4.4).*

Proof 1. Note that

$$x(1) = \int_0^1 x'(t)dt = \int_0^1 v_{\max}(x(t))dt$$

$$= \lim_{T \to \infty} \frac{1}{T} \sum_{i=1}^{T} v_{\max}(x(\frac{i}{T})).$$

Since $v_{\max}(x(\frac{i}{T})) \in P(\mathcal{M})$ and $P(\mathcal{M})$ is convex and closed, we have $x(1) \in P(\mathcal{M})$.

2. Note that

$$\frac{d}{dt} F(x(t)) = \langle x'(t), \nabla F(x(t)) \rangle = \langle v_{\max}(x(t)), \nabla F(x(t)) \rangle.$$

By Lemma 4.3.3, there exists $v \in P(\mathcal{M})$ such that

$$\langle v, \nabla F(x(t)) \rangle \geq opt - F(x(t)).$$

Therefore,

$$\langle v_{\max}(x(t)), \nabla F(x(t)) \rangle \geq opt - F(x(t)).$$

Thus,

$$\frac{d}{dt}F(x(t)) \geq opt - F(x(t)).$$

Denote $g(t) = F(x(t))$. Then we have

$$g'(t) + g(t) \geq opt \text{ and } g(0) = 0.$$

Define $h(t) = g'(t) + g(t)$. Solve this differential equation. Then, we obtain

$$g(t) = \int_0^t e^{x-t}h(x)dx.$$

Hence,

$$F(x(1)) = g(1) \geq opt \cdot \int_0^1 e^{x-1}dx = opt \cdot (1 - e^{-1}).$$

$\square$

Now, let us discuss issues about implementation of Algorithm 23. How to solve differential equation $x'(t) = v_{\max}(x(t))$ numerically? Actually, the continuous greedy (Algorithm 22) give a simple solution.

4.4 Pipage Rounding

In this section, we introduce a rounding technique, called the pipage rounding since it has application in continuous greedy algorithm.

To explain this technique, consider a specific problem:

Problem 4.4.1 (*Maximum Weight Hitting*) Given a collection C of subsets of a finite set X with nonnegative weight function w on C and a positive integer p, find a subcollection A of X with $|A| = p$ to maximize the total weight of subsets hit by A.

Assume $X = \{1, 2, \ldots, n\}$ and $C = \{S_1, S_2, \ldots, S_m\}$. Denote $w_i = w(S_i)$. Let x_i be a 0-1 variable to indicate whether element i is in subset A. Then this problem can be formulated into the following integer LP.

$$\max \quad \sum_{j=1}^{m} w_j z_j \tag{4.5}$$

$$\text{s.t.} \quad \sum_{i \in S_j} x_i \geq z_j, \ j = 1, \ldots, m,$$

$$\sum_{i=1}^{n} x_i = p$$

$$x_i \in \{0, 1\}, \ i = 1, 2, \ldots, n$$

$$z_j \in \{0, 1\}, \ j = 1, 2, \ldots, m.$$

There are two equivalent formulations as follows.

$$\max L(x) = \sum_{j=1}^{m} w_j \min \left\{ 1, \sum_{i \in S_j} x_i \right\} \tag{4.6}$$

$$\text{s.t.} \quad \sum_{i=1}^{n} x_i = p$$

$$x_i \in \{0, 1\}, \ i = 1, 2, \ldots, n$$

$$\max F(x) = \sum_{j=1}^{m} w_j \left(1 - \prod_{i \in S_j} (1 - x_i) \right) \tag{4.7}$$

$$\text{s.t.} \quad \sum_{i=1}^{n} x_i = p$$

$$x_i \in \{0, 1\}, \ i = 1, 2, \ldots, n$$

$L(x)$ and $F(x)$ has the same value when each x_i takes value 0 or 1. But, when x_i is relaxed to $0 \leq x_i \leq 1$, they may have different values. The following gives a relationship between them for real number x.

Lemma 4.4.2 $F(x) \geq (1 - 1/e)L(x)$ for $0 \leq x \leq 1$.

Proof Note that

$$1 - \prod_{i \in S_j} (1 - x_i) \geq 1 - \left(\frac{\sum_{i \in S_j} (1 - x_i)}{k} \right)^k \quad (k = |S_j|)$$

$$\geq 1 - \left(1 - \frac{\sum_{i \in S_j} x_i}{k} \right)^k .$$

Set $g(z) = 1 - (1 - \frac{z}{k})^k$. Then $g(0) = 0$ and $g(1) = 1 - (1 - \frac{1}{k})^k$. Moreover, $g(z)$ is monotone increasing and concave in $[0, 1]$ since

$$g'(z) = \left(1 - \frac{z}{k}\right)^{k-1} \geq 0$$

and

$$g''(z) = (k - 1)\left(1 - \frac{z}{k}\right)^{k-2}\left(-\frac{1}{k}\right) \leq 0.$$

Therefore,

$$g(z) \geq z\left(1 - \left(1 - \frac{1}{k}\right)^k\right) \geq z\left(1 - \frac{1}{e}\right).$$

Thus, we have

$$1 - \prod_{i \in S_j}(1 - x_i) \geq (1 - e^{-1})\sum_{i \in S_j} x_i \geq \min\left(1, \sum_{i \in S_j} x_i\right).$$

$\square$

Now, we consider LP-relaxation of (4.6).

$$\max L(x) = \sum_{j=1}^{m} w_j \min\left\{1, \sum_{i \in S_j} x_i\right\} \tag{4.8}$$

$$\text{s.t.} \quad \sum_{i=1}^{n} x_i = p$$

$$0 \leq x_i \leq 1, i = 1, 2, \ldots, n.$$

It is equivalent to the following LP.

$$\max \sum_{j=1}^{m} w_j z_j \tag{4.9}$$

$$\text{s.t.} \quad \sum_{i \in S_j} x_i \geq z_j, j = 1, \ldots, m,$$

$$\sum_{i=1}^{n} x_i = p$$

$$0 \leq x_i \leq 1, i = 1, 2, \ldots, n$$

$$0 \leq z_j \leq 1, j = 1, 2, \ldots, m.$$

Therefore, it is polynomial-time solvable. Let x^* be an optimal solution of (4.8). We will use the following rounding to find an integer solution $\bar{x}$ from x^*.

Pipage Rounding

$x \leftarrow x^*$;

while x has an non-integer component **do begin**

    choose $0 < x_k < 1$ and $0 < x_j < 1$ $(k \neq j)$;

    define $x(\varepsilon)$ by setting

$$x_i(\varepsilon) = \begin{cases} x_i & \text{if } i \neq k, j, \\ x_j + \varepsilon & \text{if } i = j, \\ x_k - \varepsilon & \text{if } i = k; \end{cases}$$

    define $\varepsilon_1 = \min(x_j, 1 - x_k)$ and $\varepsilon_2 = \min(1 - x_j, x_k)$;

    if $F(x(-\varepsilon_1)) \geq F(x(\varepsilon_2))$

        then $x \leftarrow x(-\varepsilon_1)$

        else $x \leftarrow x(\varepsilon_2)$;

end-while;

return $\bar{x} = x$.

The existence of x_k and x_j is due to the fact that when x has a non-integer component, x has at least two non-integer components since $\sum_{i=1}^{n} x_i = p$.

The following is an important property of $F(x(\varepsilon))$.

Lemma 4.4.3 $F(x(\varepsilon))$ *is convex with respect to* ε.

Proof If S_j contains only one of k and j, then the jth term of $F(x(\varepsilon))$, corresponding to $1 - \prod_{i \in S_j}(1 - x_i)$, is linear and hence convex with respect to ε. If S_j contains both k and j, then the jth term of $F(x(\varepsilon))$, corresponding to $1 - \prod_{i \in S_j}(1 - x_i)$, is in form

$$g(\varepsilon) = 1 - a(b + \varepsilon)(c - \varepsilon)$$

where a, b and c are nonnegative constants with respect to ε. If $a = 0$, then this term is a constant 1 and hence convex. For $a > 0$, since $g''(\varepsilon) = a > 0$, $g(\varepsilon)$ is convex. Finally, we note that the sum of several convex functions is convex. $\square$

By Lemma 4.4.3, the value of $F(x)$ is nondecreasing during the pipage rounding process. Therefore, $F(\bar{x}) \geq F(x^*)$.

Theorem 4.4.4 *The maximum weight hitting problem has polynomial time an* $(1 - e^{-1})$*-approximation.*

Proof $L(\bar{x}) = F(\bar{x}) \geq F(x^*) \geq (1 - 1/e)L(x^*) \geq (1 - e^{-1}) \cdot opt.$ $\square$

From above example, we may get a little impression on pipage rounding. Next, we give a general description.

Consider a bipartite graph $G = (U, V, E)$ and an integer programming with 0-1 variables x_e each associated with an edge e and each constraint is in the form

$$\sum_{e \in \delta(v)} x_e \leq p_e$$

or

$$\sum_{e \in \delta(v)} x_e = p_e$$

or

$$\sum_{e \in \delta(v)} x_e \geq p_e$$

where $\delta(v)$ is the set of all edges incident to $v \in U \cup V$ and p_e is a nonnegative integer. For example, we may consider the following integer programming.

$$\max \quad L(x) \tag{4.10}$$
$$\text{s.t.} \quad \sum_{e \in \delta(v)} x_e \leq p_v \text{ for } v \in U \cup V$$
$$x_e \in \{0, 1\} \text{ for } e \in E.$$

As shown in Fig. 4.5, suppose $L(x)$ has a company $F(x)$ such that
(A1) $L(x) = F(x)$ for $x_e \in \{0, 1\}$, and
(A2) $L(x) \leq cF(x)$ for $0 \leq x_e \leq 1$.

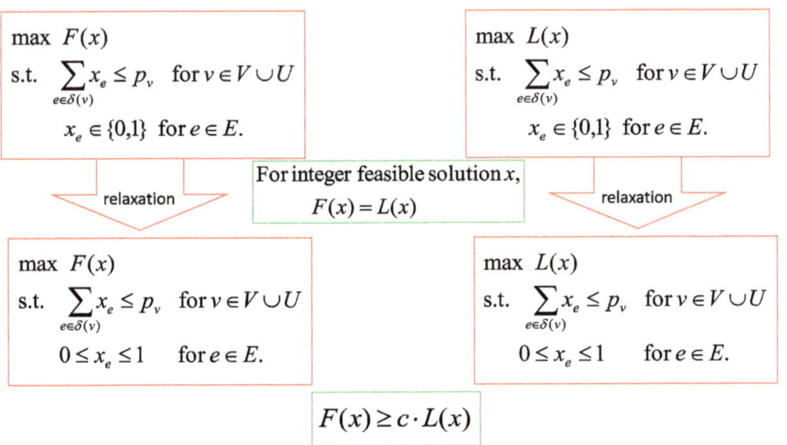

Fig. 4.5 Let x^* be a maximum solution of the right LP. By ε-convexity of $F(x)$, an integer solution $\bar{x}$ can be found with pipage rounding such that $F(\bar{x}) \geq F(x^*)$. Thus, $L(\bar{x}) = F(\bar{x}) \geq F(x^*) \geq cL(x^*) \geq c \cdot opt$

We also assume the following.

(A3) The relaxation of integer programming (4.10) is equivalent to a LP or is polynomial-time solvable.

Suppose x^* is an optimal solution of the relaxation of (4.10), i.e.,

$$\max \quad L(x) \tag{4.11}$$

$$\text{s.t.} \quad \sum_{e \in \delta(v)} x_e \le p_v \text{ for } v \in U \cup V$$

$$0 \le x_e \le 1 \text{ for } e \in E.$$

We next explain a pipage rounding procedure to obtain an integer solution $\bar{x}$ from x^*.

Initially, set $x \leftarrow x^*$. While x is not an integer solution, we will do the following:

1. Consider the subgraph H_x of G induced by all edges e with $0 < x_e < 1$. Let R be a cycle or a maximal path of H_x. Then R can be decomposed into two matchings M_1 and M_2.
2. Define $x(\varepsilon)$ by

$$x_e(\varepsilon) = \begin{cases} x_e & \text{if } e \notin R, \\ x_e + \varepsilon & \text{if } e \in M_1, \\ x_e - \varepsilon & \text{if } e \in M_2. \end{cases}$$

Define

$$\varepsilon_1 = \min(\min_{e \in M_1} x_e, \min_{e \in M_2}(1 - x_e))$$

$$\varepsilon_2 = \min(\min_{e \in M_1}(1 - x_e), \min_{e \in M_2} x_e).$$

3. If $F(x(-\varepsilon_1)) \ge F(x(\varepsilon_2))$
   then $x \leftarrow x(-\varepsilon_1)$
   else $x \leftarrow x(-\varepsilon_2)$.

Lemma 4.4.5 *For $\varepsilon \in [-\varepsilon_1, \varepsilon_2]$, $x(\varepsilon)$ is feasible for (4.11).*

Proof If R is a cycle, then $\sum_{e \in \delta(v)} x_e(\varepsilon) = \sum_{e \in \delta(v)} x_e$ and hence $x(\varepsilon)$ is feasible.

If R is a maximal path, then only for v being an endpoint of R, $\sum_{e \in \delta(v)} x_e(\varepsilon) \ne \sum_{e \in \delta(v)} x_e$. Suppose $e' \in \delta(v) \cap R$. Since R is a maximal path, we have that for $e \in \delta(v) \setminus \{e'\}$, x_e is an integer. Therefore,

$$p_v - \sum_{e \in \delta(v)} x_e(\varepsilon) = p_v - \sum_{e \in \delta(v) \setminus \{e'\}} x_e - x_{e'}(\varepsilon) \ge 1 - x_{e'}(\varepsilon) \ge 0.$$

$\square$

Assume $F(x)$ is ε-convex, i.e.,

(A4) For any R, $F(x(\varepsilon))$ is convex with respect to ε.

Then by pipage rounding, we would obtain an integer solution $\bar{x}$ such that $F(\bar{x}) \geq F(x^*)$. Therefore,

$$F(\bar{x}) \geq F(x^*) \geq cL(x^*) \geq c \cdot opt.$$

For the maximum weight hitting problem, we faced a star which is a bipartite graph $G = (U, V, E)$ where $U = \{u\}$, $V = \{v_1, v_2, \ldots, v_n\}$ and $E = \{(u, v_1), (u, v_2), \ldots, (u, v_2)\}$. Each variable x_i corresponds to an edge (u, v_i). Therefore, in each iteration of pipage rounding, we deal with a maximal path consisting of two edges.

4.5 Revenue Maximization

In viral marketing, instead of influence spread, the revenue is often considered as an objective function in the maximization problem. Suppose that each seed is realized by distributing a discount coupon. Then, the relationship between the revenue $r(S)$ and influence spread $\sigma(S)$ is as follows.

$$r(S) = \sigma(S) \cdot \alpha - |S| \cdot \beta$$

where α is the difference of selling price and the product cost for each product and β is the discount amount for each coupon.

Note that if $\sigma(S)$ is submodular, so is $r(S)$. However, even if $\sigma(S)$ is monotone nondecreasing, then $r(S)$ may not be monotone. If $\alpha \geq \beta$, then $r(S)$ is nonnegative. The following has been studied in the literature.

Problem 4.5.1 (*Revenue Maximization*) Let $f(S)$ be a nonnegative submodular function over 2^X. The problem is to maximize $f(S)$ subject to certain number of matroid constraints or/and knapsack constraints.

There are many approximation algorithms designed for the revenue optimization problem. Especially, Greedy + Max and Continuous Greedy can be modified for this purpose. However, since the objective function is non-monotone, the analysis of algorithms gets harder, and the approximation performance ratio gets worse. For example, consider a nonnegative submodular function f over 2^X and a positive integer k. Algorithm 24 is a random greedy approximation for problem $\max\{f(S \mid |S| \leq k\}$.

Algorithm 24 Random Greedy Algorithm

Input: A set function over 2^X and an integer $k > 0$.
Output: A set $S \in 2^X$ with $|S| \leq k$.
1: $S_0 \leftarrow \emptyset$;
2: **for** $i = 1$ to k **do**
3: select $M_i \subseteq X \setminus S_{i-1}$ with size at most k to maximize $\sum_{u \in M_i} \Delta_u f(S_{i-1})$;
4: choose u_i in mutually exclusive and uniformly random way from M_i with probability $1/k$ and not choose any element with probability $1 - |M_i|/k$;
5: **if** u_i is selected **then**
6:   $S_i \leftarrow S_{i-1} \cup \{u_i\}$
7: **else**
8:   $S_i \leftarrow S_{i-1}$;
9: **end if**
10: **end for**
11: **return** S_k.

To analyze Algorithm 24, let us show a lemma.

Lemma 4.5.2 *Let f be a submodular function on subsets of a finite set X. Suppose that $A(p)$ is a random subset of a set A such that each element of A appears in $A(p)$, with probability at most p. Then*

$$\mathbb{E}[f(A(p)] \geq (1-p)f(\emptyset).$$

Proof Let $A = \{u_1, u_2, \ldots, u_h\}$, where $h = |A|$. Denote $p_i = \Pr[u_i \in A(p)]$. Assume $p_1 \geq p_2 \geq \cdots p_h$. Let x_i be the indicator for the event that $u_i \in A(p)$, that is,

$$x_i = \begin{cases} 1 \text{ if } u_i \in A(p) \\ 0 \text{ otherwise.} \end{cases}$$

Let $A_i = \{u_1, u_2, \ldots, u_i\}$. Then we have

$$\mathbb{E}[f(A(p)] = \mathbb{E}\left[f(\emptyset) + \sum_{i=1}^{h} x_i \cdot \Delta_{u_i} f(A(p) \cap A_{i-1}) \right]$$

$$\geq \mathbb{E}\left[f(\emptyset) + \sum_{i=1}^{h} x_i \cdot \Delta_{u_i} f(A_{i-1}) \right]$$

$$= f(\emptyset) + \sum_{i=1}^{h} \mathbb{E}[x_i] \cdot \Delta_{u_i} f(A_{i-1})$$

$$= f(\emptyset) + \sum_{i=1}^{h} p_i \cdot \Delta_{u_i} f(A_{i-1})$$

$$= (1 - p_1) f(\emptyset) \sum_{i=1}^{h} (p_i - p_{i-1}) f(A_i) + p_h f(A)$$

$$\geq (1 - p) f(\emptyset).$$

$\square$

Now, we analyze the performance of Algorithm 24.

Theorem 4.5.3 (Buchbinder et al. [32]) *If f is a nonnegative submodular function, then Algorithm 24 has approximation performance ratio $1/e$, i.e.,*

$$\mathbb{E}[f(S_k)] \geq \frac{1}{e} \cdot opt.$$

Proof First, we claim that for $0 \leq i \leq k$,

$$\mathbb{E}[f(OPT \cup A_i)] \geq \left(1 - \frac{1}{k}\right)^i \cdot opt$$

where OPT is an optimal solution and $opt = f(OPT)$.

In fact, in each iteration i, an element of $X \setminus A_{i-1}$ lies outside of A_i, with probability at least $1 - 1/k$. Hence, each element of $X \setminus A_{i-1}$ is selected into A_i, with probability at most $1 - (1 - 1/k)^i$. Since f is submodular, $g(S) = f(S \cup OPT)$ is submodular. By Lemma 4.5.2, we have

$$\mathbb{E}[f(OPT \cup A_i)] = \mathbb{E}[g(A_i)] \geq (1 - 1/k)^i \cdot g(\emptyset) = (1 - 1/k)^i \cdot opt.$$

This completes the proof of our claim.

Fix all random processes until A_i is obtained for $1 \leq i \leq k$. By the greedy choice of A_i, we have

$$\mathbb{E}[\Delta_{u_i} f(A_{i-1})] = \frac{1}{k} \cdot \sum_{u \in M_i} \Delta_u f(A_{i-1})$$

$$\geq \frac{1}{k} \cdot \sum_{u \in OPT \setminus A_{i-1}} \Delta_u f(A_{i-1})$$

$$\geq \frac{1}{k} \cdot (f(OPT \cup A_{i-1}) - f(A_{i-1})).$$

The second inequality is due to the submodularity of f.

Release randomness of A_i and A_{i-1} and take the expectation. Using our claim, we have

$$\mathbb{E}[\Delta_{u_i} f(A_{i-1})] \geq \frac{1}{k} \cdot (\mathbb{E}[f(OPT \cup A_{i-1})] - \mathbb{E}[f(A_{i-1})])$$

$$\geq \frac{1}{k} \cdot \left(\left(1 - \frac{1}{k}\right)^{i-1} \cdot opt - \mathbb{E}[f(A_{i-1})] \right).$$

Therefore,

$$\mathbb{E}[f(A_i)] \geq \frac{1}{k} \cdot \left(1 - \frac{1}{k}\right)^{i-1} \cdot opt + \left(1 - \frac{1}{k}\right) \cdot \mathbb{E}[f(A_{i-1}]$$

$$\geq \frac{1}{k} \cdot \left(1 - \frac{1}{k}\right)^{i-1} \cdot opt.$$

Letting $i = k$, we obtain

$$\mathbb{E}[f(A_k)] \geq \left(1 - \frac{1}{k}\right)^{k-1} \cdot opt \geq e^{-1} \cdot opt.$$

$\square$

4.6   Product with Multiple Models

In this section, we study viral marketing for product with multiple models (Fig. 4.6). This is different from the viral marketing for multi-products in the following.

- For multi-products, each customer may purchase more than one product.
- For the product with multiple models, each customer would purchase at most one product with some model.

Based on this difference, viral marketing for product with multiple models may enjoy a multi-set function formulation different from the one for multi-products.

Consider a grand set V. Let $(k+1)^V$ denote the family of all k disjoint subsets of V, i.e.,

$$(k+1)^V = \{(X_1, \ldots, X_k) \mid X_i \subset V, \forall i \in [k]; X_i \cap X_j = \emptyset, \forall i \neq j\}$$

where $[k] = \{1, \ldots, k\}$ The submodularity of set functions can be generalized to multi-set functions as follows.

$$b_1|S_1| + b_2|S_2| + \cdots + b_n|S_n| \le B$$

Fig. 4.6 Product with multiple models

Definition 4.6.1 (*k-submodularity*) A function $f : (k+1)^V \to \mathbb{R}$ is said to be k-submodular if for any two k-tuples $X, Y \in (k+1)^V$,

$$f(X) + f(Y) \ge f(X \sqcap Y) + f(X \sqcup Y)$$

where

$$X \sqcap Y = (X_1 \cap Y_1, \ldots, X_k \cap Y_k)$$

and

$$X \sqcup Y = ((X_1 \cup Y_1) \setminus \cup_{i \neq 1}(X_i \cup Y_i), \ldots, (X_k \cup Y_k) \setminus \cup_{i \neq k}(X_i \cup Y_i)).$$

For each k-tuple $X \in (k+1)^V$, there exists unique set $S = \{(a, d) \mid a \in X_d, d \in [k]\}$, called the *solution* of X while each (a, d) is called an *item-dimension pair*. For simplicity, we will use X and its solution S interchangeably. For any solution S, define $U(S) = \{a \mid \exists d \in [k], (a, d) \in S\}$.

For any item-dimension pair (a, d) and a solution S, denote

$$\Delta_{a,d} f(S) = f(S \cup \{(a, d)\}) - f(S).$$

f is said to satisfy the *orthant submodularity* if for any two solutions S, S' with $S \subseteq S'$, $a \notin U(S')$, and $d \in [k]$,

$$\Delta_{a,d} f(S) \ge \Delta_{a,d} f(S').$$

For $k = 1$, it is well-known that f is 1-submodular if and only if f is orthant submodular. However, for $k \ge 2$, this is not true. In fact, we have the following.

Lemma 4.6.2 (Ward and Zivny [431]) *A function $f : (k+1)^V \to \mathbb{R}$ is k-submodular if and only if f is orthant submodular and pairwise monotone. Here, f is pairwise monotone if for any $S \in (k+1)^V$, $a \notin U(S)$, $d_1, d_2 \in [k]$, and $d_1 \neq d_2$,*

$$\Delta_{a,d_1} f(S) + \Delta_{a,d_2} f(S) \geq 0.$$

Now, let us return to viral marketing for products with multiple models. Consider a social network $G = (V, E)$ with the IC model and a product with k (≥ 2) models. A seed means to distribute a discount coupon or a free sample. Let X_i be the set of seeds with model $i \in [k]$. Let $\sigma_i(X)$ be the influence spread on model i (i.e., the expected number of nodes which adopt product with model i) under seed distribution $X = (X_1, \ldots, X_k)$. Then, the problem can be formulated as follows.

$$\max \quad f(X) = \sum_{i=1}^{k} p_i \cdot \sigma_i(X)$$

$$\text{subject to} \quad \sum_{i=1}^{k} b_i \cdot |X_i| \leq B$$

where p_i is the price of the product with model i, b_i is the cost of seed with model i, and B is the total budget limit for seed distribution.

Lemma 4.6.3 *For $i \in [k]$, $\sigma_i(X)$ is orthant submodular.*

Proof Consider two solutions S, S' with $S \subseteq S', d \in [k]$ and $a \notin U(S')$. $\Delta_{a,d} f(S') \leq \Delta_{a,d}(S)$ for the following reasons: $S \subseteq S'$ implies $U(S) \subseteq U(S')$. The influence of additional seeds in $U(S') \setminus U(S)$ will cover up the marginal influence of (a, d). $\square$

By Lemma 4.6.3, f is orthant submodular. In general, f is not pairwise monotone. This is because adding seed with the lower price model may block the spread of influence of product with the high price model. However, if all prices are the same, or our objective is the number of customers adopting the product, then it is easy to see that f is monotone nondecreasing, and hence it is k-submodular. In this special case, the problem can be formulated as follows.

Problem 4.6.4 (*Monotone k-submodular Maximization with Knapsack Constraint*) Let $f : (k+1)^V \to \mathbb{R}$ be a monotone nondecreasing k-submodular function with $f(\emptyset) = 0$. The maximization problem is as follows.

$$\max \quad f(S)$$

$$\text{subject to} \quad \sum_{x \in U(S)} b(x)x \leq B$$

$$S \in (k+1)^V$$

where V is the universe set, and $b(x)$ is the budget cost of item x.

The greedy + max algorithm can be extended to yield $\frac{1}{3}$-approximation (Algorithm 25 for Problem 4.6.4 (Tang et al. [387]).

Algorithm 25 Greedy + Max: $\frac{1}{3}$-Approximation for Problem (4.6.4)

---

input: A monotone k-submodular function $f : (k + 1)^V \to \mathbb{R}$ with $f(\emptyset) = 0$ and knapsack constraint coefficients $0 \le b(x) \le B$ for $x \in V$.
output: A subset S of V.
1: $G \leftarrow \emptyset, S \leftarrow \emptyset, Y \leftarrow V$;
2: **while** $Y \neq \emptyset$ **do**
3:    $Y' \leftarrow Y, G' \leftarrow G$;
4:    $(a', d') \leftarrow \mathrm{argmax}_{(a,d) \in Y} \frac{\Delta_{d,d} f(G)}{b(a)}$;
5:    $G \leftarrow G \cup \{(a', d')\}$;
6:    $B \leftarrow B - b(a')$;
7:    Remove all elements $a \in Y$ with $b(a) > B$;
8: **end while**
9: $(a'd') \leftarrow \mathrm{argmax}_{(a,d) \in Y'} \Delta_{a,d} f(G')$;
10: $S \leftarrow G' \cup \{(a', d')\}$
11: **return** S

For simplicity of analysis of Algorithm 25, we assume $B = 1$ and $opt = f(OPT) = 1$ by normalizing

$$b(e) \leftarrow b(e)/B, \quad f(A) \leftarrow f(A)/opt.$$

Define $b(a, d) = b(a)$ for $d \in [k]$. Let G be obtained by Algorithm 25 and $G = \{(a_1, d_1), (a_2, d_2), \ldots, (a_m, d_m)\}$ where elements are in the selection order of the algorithm. Note that S always contains one element not in G. Let (a_{m+1}, d_{m+1}) be this element. Denote $G_i = \{(a_1, d_1), (a_2, d_2), \ldots, (a_i, d_i)\}$. Next, we define two functions.

Definition 4.6.5 (*Greedy Performance Function*) For $x \in [0, 1]$, let i be the smallest index such that $b(G_i) > x$. Define greedy performance function $g(x)$ by

$$g(x) = f(G_{i-1}) + (x - b(G_{i-1})) \cdot \frac{\Delta_{a_i, d_i} f(G_{i-1})}{b(a_i)}.$$

Clearly, $g(x)$ is a monotone increasing, piecewise linear function and its right derivative exists, denoted by $g'(x)$.

Let $o_1 = \mathrm{argmax}_{o \in OPT} b(o)$. Let $b^* \ge 0$ such that there exists an index i^* satisfying

$$b(a_1) + \cdots + b(a_{i^*-1}) = 1 - b(o_1) - b^*$$

and

$$b(a_1) + \cdots + b(a_{i^*-1}) + b(a_{i^*}) > 1 - b(o_1).$$

Definition 4.6.6 (*Greedy + Max Performance Lower Bound*) For $x \in [0, 1 - b(o_1) - b^*]$, let i be the smallest index such that $b(G_i) > x$. Define

$$g_1(x) = g(x) + \max_{d \in [k]} \Delta_{o_1,d} f(G_{i-1}).$$

For preparation of analysis of Algorithm 25, let us consider a dimension selection problem.

Problem 4.6.7 (*Dimension Selection*) Consider a function $f : (k+1)^V \to \mathbb{R}$. Given ℓ items $s_1, s_2, \ldots, s_\ell$ in V, select dimensions $d_1, d_2, \ldots, d_\ell$ for these items to maximize $f(\{(s_1, d_1), (s_2, d_2), \ldots, (s_\ell, d_\ell)\})$.

Lemma 4.6.8 *Let* $T = \{(s_1, d_1^*), (s_2, d_2^*), \ldots, (s_\ell, d_\ell^*)\}$ *be the maximum solution of the dimension selection problem. Let* $S = \{(s_1, d_1), (s_2, d_2), \ldots, (s_\ell, d_\ell)\}$ *be obtained by the greedy selection*

$$(s_i, d_i) = argmax_{d_i \in [k]} f(S_{i-1})$$

where $S_{i-1} = \{(s_1, d_1), \ldots, (s_{i-1}, d_{i-1})\}$. *Define*

$$T_j = (T \setminus \{(s_1, d_1^*), \ldots, (s_j, d_j^*)\}) \cup S_j.$$

If f *is monotone nondecreasing and k-submodular, then for* $1 \le t \le \ell$,

$$f(T) \le 2f(S_t) + \sum_{(s,d) \in T_t \setminus S_t} \Delta_{s,d} f(S_t).$$

Proof For $0 \le j < t$, define

$$P_j = T_j \setminus \{(s_{j+1}, d_{j+1}^*)\} = T_{j+1} \setminus \{(s_{j+1}, d_{j+1})\}.$$

Then,

$$f(T_j) = f(P_j) + \Delta_{s_{j+1}, d_{j+1}^*} f(P_j),$$

$$f(T_{j+1}) = f(P_j) + \Delta_{s_{j+1}, d_{j+1}} f(P_j).$$

Therefore,

$$\begin{aligned}
f(T_j) - f(T_{j+1}) &= \Delta_{s_{j+1}, d_{j+1}^*} f(P_j) - \Delta_{s_{j+1}, d_{j+1}} f(P_j) \\
&\le \Delta_{s_{j+1}, d_{j+1}^*} f(S_j) \\
&\le \Delta_{s_{j+1}, d_{j+1}} f(S_j) \\
&= f(S_{j+1}) - f(S_j).
\end{aligned}$$

The first inequality is due to $S_j \subseteq P_j$ and the monotonicity of f. The second inequality is due to the greedy selection of d_{j+1}. Summing up from $j = 0$ to $t - 1$, we obtain

$$f(T_0) - f(T_t) \leq f(S_t) - f(S_0) = f(S_t).$$

Since $S_t \subseteq T_t$ and f is orthant submodular, we have

$$f(T) = f(T_0) \leq f(T_t) + f(S_t) \leq 2f(S_t) + \sum_{(a,d)\in T_t \setminus S_t} \Delta_{a,d} f(S_t).$$

$\square$

Lemma 4.6.9 (Tang et al. [387]) *For $x \in [0, 1 - b(o_1) - b^*]$,*

$$2g_1(x) + (1 - b(o_1))g'(x) \geq 1 = opt.$$

Proof Note that $g'(x)$ is a constant in $[b(G_{i-1}), b(G_i))$ and $g_1(x)$ is monotone increasing. It is sufficient to prove that the inequality holds for $x = b(G_{i-1})$. At this point,

$$g_1(x) = g(b(G_{i-1})) + \max_{d\in[k]} \Delta_{o_1,d} f(G_{i-1}) = f(G_{i-1}) + \max_{d\in[k]} \Delta_{o_1,d} f(G_{i-1}).$$

Let $V' = U(G_{i-1}) \cup U(OPT) = \{s_1, s_2, \ldots, s_\ell\}$ such that $s_1 = a_1, \ldots, s_{i-1} = a_{i-1}$ and

$$o_1 = \begin{cases} s_i & \text{if } o_1 \notin U(G_{i-1}), \\ s_{i-1} & \text{if } o_1 \in U(G_{i-1}). \end{cases}$$

Let T be an optimal solution of the dimension-selection problem on V'. Let $\{(s_1, d_1), (s_2, d_2), \ldots, (s_\ell, d_\ell)\}$ be the greedy solution.

Let us first consider the case that $o_1 \notin U(G_{i-1})$. In this case, $s_i = o_1$, and $S_i = S_{i-1} \cup \{(o_1, d_i)\} = G_{i-1} \cup \{(o_1, d_i)\}$. By Lemma 4.6.8 with $t = i$,

$$f(T) \leq 2f(S_i) + \sum_{(s,d)\in T_i \setminus S_i} \Delta_{s,d} f(S_i)$$

$$= 2f(G_{i-1} \cup \{(o_1, d_i)\}) +$$

$$\sum_{(s,d)\in T_i \setminus (G_i \cup \{(o_1,d_i)\})} \Delta_{s,d} f(G_{i-1} \cup \{(o_1, d_i)\})$$

$$= 2g_1(x) + \sum_{(s,d)\in T_i \setminus (G_{i-1} \cup \{(o_1,d_i)\})} \Delta_{s,d} f(G_{i-1} \cup \{o_1\})$$

$$= 2g_1(x) + \sum_{(s,d)\in T_i \setminus (G_{i-1} \cup \{(o_1,d_i)\})} b(s) \cdot \frac{\Delta_{s,d} f(G_{i-1} \cup \{o_1\})}{b(s)}$$

$$\leq 2g_1(x) + \sum_{(s,d)\in T_i\setminus(G_{i-1}\cup\{(o_1,d_i)\})} b(s)\cdot g'(x)$$

$$= 2g_1(x) + g'(x)\cdot b(T_i\setminus(G_{i-1}\cup\{(o_1,d_i)\}))$$

$$\leq 2g_1(x) + g'(x)(1-b(o_1)).$$

The second inequality is due to

$$\frac{\Delta_{s,d}f(G_{i-1}\cup\{(o_1,d_i)\})}{b(s)} \leq \frac{\Delta_{s,d}f(G_{i-1})}{b(s)} \leq g'(x)$$

for every $(s,d)\in T_i\setminus(G_{i-1}\cup\{(o_1,d_i)\})$.

In the case that $o_1\in U(G_{i-1})$. Note that $s_{i-1}=o_1$. Set $t=i-1$ in Lemma 4.6.8. The proof can be done similarly by following the above argument. $\qquad\square$

Theorem 4.6.10 (Tang et al. [387]) *Algorithm 25 gives a $\frac{1}{3}$-approximation for Problem 4.6.4 with running time $O(kn^2)$.*

Proof By Lemma 4.6.9 at $x=1-b(o_1)-b^*$,

$$2g_1(1-b(o_1)-b^*) + (1-b(o_1))g'(1-b(o_1)-b^*) \geq 1.$$

Consider two cases.

Case 1. $g_1(1-b(o_1)-b^*)\geq 1/3$. In this case, we have

$$\begin{aligned} f(S) &\geq g_1(b(G_{i^*-1})) \\ &= g_1(1-b(o_1)-b^*) \\ &\geq 1/3. \end{aligned}$$

Case 2. $(1-b(o_1))g'(1-b(o_1)-b^*)\geq 1/3$. By greedy rule, it can be seen that $g'(x)$ is monotone non-increasing. Since $g(0)=0$, we have

$$g(x) = \int_0^x g'(y)dy \geq \int_0^x g'(x)dy = g'(x)\cdot x.$$

Set $x=1-b(o_1)-b^*$. We obtain

$$g(1-b(o_1)-b^*) \geq g'(1-b(o_1)-b^*)\cdot(1-b(o_1)-b^*). \tag{4.12}$$

Note that $f(G_{i^*-1})=g(1-b(o_1)-b^*)$ and $b^*\leq b(s_{i^*})$. Thus,

$$\begin{aligned} f(S) &\geq f(G_{i^*}) \\ &= f(G_{i^*-1}) + b(s_{i^*})\cdot\frac{\Delta_{s_{i^*}}f(G_{i^*-1})}{b(s_{i^*}} \\ &= g(1-b(o_1)-b^*) + b(s_{i^*})\cdot g'(1-b(o_1)-b^*). \end{aligned}$$

By (4.12),

$$
\begin{aligned}
f(S) &\geq g(1 - b(o_1) - b^*) + b(s_{i*}) \cdot g'(1 - b(o_1) - b^*) \\
&\geq g'(1 - b(o_1) - b^*) \cdot (1 - b(o_1) - b^*) + b^* \cdot g'(1 - b(o_1) - b^*) \\
&= g'(1 - b(o_1) - b^*)(1 - b(o_1)) \\
&\geq 1/3.
\end{aligned}
$$

For the estimation of running time, we note that there exist at most n iterations and in each iteration, the running time is $O(n)$. Therefore, the total time is $O(n^2)$. $\qquad\square$

Exercises

1. Consider a set function f over 2^X and a positive integer k. Algorithm 24 is a random greedy approximation for problem $\max\{f(S) \mid |S| \leq k\}$. Prove that if f is monotone nondecreasing and submodular, then Algorithm 24 has approximation performance ratio $1 - 1/e$, i.e.,

$$
\mathbb{E}[f(S_k)] \geq (1 - 1/e) \cdot opt.
$$

2. Let $\mathcal{I}$ be a collection of subsets of a finite set E. $(E, \mathcal{I})$ is called an *independent system* if for any $A \in \mathcal{I}$ and $B \subset A$, we have $B \in \mathcal{I}$. Consider an independent system $(E, \mathcal{I})$. A subset A of E is said to be *independent* if $A \in \mathcal{I}$. Consider the following problem:

$$
\max\{f(A) \mid A \in \mathcal{I}\}
$$

   where $f(A) = \sum_{x \in A} f(x)$ and f is a nonnegative function over E. Please design a greedy algorithm with performance ratio at least

$$
\max_{F \subseteq E} \frac{u(F)}{v(F)}
$$

   where $u(F)$ is the smallest size of maximal independent set contained in F and $v(F)$ is the largest size of maximal independent set contained in F.

3. Note that an independent system $(\mathcal{C}, E)$ is a matroid if and only if for any $F \subseteq E$, $u(F) = v(F)$. Please show that every independent system is the intersection of finitely many matroids.

4. A linear function is monotone nondecreasing and submodular if and only if all coefficients are nonnegative. Show that for Problem 4.2.3, if the objective function f is linear, then there is a greedy algorithm with a better performance ratio $1/h$. (Hint: Use the result in Problem 2.)

5. (Budgeted Maximum Coverage [213]) Consider a collection $\mathcal{C}$ of subsets of a finite set X. Every subset S in $\mathcal{C}$ has a nonnegative cost $c(S)$ and every element x in X has a nonnegative weight $w(x)$. Given a budget L, the problem is to find a subcollection $\mathcal{C}'$ of $\mathcal{C}$ with total cost not exceed L, i.e., $\sum_{S \in \mathcal{C}'} c(S) \leq L$, to maximize the total weight of elements covered by $\mathcal{C}'$. Please prove that this problem is a special case of monotone nondecreasing submodular function maximization with knapsack constraint.

6. ([468]) Let ξ be a continuous and piecewise smooth function $[u, v] \to \mathbb{R}^+$. If for some $\alpha, \beta > 0$, we have $\xi(x) + \alpha\xi'(x) \geq \beta$ for $x \in [u, v]$, then $\xi(v) \geq \beta + (\xi(u) - \beta)e^{\frac{u-v}{\alpha}}$, where $\xi(x)$ represents the right derivative of $\xi(x)$.

7. (Standard Greedy Inequality [468]) Let g be the greedy performance function defined in Definition 4.1.6 and o_1 the largest cost item in OPT. Show the following.

   (a) For $x \in [0, 1 - c(o_1)]$,

$$g(x) + g'(x) \geq 1$$

      where $g'(x)$ represents the right derivative of $g(x)$.

   (b) For $x \in [0, 1 - c(o_1)]$, $g(x) \geq 1 - e^{-x}$.

8. Suppose $f : 2^X \to \mathbb{R}$ is a submodular function. Show that

$$f(A \cup C) - f(A) \leq \sum_{c \in C}[f(A \cup \{c\}) - f(A)],$$

and

$$f(A) - f(A \cap C) \geq \sum_{a \in A \setminus C}[f(A) - f(A \setminus \{a\})].$$

9. Consider a graph $G = (V, E)$. For any node subset $A \subset V$, define the cut of A by

$$\partial A = \{(u, v) \mid u \in A, v \notin A\}.$$

   Show that $f(A) = |\partial A|$ is a submodular function, but not monotone nondecreasing.

10. Consider a collection $\mathcal{C}$ of subsets of a finite set X. Suppose $f : 2^X \to \mathbb{R}$ is a submodular function. Show that $g : 2^{\mathcal{C}} \to \mathbb{R}$ is also a submodular function.

11. ([19]) Consider s graph $G = (V, E)$. For any node subset A, let $f(A)$ be the number of edges in the subgraph induced by A. Show that f is supermodular.

12. Consider function $f : 2^V \to \mathbb{R}$. Show that each of the following is necessary and sufficient condition for f to be submodular.

   (a) For any sunsets S and T of V,

$$f(T) \leq f(S) + \sum_{v \in T \setminus S} \Delta_v f(S) - \sum_{v \in S \setminus T} \Delta_v f((S \cup T) \setminus \{v\}).$$

(b) For any $S \subseteq T \subseteq V$,

$$f(T) \leq f(S) + \sum_{v \in T \setminus S} \Delta_v f(S).$$

(c) For any $S, T \subseteq V$,

$$f(T) \leq f(S) + \sum_{v \in T \setminus S} \Delta_v f(S \cap T) - \sum_{v \in S \setminus T} \Delta_v f(S \setminus \{v\}).$$

(d) For any $T \subseteq S \subseteq V$,

$$f(T) \leq f(S) - sum_{v \in S \setminus T} \Delta_v f(S \setminus \{v\}).$$

(e) For any $S, T \subseteq V$,

$$f(T) \leq f(S) + \sum_{v \in T \setminus S} \Delta_v f(T \cap S) - \sum_{v \in S \setminus T} \Delta_v f((S \cup T) \setminus \{v\}).$$

13. Consider function $f : (k+1)^V \to \mathbb{R}$. Show that each of conditions in above problem is necessary for f to be orthant submoular. Please determine whether anyone of them is also sufficient.

14. (Multi-Grade IC Model) Consider a social network $G = (V, E)$ with the IC model and a product with n grades $g_1, \ldots, g_n$. To address the viral marketing of this product in the social network G, the IC model is modified as follows: Each node x is associated with an inherent valuation vector $\vec{v}_x = (v_{x,1}, \ldots, v_{x,n})$ which records the valuation for purchase of the grade of the product. Each edge (x, y) is associated with a weight matrix W_{xy} which records valuation increments obtained by influence between social relationship of x and y and the promotion relationship between g_i and g_j. Each node has three states, inactive, active, and purchased. Initially, all seeds are in purchased state and all others are inactive. In each subsequent step, if x is in purchased state and y is inactive, then y becomes active with probability p_{xy}. (This is similar to the IC model.) Each node y updates its valuation vector by

$$v'_{y,j} \leftarrow v_{y,j} + \max_{(x,i) \in V_{act}(y)} W_{xy}(i, j)$$

where $V_{act}(y)$ is the activation set of y; its element (x, i) means that x, as an in-neighbor of y, already purchased grade g_i and tries to activate y. Let p_j be the price of grade g_j. If $v'_{y,j} < p_j$ for every $j = 1, \ldots, n$, then y will not change its state to purchased. Otherwise, select j such that $v'_{y,j} - p_j = \max_{1 \leq j' \leq n}(v'_{y,j'} - p_{j'})$. Node y would purchase grade g_j. Similar to the IC model. Only freshly purchased-state node have influence power. Define

$$f(S) = \sum_{j=1}^{n} p_j \cdot m_j(S)$$

where S is the seed set and $m_j(S)$ is the number of nodes who purchase grade g_j. Please show that $f(S)$ is neither submodular nor supermodular.

15. Show, by the following modifications, that Algorithm 25 is a $\frac{1}{4}$-approximation algorithm for non-monotone k-submodular maximization with knapsack constraint.

(a) In Lemma 4.6.8, add statement: If f is non-monotone k-submodular, then

$$f(T) \leq 3f(S_t) + \sum_{(s,d)\in T_t \setminus S_t} \Delta_{s,d} f(S_t).$$

(b) In Lemma 4.6.9, add statement: If f is non-monotone k-submodular, then for $x \in [0, 1 - b(o_1) - b^*]$,

$$3g_1(x) + (1 - b(o_1))g'(x) \geq 1 = opt.$$

16. (Feldman et al. [108]) The modification of continuous greedy algorithm is as shown in Algorithm 26. The outline of analysis is as follows. Please check the reference and complete all proofs.

Algorithm 26 Measured Continuous Greedy

---

input $F(x)$, $P(\mathcal{M})$, and T.
output $x(1)$.
1: $\alpha \leftarrow \frac{1}{T}$.
2: $x(0) \leftarrow 0 \in R^n$.
3: **for** $t \leftarrow 1$ to T **do**
4:     $v \leftarrow v_{\max}(x((t-1)\alpha))$
5:     where $v_{\max}(x) = \mathrm{argmax}_{v \in P(\mathcal{M})} \langle v, \nabla F(x) \rangle$.
6:     **for** each $e \in X$ **do**
7:         $x_e(t\alpha) \leftarrow x_e((t-1)\alpha) + \alpha v_e \cdot (1 - x_e((t-1)\alpha))$
8:     **end for**
9: **end for**
10: **return** $x(1)$.

(a) $x(1) \in P(\mathcal{M})$.
(b) For any $0 \leq t \leq T$,

$$\sum_{e \in X} (1 - x_e(t\alpha)) \cdot v_e \cdot \partial_e F(x(t\alpha)) \geq F(x(t\alpha) \vee 1_{OPT}) - F(x(t\alpha)).$$

$$(4.13)$$

(c) For any $0 \leq t \leq T$,

$$F(x((t+1)\alpha)) - F(x(t\alpha)) \geq \alpha \cdot [F(x(t\alpha) \vee \mathbf{1}_{OPT})) - F(x(t\alpha))] - O(n^3\alpha^2) \cdot opt$$

where $\mathbf{1}_{OPT}$ is the characteristic vector of the set OPT and the set OPT is the optimal solution.

(d) Consider a vector $x \in [0, 1]^X$ with $x_e \leq a$ for every $e \in X$. Then for very set $S \subseteq X$,

$$F(x \vee \mathbf{1}_S) \geq (1-a)f(S).$$

(e) For any $0 \leq t \leq T$,

$$F(x((t+1)\alpha)) - F(x(t\alpha)) \geq \alpha \cdot [e^t \cdot opt - F(x(t\alpha))] - O(n^3\alpha^2) \cdot opt.$$

(f) Define $g(t)$ by letting $g(0) = 0$ and

$$g((t+1)) = g(t) + \alpha[e^{-t} \cdot opt - g(t)].$$

By using above results, prove the following.

- For any $0 \leq t \leq T$,

$$g(t) \leq F(x(t\alpha)) + O(n^3\alpha) \cdot t \cdot opt. \tag{4.14}$$

- For any $0 \leq t \leq T$,
$$g(t) \geq te^{-t} \cdot opt. \tag{4.15}$$

(g) Consider a nonnegative submodular function $f : 2^X \to \mathbb{R}^+$ and the problem of maximizing f with a matroid constraint. The measured continuous greedy algorithm computes a point $x(1)$ such that

$$F(x(1)) \geq [e^{-1} - o(1)] \cdot opt,$$

where F is the multilinear extension of f.

(h) For nonnegative submodular function maximization, there exists a polynomial-time algorithm generating an approximation $\bar{x}$ such that

$$f(\bar{x}) \geq (e^{-1} - o(1)) \cdot opt.$$

Historical Notes

One of the important applications of social influence is the viral marketing
[77, 138, 139, 215, 234, 281, 347, 351]. The viral marketing for multiple products
have been studied extensively in various models [43, 89, 309, 312, 482]. Solutions
for those problems are closely related to the monotone submodular maximization
with a knapsack constraint, which has two well-known approximation algorithms
with performance ratio $(1 - e^{-1})/2$ [238] and $1 - e^{-1}$ [374], respectively. The budgeted maximization coverage problem [213] is a special case of the monotone submodular maximization with knapsack constraint. Nguyen et al. [307] and Pham et
al. [331] studied the cost-aware targeted viral marketing based on a greedy algorithm
in [213] for the budgeted maximum coverage problem. This greedy algorithm has
approximation performance ratio $1 - e^{-1/2} > (1 - e^{-1})/2$, which leaves a potential to improve the faster approximation algorithm for the monotone submodular
maximization with knapsack constraint.

Balanced seed-distribution is one of considerations in problem settings of
[167, 168]. This consideration leads to the monotone submodular maximization with
matroid constraints, which has been studied extensively in the literature [34, 64, 96,
113, 183, 200, 221, 301, 302].

The continuous greedy was proposed by Calinescu et al. [34] for design of
$(1 - e^{-1})$-approximation for the monotone submodular maximization with a matroid
constraint. This algorithm employs the pipage rounding which was initiated by Ageev
and Sviridenko [1]. The pipage rounding is an important technique rounding a fractional solution to an integer solution in design of approximation algorithms, which
received a lot of applications [38, 127].

The approximation design for non-monotone submodular maximization problems was initiated by Feige et al. [104] with local search technique. This work was
followed up by a sequence of research efforts [107, 108, 133, 231, 412]. In particular, Feldman et al. [107, 108] employed the continuous greedy in this research
direction. The revenue maximization problem was introduced by Zhu et al. [504].
In the literature, the revenue maximization is also called the profit maximization [126, 155, 263, 343, 463].

Chapter 5
Nonsubmodular Optimization

> Submodularity is an ideal model for cooperation, complexity, and attractiveness as well as for diversity, coverage, and information.
> — Jeff Bilmes

In recent developments of information technologies, many new types of optimization problems appear which motivate many new research subjects in optimization theory. Especially, in viral marketing, there exist many problems whose objective functions are neither submodular nor supermodular. Those problems encourage us to extend our study from the submodular optimization to the nonsubmodular optimization. This extension requires new methodologies. In this chapter, we discuss two of them, the sandwich method and the parameterized method.

5.1 Sandwich Method

In promotion of online game, the profit depends on the interactions among users besides the number of users. This motivates interaction-aware influence maximization problem. In this problem, the objective for maximization has to take both the number of active users and the user-to-user interactions into account.

Specifically, let us consider the interaction-aware viral marketing problem on social network $G = (V, E)$ with independent cascade model. This problem can be formulated to the following set function maximization:

$$\max \quad f(S)$$
$$\text{subject to} \quad |S| \leq k, S \subseteq V,$$

W. Wu et al., *Computational Aspects of Social Networks*, Springer Optimization and Its Applications 234, https://doi.org/10.1007/978-3-032-14833-9_5

where S is called the seed set. The objective function $f(S)$ can be represented as

$$f(S) = \mathbb{E}_{g \sim D}[\alpha \cdot |I_g(S)| + \beta \cdot \sum_{\{u,v\} \subseteq I_g(S)} b(u, v)]$$

where g is the subgraph of G consisting of all alive edges at end of cascade process, D is the distribution of g, $I_g(S)$ is the set of active nodes, and $b(u, v)$ is benefit generated by interaction between nodes u and v. α and β are used to balance two types of profits.

Counterexample 5.1.1 ($[125]$) $f(S)$ is neither submodular nor supermodular.

Construction. Consider a counterexample in Fig. 5.1. Each edge (u, v) is associated with a pair of labels, $(p_{uv}, b(u, v))$, i.e., the first label is the propagation probability and the second label is the activity benefit between two nodes. Set $\alpha > 0$ and $\beta > 0$. Then,

$$\begin{aligned}
f(\{a, d\}) - f(\{a\}) &= (2\alpha + \beta) - \alpha \\
&< f(\{a, b, d\}) - f(\{a, b\}) \\
&= (3\alpha + 3\beta) - 2\alpha,
\end{aligned}$$

which means that f is not submodular. Moreover,

$$\begin{aligned}
f(\{c\}) - f(\emptyset) &= 2\alpha + 2\beta \\
&> f(\{c, d\}) - f\{d\}) \\
&= (2\alpha + 2\beta) - \alpha,
\end{aligned}$$

which means that f is not supermodular. $\qquad\square$

Define two set functions $B_1(S)$ and $B_2(S)$ on 2^X as follows:

$$B_1(S) = \mathbb{E}_{g \sim D}[\alpha \cdot |I_g(S)| + \beta \cdot \sum_{u \in I_g(S), v \in V} b(u, v)],$$

$$B_2(S) = \mathbb{E}_{g \sim D}[\alpha \cdot |I_g(S)|].$$

Fig. 5.1 A counterexample

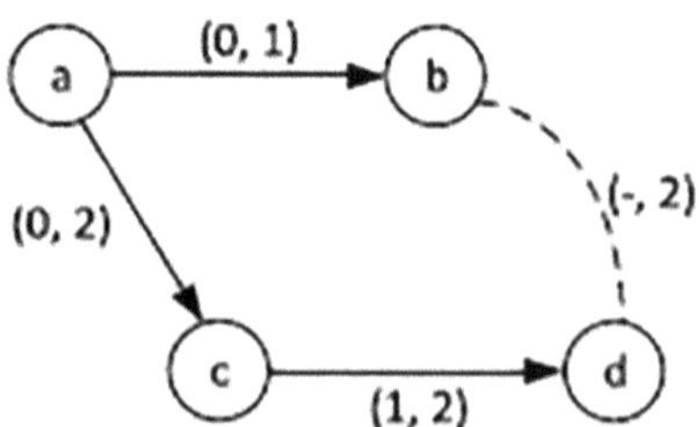

Lemma 5.1.2 *Both $B_1(S)$ and $B_2(S)$ are monotone nondecreasing and submodular such that for any $S \in 2^X$,*

$$B_1(S) \geq f(S) \geq B_2(S).$$

Therefore, there is a $(1 - 1/e)$-approximation solution S_1 for $\max\{B_1(S) \mid |S| \leq k\}$ and there is a $(1 - 1/e)$-approximation solution S_2 for $\max\{B_2(S) \mid |S| \leq k\}$. Choose

$$S_3 = \operatorname{argmax}(f(S_1), f(S_2)).$$

Above method is called the sandwich method. This method has been used quite often for solving nonsubmodular optimization problems in the literature. In general, it runs as follows.

Suppose that we face a problem $\max_{A \in \Omega} f(A)$ where Ω is a collection of subsets of a finite set X.

Sandwich Method:

- **Input** a set function $f : \Omega \to \mathbb{R}$.
- Initially, find two submodular functions u and l such that $l(A) \leq f(A) \leq u(A)$ for $A \in \Omega$. Then carry out the following operations:

  - Compute an α-approximation solution S_u for $\max_{A \in \Omega} u(A)$ and a β-approximation solution S_l for $\max_{A \in \Omega} l(A)$.
  - Compute a feasible solution S_o for $\max_{A \in \Omega} f(A)$.
  - Set $S = \operatorname{argmax}(f(S_u), f(S_o), f(S_l))$.

- **Output** S.

The performance ratio of this algorithm is data-dependent as follows. Hence, this algorithm is also called a *data-dependent* approximation algorithm.

Theorem 5.1.3 (Lu et al. [271]) *The solution S produced by the sandwich method satisfies the following:*

$$f(S) \geq \max\left\{ \frac{f(S_u)}{u(S_u)} \cdot \alpha, \ \frac{opt_l}{opt_f} \cdot \beta \right\} \cdot opt_f,$$

where opt_f (opt_l) is the objective function value of the minimum solution for $\max_{A \in \Omega} f(A)$ $(\max_{A \in \Omega} l(A))$.

Proof Since S_u is a β-approximation solution for $\max_{A \in \Omega} u(A)$, we have

$$f(S_u) = \frac{f(S_u)}{u(S_u)} \cdot u(S_u) \geq \frac{f(S_u)}{u(S_u)} \cdot \alpha \cdot opt_u \geq \frac{f(S_u)}{u(S_u)} \cdot \alpha \cdot u(OPT_f) \geq \frac{f(S_u)}{u(S_u)} \cdot \alpha \cdot opt_f,$$

where OPT_f is an optimal solution for $\min_{A \in \Omega} f(A)$. Since S_l is an β-approximation solution for $\max_{A \in \Omega} l(A)$, we have

$$f(S_l) \geq l(S_l) \geq \beta \cdot opt_l = \beta \cdot \frac{opt_l}{opt_f} \cdot opt_f.$$

Therefore, the theorem holds. $\square$

Now, we come back to the interaction-aware marketing problem.

Corollary 5.1.4 S_3 *is an approximation solution for* $\max\{f(S) \mid |S| \leq k\}$ *with performance*

$$f(S) \geq \max\left\{\frac{f(S_1)}{B_1(S_1)}, \frac{opt_2}{opt_f}\right\} \cdot (1 - 1/e) \cdot opt_f,$$

where opt_f *(opt_2) is the objective function value of the optimal solution for* $\max\{f(S) \mid |S| \leq k\}$ *($\max\{B_2(S) \mid |S| \leq k\}$).*

This performance measure has the following disadvantage: There exist many pairs of monotone nondecreasing submodular upper and lower bounds. This performance measure cannot tell which pair is better. Therefore, it is not a satisfied performance measure.

5.2 Complementary Product

When you purchase a product A in an online store, a recommendation could come out "One who buys A may also like to buy B." When you purchase both A and B, another recommendation could jump out "One who buy both A and B may also like to buy C." Both recommendations are generated from analysis on historical data. Complementary products are those that tend to be purchased together. For example, products A, B, and C are complementary.

Consider a directed social network $G = (V, E)$ with IC model, i.e., each arc (u, v) is associated with a number p_{uv} which is the probability that node v accepts influence of u. Let $g_1, g_2, ..., g_k$ be k products with prices $c_1, c_2, ..., c_k$, respectively. Suppose that a customer u who buys product g_i would also buy product g_j with probability p_{ij}^g and who buys products g_i and g_j would also buy product g_h with probability p_{ijh}^g. In this section, we study the following problem.

Problem 5.2.1 (*Viral Marketing for Complementary Products [169]*) Consider a social network and a set of products with complementary relationships. Given a budget B, find a set of customers for giving discounts (e.g., free samples) within the budget B to maximize the expected total sales of the product.

Let us first consider "One who buys A may also like to buy B," i.e., the recommendation from one product. In this case, at each step of information diffusion process, there are two types of influences to each customer v for each product g_j:

Fig. 5.2 Netwok $\hat{G}$

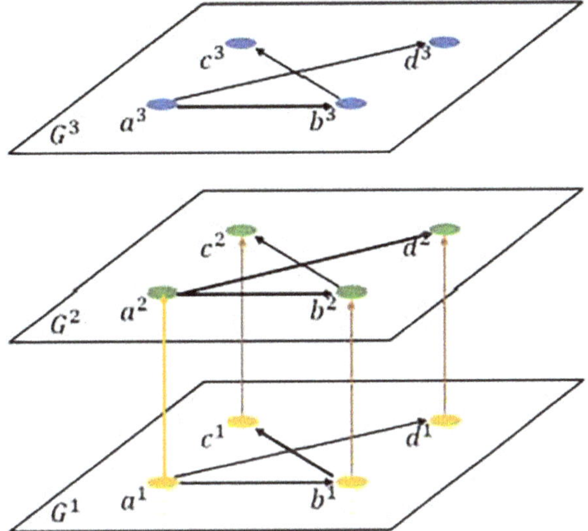

- The influence from a customer u who already purchased product g_j. The success probability of this influence is p_{uv}.
- The influence from a product g_i which comes from a recommendation when the customer v purchased g_i at a previous step. The success of this influence has probability p_{ij}^v.

These influences are naturally assumed to be independent.

Let G be the social network in the problem of viral marketing for complementary products. As shown in Fig. 5.2, for each product g_i, make a copy G^i of G. Denote by u^i the copy of node u in G^i. Consider network $\hat{G} = G^1 \cup G^2 \cup \cdots \cup G^k$. For every two nodes u^i and u^j, if there exists a recommendation "a customer who purchased g_i would also like to purchase g_j," then add an arc (u^i, u^j) associated with a probability p_{ij}^g. The problem of viral marketing for complementary products would be equivalent to the influence maximization on $\hat{G}$ with cost function $c(u^i) = c_i$.

Thus, we have the following.

Theorem 5.2.2 (Guo and Wu [169]) *Suppose that influences from recommendations and from social relations at each customer are independent and the influence from recommendation can be successful only at fresh (i.e., when the customer made a purchase and the recommendation just pop up). Then the problem of viral marketing for complementary products with recommendations each from only one product is the influence maximization problem on network $\hat{G}$ with knapsack constraint, i.e., a monotone nondecreasing submodular set function maximization with knapsack constraint.*

Next, we consider recommendations in form "A customer who purchased A and B may also like to purchase C." Let us call this type of recommendation as the *composed recommendation*.

Counterexample 5.2.3 When composed recommendations are considered, the expected total sales (we may still called it as the influence spread for simplicity) is neither submodular nor supermodular

Construction. Let us use $i \to j$ denote the recommendation that buying g_i influences buying g_j and $(i, j) \to h$ denote the recommendation that buying g_i and g_j influence buying g_h. Consider following recommendations:

$$1 \to 2, 1 \to 3, 3 \to 4,$$

$$(1, 2) \to 5, (2, 4) \to 6, (2, 3) \to 7, (3, 4) \to 8.$$

Assume $c_1 = c_2 = c_3 = c_4 = c_5 = 1$ and $c_6 = c_7 = c_8 = 2$ and assume that only one customer u exists. Then, we have

$$\sigma(\{u^1, u^2\}) - \sigma(\{u^1\}) = 2 < \sigma(\{u^1, u^2, u^4\}) - \sigma(\{u^1, u^2\}) = 5,$$

which means that the influence spread σ is not submodular. Moreover,

$$\sigma(\{u^3\}) - \sigma(\emptyset) = 4 > \sigma(\{u^3, u^4\}) - \sigma(\{u^4\}) = 3,$$

which means that σ is not supermodular. $\qquad\qquad\square$

To employ the sandwich method, let us establish upper and lower bounds in the following.

Upper Bound (Guo and Wu [164]): For each composed recommendation $(i, j) \to h$, replace p_{ih}^g and p_{jh}^g by

$$p_{ih}^{\prime g} = 1 - (1 - p_{ih}^g)(1 - p_{ijh}^g)$$

and

$$p_{jh}^{\prime g} = 1 - (1 - p_{jh}^g)(1 - p_{ijh}^g),$$

respectively. This is equivalent to relax the activate composed recommendation so that as long as one of g_i and g_j is activated, the composed recommendation can act. The influence spread in obtained network is monotone nondecreasing and submodular, which is an upper bound for original one.

Lower Bound: We may simply remove all composed recommendations. Could we keep some composed recommendations? The following counterexample indicates that the existence of only one composed recommendation can make the influence spread not submodular.

Consider a network with only one customer and three products g_1, g_2, g_3. There is a composed recommendation $(1, 2) \rightarrow 3$. Assume $c_1 = c_2 = c_3 = 1$ and all influences are deterministic. Then, we have

$$\sigma(\{1\}) - \sigma(\emptyset) = 1 - 0 = 1$$

and
$$\sigma(\{1, 2\}) - \sigma(\{2\}) = 3 - 1 = 2.$$

Thus,
$$\sigma(\{1\}) - \sigma(\emptyset) < \sigma(\{1, 2\}) - \sigma(\{2\}).$$

This means that σ is not submodular.

5.3 Multi-feature Product

There may exist various situations about the product and the company in market, which lead to different problem formulations. Usually, a product may have multiple features considered by a customer to make a decision "buy or not." For example, except the quality, the price and the reputation (or brand image) may also be important factors. Each person has different levels of expertise on various features and hence has different levels of influence on others. When consider multiple features of a product, different persons have various emphasis points. For example, a wealthy one may like a product with high reputation and a less wealthy one may prefer to have a product with high quality and less price. Therefore, in study of viral marketing, a model based on multi-feature influence should be considered.

Definition 5.3.1 (*Multi-dimensional Mixed (MDM) Model*) Consider a product with r features and a directed online social network $G = (V, E)$. In the following, we list important facts about this model consisting of discrete steps:

- Each node represents a customer with two states, active and inactive. An active node represents a customer who has already purchased the product.
- Every edge (u, v) is assigned with k probabilities p_{uv}^1, ..., p_{uv}^r. Those probabilities represent the expertise of u from the view point of v. When u just becomes active, v will be influenced by u and accepts the feature ϕ with probability p_{uv}^ϕ.
- When v receives influence from more than one active in-neighbors, v treats them as independent events.
- Each customer v has a threshold θ_v and a weight w_v^ϕ for each feature ϕ with $\sum_{\phi=1}^r w_v^\phi = 1$. The threshold θ_v is selected randomly and uniformly from $[0, 1]$ at the initial step. Customer v decides to purchase the product if and only if the total weight of accepted factors reaches at least θ_v. The weight w_v^ϕ represents the preference of customer v on feature ϕ.

- Initially, a set of seeds are selected and activated. At each subsequent step, every node check if the activation condition is satisfied. This process ends if no node becomes active at current step.

If θ_v is not randomly selected, then the influence spread is neither submodular nor supermodula since it is similar to the positive influence in Sect. 3.5. What would occur as above, i.e., θ_v is randomly uniformly selected from $[0, 1]$? The influence spread is till not submodular. The following is a counterexample.

Counterexample 5.3.2 With the MDM model, the influence spread is neither submodular nor supermodular.

Construction. Consider a social network G with four nodes a, b, c, d and a product with two features $1, 2$ as shown in Fig. 5.3. Set $p_{ac}^1 = 1$, $p_{ac}^2 = 0$, $p_{bc}^1 = 0$, $p_{bc}^2 = 1$, $p_{cd}^1 = 1$, $p_{cd}^2 = 0$. Let $w_c^1 = w_c^2 0.5$ and $w_d^1 = 1$, $w_d^2 = 0$. The, we have

$$\sigma(\{a\}) = 1 + 0.5 \cdot 2 = 2, \quad \sigma(\{b\}) = 1 + 0.5 = 1.5, \quad \sigma(\{a, b\}) = 2 + 2 = 4.$$

Thus,

$$\sigma(\{a\}) - \sigma(\emptyset) = 2 < 2.5 = \sigma(\{a, b\}) - \sigma(\{b\}).$$

This means that σ is not submodular.

It is quite easy to construct a counterexample for supermodularity. In fact, consider a product with only one feature and a social network with only one directed edge (a, b) and deterministic model (note: this is a special case of the IC mode.) Assume $c_a = c_b = 1$. Then, we have

$$\sigma(\{a\}) - \sigma(\emptyset) = 2 > 1 = \sigma(\{a, b\}) - \sigma(\{b\}).$$

This means that σ is not supermodular.   □

This counterexample is designed based on a key fact that only an active node has ability to diffuse information. This means that a customer has ability to spread the information of a product only in case that the customer has purchase the product. This

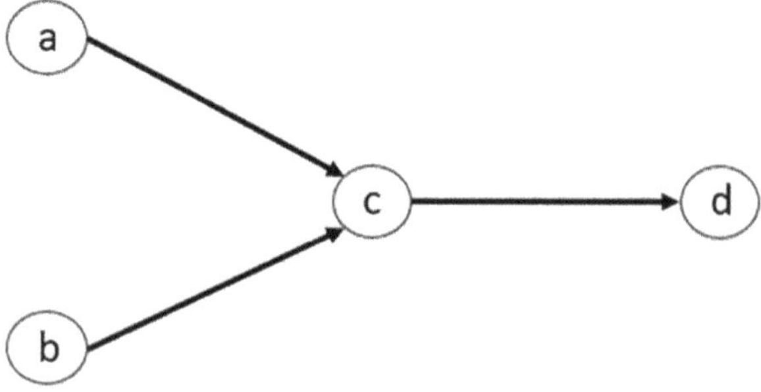

Fig. 5.3 Counterexample 5.3.2

constraint may be too strong for customers in the viral marketing. In fact, a customer who accepts a feature of a product may diffuse his opinion to his/her friends. For example, for products of Hermes, a lot of ones do not have them, but agree that they have high quality. In the following model, we relax this constraint.

Definition 5.3.3 (*Multi-feature (MF) Model [161]*) Consider a product with r features and a social network G.

- For each feature ϕ, make a copy G^ϕ of G. Let u^ϕ denote the copy of node u (Fig. 5.4). Let p_{uv}^ϕ be the probability associated with edge (u^ϕ, v^ϕ), which represents the probability that customer v accepts the influence of customer u on feature ϕ. Every customer u has a weight w_u^ϕ for each feature ϕ and $\sum_{\phi=1}^{r} w_u^\phi = 1$.
- each node u has two states for each feature ϕ, active and inactive for feature ϕ. If u is active for feature ϕ, then u has ability to influence on its outgoing neighbors on feature ϕ.
- Initially, every node is inactive for every feature. Then, k customers are selected to receive free samples so that these k node become active for any feature. Then, on each G^ϕ, the information on feature ϕ is diffused with IC model.
- After in all G^ϕ the diffusion process ends, every node u select a threshold θ_u randomly and uniformly from $[0, 1]$ and check if

$$\sum_{u:\text{active for } \phi} w_u^\phi \geq \theta_u.$$

If answer is yes, then u purchases the product; otherwise, not.

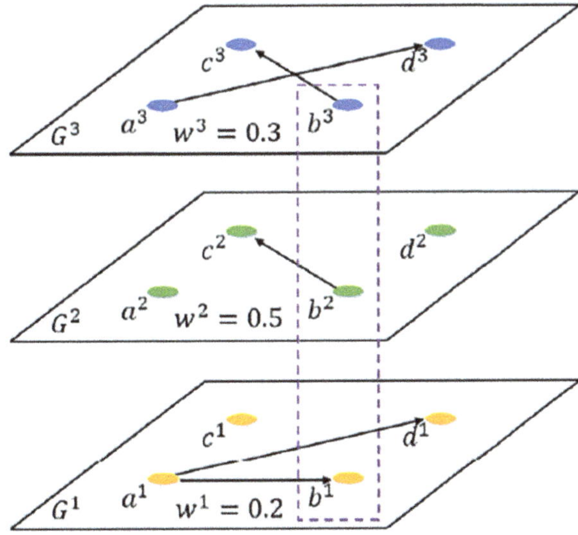

Fig. 5.4 For each feature, make a copy of G

Remark: In above definition for the MF model, the last rule can also be replaced by the following: In the initial step, the very node u selects a threshold θ_u randomly and uniformly from $[0, 1]$ and in each subsequent step, every one u who did not buy the product checks if

$$\sum_{u:\text{active for } \phi} w_u^\phi \geq \theta_u.$$

If answer is yes, then u purchases the product; otherwise, not. This change does not change the set of customers who purchased the product at the end of process.

The following can be proved easily.

Theorem 5.3.4 (Guo et al. [161]) *The influence spread $\sigma_{MF}(S)$ is monotone nondecreasing and submodular. Moreover, for every seed set S,*

$$\sigma_{MDM}(S) \leq \sigma_{MF}(S).$$

In the literature, we have not found a nontrivial lower bound for $\sigma_{MDM}(S)$, which is monotone nondecreasing and submodular with respect to S.

5.4 Influence in Hypergraph

In the real world, individual behavior is heavily influenced by the crowd [95]. The crowd influence is different from the combination of individual influence. For example, suppose that a crowd consists of two persons A and B, they have influence on third person C. A has influence on C. B has influence on C. Except these two individual influences, A and B may also have a crowd influence (or composed influence) on C. In order to formulate problems about crowd influence, Zhu et al. [499] employ hypergraph and study the influence maximization in hypergraphs, a special type of hypergraphs.

Consider a directed hypergraph $H = (V, E)$. Each hyperedge $e \in E$ can be partitioned into two parts, head and tail. The head $head(e)$ is a subset of V and the tail $tail(e)$ is a singleton, or for simplicity to say, a node in V. The IC model can be extended to this type of hypergraphs as follows.

Definition 5.4.1 (*IC Model in Hypergraph [499]*) It is an extension of the IC model from graph to hypergraph. There are the following rules for the hyperedge.

- Each hyperedge e is associated with a probability p_e. Set $H_e = head(e)$ and $v = tail(e)$. Then crowd H_e influences v successfully with probability p_e.
- When a node v receives influences from multiple crowds and/or nodes, those influences are independent.
- A crowd H_e is said to be freshly active only if every node in H_e is freshly active. Only freshly active crowd or node can have ability to influence a node v.

Definition 5.4.2 (*Influence Maximization in Hypergraphs [499]*) Give a hypergraph $H = (V, E)$ with the IC model, and a positive integer k, find a set S of at most k seeds to maximize the influence spread $\sigma(S)$ in the hypergraph H.

Next, we study properties of objective function $\sigma(\cdot)$.

Theorem 5.4.3 (Zhu et al. [499]) *The influence spread $\sigma(\cdot)$ in hypergraph is neither submodular nor supermodular.*

Proof We will present a counterexample to prove this theorem. Consider a hypergraph $H = (V, E)$ withe the IC model, where $V = \{v_1, v_2, v_3, v_4\}$, $E = \{(v_1, v_4), (v_3, v_4), (\{v_1, v_2\}), v_3)\}$, and $p_{(v_1, v_4)} = p_{(v_3, v_4)} = p_{(\{v_1, v_2\}, v_3)} = 1$ (Fig. 5.5).

First, set $A = \emptyset$ and $B = \{v_2\}$. Then $\sigma(A) = 0$ and $\sigma(B) = 1$. Putting v_1 into A and B, we obtain $\sigma(A \cup \{v_1\}) = \sigma(\{v_1\}) = 2$ and $\sigma(B \cup \{v_1\}) = \sigma(\{v_1, v_2\}) = 4$. Therefore,

$$\sigma(A \cup \{v_1\}) - \sigma(A) = 2 < \sigma(B \cup \{v_1\}) - f(B) = 3,$$

which indicates that $\sigma(\cdot)$ is not submodular.

Next, set $A = \emptyset$ and $B = \{v_1\}$. Then $\sigma(A) = 0$ and $\sigma(B) = 2$. Putting v_3 into A and B, we obtain $\sigma(A \cup \{v_3\}) = \sigma(\{v_3\}) = 2$ and $\sigma(B \cup \{v_3\}) = \sigma(\{v_1, v_3\}) = 3\}$. Therefore,

$$\sigma(A \cup \{v_1\}) - \sigma(A) = 2 < \sigma(B \cup \{v_1\}) - f(B) = 1.$$

This means that $\sigma(\cdot)$ is not supermodular. $\square$

Zhu et al. [499] employ the sandwich method to find approximation solutions for the influence maximization in hypergraphs. In order to establish a monotone non-decreasing and submodular upper bound for $\sigma(\cdot)$, they construct a graph $G_u(V, E_u)$ based on hypergraph $H = (V, E)$, by replacing each hyperedge (H_e, v) by a group of edges as follows.

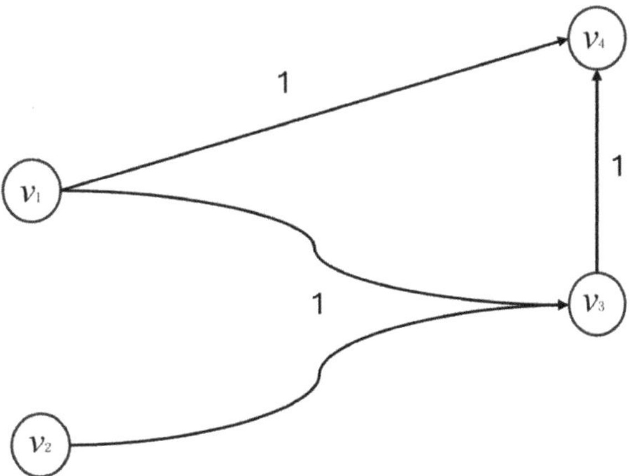

Fig. 5.5 A counterexample

- Each hyperedge (H_e, v) is replaced by a group of edges, (u, v) for all $u \in H_e$.
- Suppose that node u is contained in h hyperedges, $(H_1, v), (H_2, v), ..., (H_h, v)$. Then (u, v) is associated with the probability

$$1 - (1 - p_{(H_1,v)})(1 - p_{(H_2,v)}) \cdots (1 - p_{(H_h,v)}).$$

Let $\sigma_u(S)$ denote the influence spread of seed set S in graph G_u. Then we have

Theorem 5.4.4 (Zhu et al. [499]) *For any seed set S, $\sigma(S) \leq \sigma_u(S)$.*

Proof Break down the construction of graph G_u into two stages.

- Stage 1. Replace each hyperedge (H_e, v) by edges (u, v) for $u \in H_e$, each associated with probability $p_{(H_e,v)}$. Result in a graph $G_m(V, E_m)$ with possible multiple edges between each pair of nodes u and v.
- Stage 2. Merge multiple edges, with probabilities $p_{(H_1,v)}$, $p_{(H_2,v)}$, $\ldots$, $p_{(H_h,v)}$, respectively, into one with probability

$$1 - (1 - p_{(H_1,v)})(1 - p_{(H_2,v)}) \cdots (1 - p_{(H_h,v)}).$$

Let $\sigma_m(S)$ denote the influence spread of seed set S in multi-graph G_m. Note that H_e influences v successfully with probability $p_{(H_e,v)}$. Moreover, the probability that one of the edges (u, v) for $u \in H_e$, each associated with probability $p_{(H_e,v)}$, influences v successfully is

$$1 - (1 - p_{(H_e,v)})^{|H_e|}.$$

Since

$$1 - (1 - p_{(H_e,v)})^{|H_e|} \geq p_{(H_e,v)},$$

we have $\sigma(S) \leq \sigma_m(S)$. Furthermore, note that $\sigma_u(S) = \sigma_m(S)$. Then we have

$$\sigma(S) \leq \sigma_u(S).$$

$\square$

For a monotone nondecreasing and submodular lower bound of $\sigma(S)$, a simple one can be obtained by deleting all hyperedges (H_e, v) with $|H_e| \geq 2$. Let $\sigma_l(S)$ denote the influence spread of seed set S in remaining graph. Then $\sigma(S) \geq \sigma_l(S)$.

An improved one is given in [499]. First, construct graph G by modifying H as follows.

- For each hyperedge (H_e, v), add a new node $v(H_e)$.
- For two nodes $v(H_e)$ and $v(H_{e'})$, if for every $x \in H_{e'}$, (H_e, x) is a hyperedge in H, then add edge $(v(H_e), v(H_{e'})$ associated with influence probability $\prod_{x \in H_{e'}} p_{(H_e,x)}$.

- For each hyperedge (H_e, v), delete hyperedge (H_e, v) and add edge $(v(H_e), v)$ associated with influence probability $p_{(H_e,v)}$.

Let $\sigma_{ll}(S)$ denote the influence spread of seed set S in graph G. Then we have

Theorem 5.4.5 *For any seed set S, $\sigma_{ll}(S) \leq \sigma(S)$.*

Proof Before delete all hyperedges, the modification keeps the influence power the same as before. When a hyperedge is deleted, added edge only keep influence power partially. $\qquad\square$

5.5 Budget Saving

For the influence maximization, an inverse problem is as follows.

Definition 5.5.1 *(Budget Minimization)* Consider a social network $G = (V, E)$ with the information diffusion model m. Given a positive number $\bar{\sigma}$, find a seed set S to minimize its cardinality $|S|$ under constraint $\sigma_m(S) \geq \bar{\sigma}$, where $\sigma_m(S)$ is the influence spread of S in G.

If m is the deterministic model, then $\sigma_m(S)$ is a monotone nondecreasing and submodular integer function. The budget minimization is a sample of the problem of minimum submodular set cover. Hence, a good approximation is given in Algorithm 7 and its performance is stated in Theorem 1.6.4.

If m is the IC model or the LT model, then $\sigma_m(S)$ may not be an integer. Therefore, we have to make use of a different theorem to analyze the performance of greedy algorithm as follows.

Consider a monotone nondecreasing and submodular function $f : 2^X \to \mathbb{R}$ with $f(\emptyset) = 0$. Let α be any positive number, consider the following problem:

$$\min \quad c(A) = \sum_{x \in A} c(x)$$
$$\text{subject to} \quad f(A) \geq \alpha, \tag{5.1}$$

where $c : X \to \mathbb{R}_{\geq 0}$ is a nonnegative cost function.

Algorithm 27 is a greedy algorithm for solving this problem. Theorem 5.5.2 states the performance of this algorithm.

Algorithm 27 Greedy Algorithm for Problem (5.1).

Input: A polymotroid function f over 2^X, a positive number α and a nonnegative cost function c on X.

Output: A subset A of X such that $f(A) \geq \alpha$.

1: $A \leftarrow \emptyset$;
2: **while** $f(A) < \alpha$ **do**
3:     choose $x \in X \setminus A$ to maximize $\frac{\Delta_x f(A)}{c(x)}$ and
4:     $A \leftarrow A \cup \{x\}$;
5: **end while**
6: **return** A.

Theorem 5.5.2 *Let A_{out} denote the output of A in Algorithm 27. Suppose that in each iteration of Algorithm 27, selected x always satisfies*

$$\Delta_x f(A) \geq c(x).$$

Then

$$f(A_{out}) \leq (1 + \ln(\alpha/opt)) \cdot opt.$$

Proof We first assume $\alpha = \max_{A \in 2^X} f(A)$, so that for any feasible solution A, $f(A) = \alpha$. At the end of the proof, we will explain how to deal with the general case.

Let $A_{out} = \{x_1, x_2, ..., x_g\}$. Denote $A_i = \{x_1, x_2, ..., x_i\}$. Let $OPT = \{y_1, y_2, ..., y_h\}$ be an optimal solution. Denote $a_i = \Delta_{OPT} f(A_i)$. Then we have

$$
\begin{aligned}
a_{i-1} - a_i &= \Delta_{OPT} f(A_{i-1}) - \Delta_{OPT} f(A_i) \\
&= f(A_{i-1} \cup OPT) - f(A_{i-1}) - (f(A_i \cup OPT) - f(A_i)) \\
&= \Delta_{x_i} f(A_{i-1}).
\end{aligned}
$$

Thus,

$$
\begin{aligned}
\frac{a_i}{opt} &\leq \frac{\sum_{j=1}^{h} \delta_{y_j} f(A_{i-1})}{\sum_{j=1}^{h} c(y_j)} \\
&\leq \max_{1 \leq j \leq h} \frac{\Delta_{y_j} f(A_{i-1})}{c(y_j)} \\
&\leq \frac{\Delta_{x_i} f(A_{i-1})}{c(x_i)} \\
&= \frac{a_i - a_{i-1}}{c(x_i)}.
\end{aligned}
$$

Therefore,

$$a_i \leq a_{i-1}\left(1 - \frac{c(x_i)}{opt}\right)$$

$$\leq a_{i-1}\exp\left(-\frac{c(x_i)}{opt}\right) \quad (\text{note}: 1 + x < e^x)$$

$$\leq \cdots$$

$$\leq a_0 \exp\left(-\frac{c(x_i) + c(x_{i-1}) + \cdots + c(x_1)}{opt}\right),$$

where $a_0 = f(OPT \cup A_0) - f(A_0) = f(OPT)$.

Note that by assumption made at the begging of the proof,

$$a_g = \Delta_{OPT} f(A_g) = f(OPT \cup A_{out}) - f(A_{out}) = 0$$

$$a_0 = f(OPT) = f(A_{out}) \geq \sum_{i=1}^{g} \Delta_{x_i} f(A_{i-1})$$

$$\geq \sum_{i=1}^{g} c(x_i) = c(A_{out}) \geq opt > 0.$$

Therefore, there exists $0 \leq i < g$ such that $a_{i+1} < opt \leq a_i$.

Denote $a' = a_i - opt, a'' = opt - a_{i+1}$. Then $a_i - a_{i+1} = a' + a''$. Determine c' and c'' by

$$\frac{a'}{c'} = \frac{a''}{c''} = \frac{\Delta_{x_{i+1}} f(A_i)}{c(x_{i+1})} \geq 1.$$

Then $c(x_{i+1}) = c' + c''$. Note that

$$\frac{a_i - opt}{c'} = \frac{a'}{c'} = \frac{\Delta_{x_{i+1}} f(A_i)}{c(x_{i+1})} \geq \frac{a_i}{opt}.$$

Therefore,

$$opt \leq a_i(1 - c'/opt) \leq a_0 \exp(c' + c(x_i) + \cdots + c(x_1)).$$

Hence,

$$c' + c(x_i) + \cdots + c(x_1) \leq opt \cdot \ln\frac{a_0}{opt}.$$

Note that

$$c'' + c(x_{i+2}) + \cdots + c(x_g)$$
$$\leq a'' + \Delta_{x_{i+2}} f(A_{i+1}) + \cdots + \Delta_{x_g} f(A_{g-1})$$
$$= opt - a_{i+1} + a_{i+1} - a_{I+2} + \cdots + a_{g-1} - a_g$$
$$= opt.$$

Thus,

$$c(A_{out}) = c(A_g) \leq opt \cdot (1 + \ln \frac{a_0}{opt}) = opt \cdot (1 + \ln \frac{\alpha}{opt}).$$

Now, we consider the general case, that is, it may not be true that $\alpha = \max_{A \in 2^X} f(A)$. In this case, define $\phi(A) = \min(\alpha, f(A))$. Then $\phi(A)$ is a monotone nondecreasing and submodular function with $\phi(\emptyset) = 0$ and $\max_{A \in 2^X} \phi(A) = \alpha$. Instead of $f(A)$, use $\phi(A)$ in above proof, we will obtain required result. $\qquad\square$

Next, we consider an inverse problem of the revenue maximization.

Definition 5.5.3 (*Revenue-Constrained Budget Minimization*) Consider a social network $G = (V, E)$ with the information diffusion model m. Given a positive number β, find a seed set S to minimize its cardinality $|S|$ under constraint $r_m(S) \geq \beta$, where $r_m(S)$ is the revenue function of S in G.

In this problem, $r(S)$ is submodular, but non-monotone. So far, no theoretical research effort has been found to deal with this problem. It looks that the sandwich method is not suitable.

5.6 DS Decomposition

In the last three sections, we see three examples of viral marketing. In the first two examples, our expected lower and upper bounds for the sandwich method were found. However, for the third example, the expected lower bound does not seem easy to find. In general, what is the existence of those upper and lower bounds? In this section, let us study their theoretical foundation.

We start by introducing the following fundamental property of set functions.

Theorem 5.6.1 (DS Decomposition [196, 300]) *Every set function* $f : 2^X \rightarrow \mathbb{R}$ *can be expressed as the difference between two nondecreasing submodular functions monotone* g *and* h, *i.e.,* $f = g - h$, *where* X *is a finite set.*

Proof Define $\zeta(f) = \min_{A \subset B \subseteq X, x \in X \setminus A} \{\Delta_x f(A) - \Delta_x f(B)\}$. Note that $\zeta(f) \geq 0$ if and only if f is monotone nondecreasing and submodular. Consider set function $p(A) = \sqrt{|A|}$. Then $\zeta(p) > 0$ since

$$\min_{A\subset B\subseteq X, x\in B\setminus A}\{\Delta_x p(A)-\Delta_x p(B)\} = \min_{A\subset B\subseteq X, x\in B\setminus A}\{\Delta_x p(A)\}$$

$$= \min_{A\subset X}\{\sqrt{|A|+1}-\sqrt{|A|}\}$$

$$\geq \sqrt{|X|+1}-\sqrt{|X|}$$

$$> 0$$

and

$$\min_{A\subset B\subset X, x\in X\setminus B}\{\Delta_x p(A)-\Delta_x p(B)\}$$

$$= \min_{A\subset B\subset X}\{(\sqrt{|A|+1}-\sqrt{|A|})-(\sqrt{|B|+1}-\sqrt{|B|})\}$$

$$= \min_{A\subset X}\{(\sqrt{|A|+1}-\sqrt{|A|})-(\sqrt{|A|+2}-\sqrt{|A|+1})\}$$

$$> 0,$$

since $\sqrt{n}$ is a strictly concave function. If $\zeta(f)\geq 0$, then f is monotone, nondecreasing, and submodular, and hence $f = f - 0$ is a trivial decomposition meeting the requirement.

Now, assume $\zeta(f) < 0$. Define the set functions $h = 2\cdot\frac{-\zeta(f)}{\zeta(p)}\cdot p$ and $g = f + h$. Then $\zeta(h) = 2\cdot\frac{-\zeta(f)}{\zeta(p)}\cdot\zeta(p) = -2\zeta(f) > 0$ and $\zeta(g) = \zeta(f)+\zeta(h) = -\zeta(f) > 0$. $\qquad\square$

In the revenue maximization problem, the objective function is

$$r(S) = \sigma(S)\cdot\alpha - |S|\cdot\beta.$$

This gives a quite natural DS decomposition when the influence spread $\sigma(S)$ is monotone nondecreasing and submodular.

For a non-trivial example, consider the set function $f : 2^X \to \setminus$ defined by

$$f(S) = \sum_{\{u,v\}\subseteq S} b(u,v),$$

where $b(u,v)$ is nonnegative and symmetric, i.e., $b(u,v) = b(v,u)$, and moreover, $b(u,u) = 0$. Let

$$g(S) = \sum_{u\in S}\sum_{v\in X} b(u,v),$$

$$h(S) = \sum_{\{u,v\}\subseteq S} b(u,v) + \sum_{u\in S, v\in X\setminus S} b(u,v).$$

Since

$$\sum_{u\in S}\sum_{v\in S} b(u,v) = 2 \sum_{\{u,v\}\subseteq S} b(u,v),$$

we have $f(S) = g(S) - h(S)$. Moreover, we have the following.

Claim 5.6.2 (Gao et al. [125]) *g and h are monotone nondecreasing and submodular.*

Proof Note that

$$g(S) = \sum_{u\in S} w(u),$$

where $w(u) = \sum_{v\in X} b(u,v)$. Thus, $g(S)$ is nonnegaive-weighted cardinality of S and hence monotone nondecreasing and submodular.

Next, consider $h(S)$. For $A \subset B$, we have

$$
\begin{aligned}
h(A) &= \sum_{\{u,v\}\subseteq A} b(u,v) + \sum_{u\in A, v\in X\setminus A} b(u,v) \\
&= \sum_{\{u,v\}\subseteq A} b(u,v) + \sum_{u\in A}\sum_{v\in B\setminus A} b(u,v) + \sum_{u\in A}\sum_{v\in X\setminus B} b(u,v) \\
&\leq \sum_{\{u,v\}\subseteq A} b(u,v) + \sum_{u\in A}\sum_{v\in B\setminus A} b(u,v) + \sum_{u\in B}\sum_{v\in X\setminus B} b(u,v) \\
&\leq \sum_{\{u,v\}\subseteq B} b(u,v) + \sum_{u\in B}\sum_{v\in X\setminus B} b(u,v) \\
&= h(B).
\end{aligned}
$$

For $A \subset B$ and $x \in X \setminus B$, we have

$$
\begin{aligned}
\Delta_x h(A) &= \sum_{v\in X\setminus(A\cup\{x\})} b(x,v) \\
&\geq \sum_{v\in X\setminus(B\cup\{x\})} b(x,v) \\
&= \Delta_x h(B).
\end{aligned}
$$

$\square$

For each specific set function, finding a DS decomposition is possible, but may not be so easy. In fact, no efficient approach has been found so far. The following is an important open problem.

Open Problem 1 *Is there an efficient algorithm to produce a DS decomposition for any give set function?*

The following corollary gives a theoretical foundation of sandwich method.

Corollary 5.6.3 *For any set function $f : 2^X \to \mathbb{R}$, there exist two monotone nondecreasing submodular functions μ and ℓ on 2^X such that $\mu \geq f \geq \ell$.*

Proof Let $f = g - h$ where g and h are monotone nondecreasing submodular. For any set $S \in 2^X$, set $\mu(S) = g(S) - h(\emptyset)$ and $\ell(S) = g(S) - h(X)$. Then, for any set $S \in 2^X$,

$$\mu(S) \geq f(S) \geq \ell(S)$$

and μ and ℓ are monotone nondecreasing submodular. $\qquad\square$

5.7 Positive Influence

The positive influence is one of information diffusion models for viral marketing, which works as follows.

Consider a directed graph $G = (V, E)$. The positive influence process is defined as follows: Every node has two states, active and in active. Before process, every node is inactive. Initially, select at most k nodes (called seeds) and activate them. In each subsequent step, every inactive node v checks whether the number of active incoming neighbors is not less than the number of inactive incoming neighbors. If yes, then v becomes active. The process will end if no new inactive node becomes active. The influence spread is the number of active nodes at the end of process.

Counterexample 5.7.1 With positive influence, the influence spread is neither submodular nor supermodular as a set function with respect to the seed set.

Construction. Consider a counterexample in Fig. 5.6, in which a node u has seven coming neighbors. Let A and B be two subsets of incoming neighbors with $|A| = 2$, $|B| = 3$, and $A \subseteq B$. Let x be an incoming neighbor outside of B. Then, we have

$$\Delta_x \sigma(A) = 0 < 1 = \Delta_x \sigma(B).$$

This means that the influence spread σ is not submodular.

Fig. 5.6 $\Delta_x \sigma(A) < \Delta_x \sigma(B)$

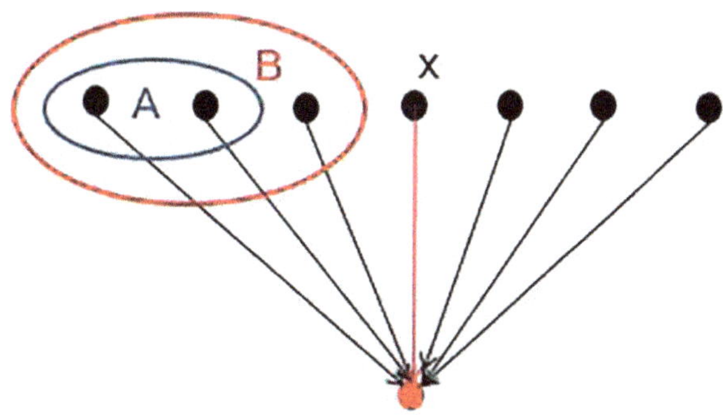

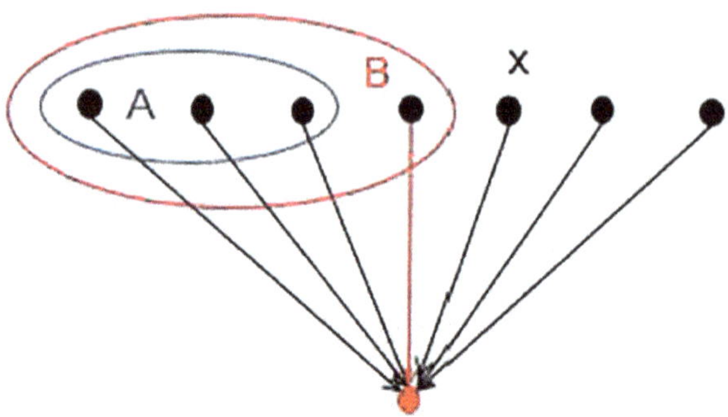

Fig. 5.7 $\Delta_x \sigma(A) > \Delta_x \sigma(B)$

Next, as shown in Fig. 5.7, let A and B be two subsets of incoming neighbors with $|A| = 3$, $|B| = 4$, and $A \subseteq B$. Let x be an incoming neighbor outside of B. Then, we have

$$\Delta_x \sigma(A) = 1 < 0 = \Delta_x \sigma(B).$$

This means that the influence spread σ is not supermodular. $\square$

To give an upper bound, we consider a general threshold model in which, at each node u, its threshold function is defined by

$$f_u(S) = \min(1, 2\frac{|S|}{|N^-(u)|}),$$

where $N^-(u)$ is the set of incoming neighbors of u. Since $|S| \geq \frac{1}{2} \cdot |N^-(u)|$ implies $f_u(S) = 1$, the influence spread of this general threshold model is at least the influence spread of the positive influence. Note that f_u is monotone nondecreasing and submodular. Thus, the influence spread of this general threshold model is monotone nondecreasing and submodular.

What is a lower bound? So far, no nontrivial lower bound has been found for the influence spread of positive influence. Especially, the following is still open.

Open Problem 2 *Please find a DS decomposition for the influence spread of the positive influence.*

5.8 Friending

There is a variation of DS decomposition as follows.

Theorem 5.8.1 (Du et al. [248]) *Any set function f can be expressed as the difference of two monotone nondecreasing and supermodular functions g and h, i.e., $f = g - h$.*

This variation has an interesting application in the study of social friending. Let us first introduce this application and then give the proof of this theorem.

The social friending is an activity frequently appearing in online social networks. For example, if you have a Facebook or LinkedIn account, you may frequently receive a message from LinkedIn or FaceBook, which informs you that someone added you as his/her friend. The aim of this message is to let you add new friends so that the connections of the network is expended.

Several optimization problem raised from social friending. One of them is proposed in Yang et al. [458] as follows.

Problem 5.8.2 (*Active Friending [458]*) Consider a social network $G = (V, E)$ with information diffusion model m. Let S be the list of existing friends of a node s and t a node which is a target of s, i.e., s wants to include t in his friend list. Given an integer $r > 0$, the problem is to find a subset R of at most r nodes to maximize the success probability $\Pr_m(s, S, R, t)$, i.e., the probability that node t is activated through subgraph induced by $R \cup S \cup \{s\}$. (Fig. 5.8).

It is interesting to note the following.

Lemma 5.8.3 (Yuan and Wu [478]) *The success probability* $\Pr_m(s, S, R, t)$ *is monotone nondecreasing upermodular with respect to* R *with the linear threshold model.*

Proof Under the LT model,

$$\Pr_{LT}(s, S, R, t) = \sum_{P \in \mathcal{P}_R} \Pr(P),$$

where $\mathcal{P}_R$ is the set of paths from $S \cup \{s\}$ to t and $\Pr(P)$ is the probability that t accepts the influence from $S \cup \{s\}$ through path P. Clearly, $\Pr_{LT}(s, S, R, t)$ is monotone nondecreasing. Next, we show its supermodularity, i.e.,

$$\Pr_{LT}(s, S, R_1, t) + \Pr_{LT}(s, S, R_2, t) \leq \Pr(s, S, R_1 \cup R_2, t) + \Pr(s, S, R_1 \cap R_2, t).$$

Note that that

$$\mathcal{P}_{R_1} \cup \mathcal{P}_{R_2} \subseteq \mathcal{P}_{R_1 \cup R_2}.$$

Fig. 5.8 Active friending

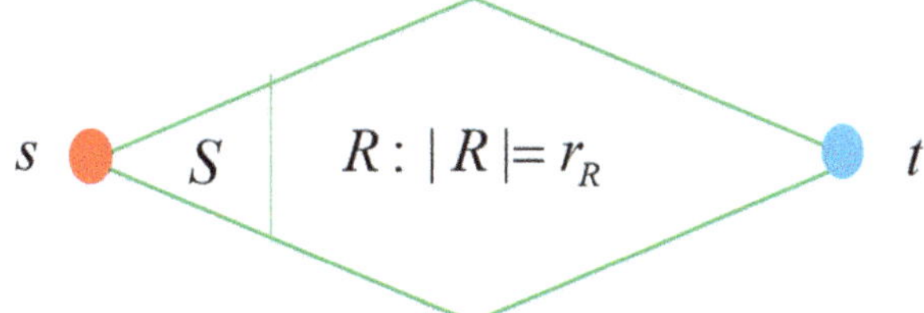

Moreover, a path P appearing in $\mathcal{P}_{R_1}$ and $\mathcal{P}_{R_2}$ must appear in $\mathcal{P}_{R_1 \cap R_2}$. Therefore,

$$\Pr_{LT}(s, S, R_1, t) + \Pr_{LT}(s, S, R_2, t) \leq \Pr(s, S, R_1 \cup R_2, t) + \Pr(s, S, R_1 \cap R_2, t).$$

$\square$

Lemma 5.8.4 (Gu et al. [156]) *With the IC model,*

$$\Pr(s, S, R, t) = (\Sigma_1 + \Sigma_3 + \cdots) - (\Sigma_2 + \Sigma_4 + \cdots),$$

where

$$\Sigma_k = \sum_{\{P_1,\ldots,P_k\} \subseteq \mathcal{P}} \Pr(P_1 \cup \cdots P_k)$$

is monotone nondecreasing and supermodular.

Proof By the inclusive-exclusive principal, we have

$$\begin{aligned}
\Pr(s, S, R, t) &= \Sigma_1 - \Sigma_2 + \Sigma_3 - \Sigma_4 + \cdots \\
&= (\Sigma_1 + \Sigma_3 + \cdots) - (\Sigma_2 + \Sigma_4 + \cdots).
\end{aligned}$$

By the argument similar to the proof of Lemma 5.8.3, we can prove that Σ_k is monotone nondecreasing and submodular. $\square$

Based on above two lemmas, Gu et al. [156, 157] designed approximation algorithms for solving the Active Friending problem.

Now, we come back to prove Theorem 5.8.1. Before doing so, we first state some properties of modular functions.

Recall that a set function f is said to be modular if for any two sets A and B,

$$f(A) + f(B) = f(A \cup B) + f(A \cap B).$$

In the following, we state three properties of modular functions. Their proofs are left to the reader as exercises.

Lemma 5.8.5 *Every supermodular function $f : 2^X \to \mathbb{R}$ can be expressed as $f = p_f + m_f$ where p_f is a monotone nondecreasing supermodular function with $p_f(\emptyset) = 0$ and m_f is a modular function.*

Lemma 5.8.6 *Let m is a modular function. Then, for any set Y,*

$$m(Y) = m(\emptyset) + \sum_{y \in Y} (m(\{y\}) - m(\emptyset)).$$

Now, let us prove Theorem 5.8.1.

Proof (Theorem 5.8.1) By Theorem 5.6.1, f can be decomposed into the difference of two monotone nondecreasing submodular functions g and h, i.e.,

$$f = g - h = (-h) - (-g).$$

Note that $-h$ and $-g$ are supermodular. By Lemma 5.8.5,

$$-h = p_{-h} + m_{-h},$$
$$-g = p_{-g} + m_{-g},$$

where p_{-h} and p_{-g} are monotone nondecreasing supermodular functions and m_{-h} and m_{-g} are modular functions. Therefore, for any set Y,

$$f(Y) = (p_{-h}(Y) + m_{-h}(\emptyset)) - (p_{-g}(Y) + m_{-g}(\emptyset)) + m(Y),$$

where

$$m(Y) = (m_{-h}(Y) - m_{-h}(\emptyset)) - (m_{-g}(Y) - m_{-g}(\emptyset).$$

Note that $m(\emptyset) = 0$. By Lemma 5.8.6,

$$m(Y) = \sum_{y \in Y} m(\{y\}).$$

Denote $X^+ = \{x in X \mid m(\{x\}) \geq 0\}$ and $X^- = X \setminus X^+$. Define $m^+(Y) = m(Y \cap X^+)$ and $m^-(Y) = m(Y \cap X^-)$. Then, m^+ and $-m^-$ are monotone nondecreasing modular functions and $m = m^+ + m^-$. Hence,

$$f(Y) = (p_{-h}(Y) + m_{-h}(\emptyset) + m^+(Y)) - (p_{-g}(Y) + m_{-g}(\emptyset) - m^-(Y)),$$

where $p_{-h}(Y) + m_{-h}(\emptyset) + m^+(Y)$ an $p_{-g}(Y) + m_{-g}(\emptyset) - m^-(Y))$ and monotone nondecreasing supermodular. $\qquad\square$

5.9 Parameterized Method

There is a family of algorithms appearing in the earlier stage of study for nonsubmodular optimizations. Each algorithm in this family often involves one or more parameters. Solutions generated by them are evaluated with approximation performance ratio containing those parameters, which usually give pretty nice theorems. However, when apply to specific problems in the real world. The parameter tends to the value which makes the performance ratio in nonsense. Therefore, recently, one does not make many efforts in the study of parameterized method. To give the reader a specific feeling, we present an example.

To deal with nonsubmodular optimization, one intends to measure how far the function differs from the submodularity. Motivated from this intension, several parameters are introduced and theoretical results for submodular optimization are extended to nonsubmodular optimization, usually with parameter involving in performance analysis. Let us give two examples in the following.

Consider a set function $f : 2^X \to R$. The *supermodular degree of an element* $u \in X$ by a function f is defined to be $|\mathcal{D}^+(u)|$ where

$$\mathcal{D}_f^+(u) = \{v \in X \mid \exists A \subseteq X : \Delta_u f(A \cup \{v\}) > \Delta_u f(A)\}.$$

The *supermodular degree of function* f is defined by

$$\mathcal{D}_f^+ = \max_{u \in X} |\mathcal{D}^+(u)|.$$

When only function f is studied on supmodular degree, we may simply write $\mathcal{D}^+ = \mathcal{D}_f^+$.

With the supermodular degree, a nice theoretical result can be established.

Consider the monotone nonsubmodular maximization with matroid constraints as follows.

Problem 5.9.1 (*Monotone Nonsubmodular Maximization*) Let f be a nonnegative and monotone nondecreasing function over 2^X where X is a finite set. Let $(X, \mathcal{C}_i)$ be a matroid for $i = 1, 2, ..., k$. Consider the following problem.

$$\max \quad f(A)$$
$$\text{subject to} \quad A \in \mathcal{C}_i \text{ for } i = 1, 2, ..., k,$$

Algorithm 28 is an extension of a greedy algorithm for submodular maximization with matroid constraints.

Algorithm 28 Greedy Approximation for Problem 5.9.1.

Input: a nonnegative monotone set function f on 2^X and k matroids $(X, \mathcal{C}_i)$ for $1 \leq i \leq k$. Let $\mathcal{C} = \cap_{i=1}^k \mathcal{C}_i$.
Output: A subset S_i of X.
1: $S_0 \leftarrow \emptyset$;
2: $i \leftarrow 0$;
3: **while** S_i is not a maximal independent set **do**
4:     $i \leftarrow i+1$;
5:     choose $u_i \in X \setminus S_{i-1}$ and $D_{i-1} \in \mathcal{D}^+(u_i)$ to maximize
6:     $f(D_i \cup \{u_i\} \cup S_{i-1}) - f(S_{i-1})$ subject to $D_i \cup \{u_i\} \cup S_{i-1} \in \mathcal{C}$;
7:     set $S_i \leftarrow D_i \cup \{u_i\} \cup S_{i-1}$;
8: **end while**
9: **return** S_i.

Theorem 5.9.2 (Feldman and Izsak [109]) *Greedy Algorithm 28 produces a* $\frac{1}{k(\mathcal{D}^+ +1)+1}$*-approximation solution for the maximization of monotone nondecreasing nonnegative set function with k matroid constraints (Problem 5.9.1).*

Proof Let $S_0, S_1, ..., S_g$ be produced by Greedy Algorithm 28. Let H_0 be an optimal solution. For $1 \le i \le g$, let H_i be a maximal independent set of $H_{i-1} \cup S_i$, containing S_i. It is easy to show that

$$|H_{i-1} \setminus H_i| \le k|S_i \setminus H_{i-1}| \le k|S_i \setminus S_{i-1}| \le k(\mathcal{D}^+ + 1).$$

Suppose $H_{i-1} \setminus H_i = \{v_1, ..., v_r\}$. Denote $H_{i-1}^j = H_{i-1} \setminus \{v_1, ..., v_j\}$. Then

$$
\begin{aligned}
&f((\mathcal{D}^+(v_j) \cap H_{i-1}^j) \cup \{v_j\} \cup S_{i-1}) - f(S_{i-1}) \\
&= \Delta_{v_j} f((\mathcal{D}^+(v_j) \cap H_{i-1}^j) \cup S_{i-1}) + f((\mathcal{D}^+(v_j) \cap H_{i-1}^j) \cup S_{i-1}) - f(S_{i-1}) \\
&\ge \Delta_{v_j} f((\mathcal{D}^+(v_j) \cap H_{i-1}^j) \cup S_{i-1}) \\
&\ge \Delta_{v_j} f(H_{i-1}^j \cup S_{i-1}).
\end{aligned}
$$

Since

$$((\mathcal{D}^+(v_j) \cap H_{i-1}^j) \setminus S_{i-1}) \cup \{v_j\} \cup S_{i-1} = (\mathcal{D}^+(v_j) \cap H_{i-1}^j) \cup \{v_j\} \cup S_{i-1}$$

and $(v_j, (\mathcal{D}^+(v_j) \cap H_{i-1}^j) \setminus S_{i-1})$ is a candidate pair for the ith step of the algorithm, we have

$$
\begin{aligned}
&f(S_i) - f(S_{i-1}) \\
&\ge f(((\mathcal{D}^+(v_j) \cap H_{i-1}^j) \setminus S_{i-1}) \cup \{v_j\} \cup S_{i-1}) - f(S_{i-1}) \\
&= f((\mathcal{D}^+(v_j) \cap H_{i-1}^j) \cup \{v_j\} \cup S_{i-1}) - f(S_{i-1}) \\
&\ge \Delta_{v_j} f((\mathcal{D}^+(v_j) \cap H_{i-1}^j) \cup S_{i-1}) \\
&\ge \Delta_{v_j} f(H_{i-1}^j \cup S_{i-1}).
\end{aligned}
$$

Thus,

$$
\begin{aligned}
r(f(S_i) - f(S_{i-1})) &\ge \sum_{j=1}^{r} \Delta_{v_j} f(H_{i-1}^j \cup S_{i-1}) \\
&= f(H_{i-1}^0 \cup S_{i-1}) - f(H_{i-1}^r \cup S_{i-1}) \\
&= f(H_{i-1}) - f(H_i).
\end{aligned}
$$

Note that $r \leq k(\mathcal{D}^+ + 1)$. Thus,

$$k(\mathcal{D}^+ + 1)[f(S_i) - f(S_0)] \geq r \sum_{i=1}^{g} [f(S_i) - f(S_{i-1})]$$

$$\geq \sum_{i=1}^{g} [f(H_{i-1}) - f(H_i)]$$

$$= f(H_0) - f(H_i).$$

Since $S_g = H_g$, $f(H_0) = opt$, and $f(S_0) \geq 0$, we finally have

$$(k(\mathcal{D}^+ + 1) + 1) f(S_g) \geq opt.$$

$\square$

The supermodular degree has been successfully applied to a few optimization problems, such as the secretary problem and committee member selection.

We next introduce other two parameters, submodularity ratio and curvature.

Definition 5.9.3 (*Submodularity Ratio [70]*) The submodularity ratio of a nonnegative set function f on 2^V is the largest scalar γ such that for any $\Omega, S \subset V$,

$$\sum_{omega \in \Omega \setminus S} \Delta_\omega f(S) \geq \gamma \cdot \Delta_\Omega f(S).$$

Definition 5.9.4 (*Curvature [64]*) The curvature of a nonnegative set function f on 2^V is the smallest scalar α such that for any $\Omega, S \subset V$ and $i \in S \setminus \Omega$,

$$\Delta_i f((S \setminus \{i\}) \cup \Omega) \geq \alpha \cdot \Delta_i f(S \setminus \{i\}).$$

Consider the following problem and Algorithm 29:

$$\max \quad f(S)$$

$$\text{subject to} \quad |S| \leq k, S \subseteq V,$$

where $f : 2^V \to \mathbb{R}^+$ is nonnegative and monotone nondecreasing.

Algorithm 29 Greedy Approximation.

Input: a nonnegative monotone nondecreasing set function $f : 2^V \to \mathbb{R}^+$ and a positive integer k.
Output: A subset S of V with $|S| \leq k$.
1: $S \leftarrow \emptyset;\ i \leftarrow 0$;
2: **for** $i = 1$ to k **do**
3:    $u_i \leftarrow \text{argmax}_{v \in V \setminus S_{i-1}} \Delta_v f(S_{i-1})$;
4:    $S_i \leftarrow S_{i-1} \cup \{u_i\}$;
5: **end for**
6: **return** S_k.

The following result holds.

Theorem 5.9.5 (Bian et al. [18]) *Suppose f is nonnegative monotone nondecreasing set function with submodularity ration $\gamma \in [0, 1]$ and curvature $\alpha \in [0, 1]$. Then, Algorithm 29 gives a $\frac{1}{\alpha}(1 - e^{\alpha\gamma})$-approximation for maximizing $f(S)$ subject to $|S| \leq k$.*

Exercises

1. Show that any submodular function f can be expressed as the sum of a polymatroid function p_f and a modular function m_f, i.e., $f = p_f + m_f$.
2. Show that any modular function m can be expressed as the difference of two monotone nondecreasing modular functions.
3. Prove or disprove that any monotone nondecreaing set function can be expressed as the sum of a monotone nondecreasing submodular function and a monotone nondecreasing supermodular function.
4. Consider a set function f defined on subsets of V. Define

$$\alpha(f) = \min_{A \subset B \subseteq V \setminus \{x\}} \{\Delta_x f(A) - \Delta_x(B)\}.$$

   Show that if a lower bound of $\alpha(f)$ can be computed in polynomial time, then two submodular functions g and h can be found in polynomial time such that $f = g - h$.
5. (Lovasz Extension) Let $f : 2^V \to \mathbb{R}$. Assume $V = \{1, 2, ..., n\}$ where $n = |V|$. For each $z \in \mathbb{R}^n$, suppose z's components in ordering $z_1 \geq z_2 \geq \cdots \geq z_n$. Define $S_i = \{1, 2, ..., i\}$. The Lovasz extension of f at z is defined by

$$\hat{f}(z) = \sum_{i=1}^{n-1} f(S_i) + z_n f(S_n).$$

   Please prove the following property of $\hat{f}$.

(a) Let $f : 2^V \to \mathbb{R}$ be a submodular function with $f(\emptyset) = 0$. Then for any $z \in [0, 1]^n$ where $n = |V|$,

$$\hat{f}(z) = \max_{x \in P_f} z^T x,$$

where
$$P_f = \{x \mid x(S) \le f(X), \forall S \subset V; x(V) = f(V)\}.$$

(b) $\hat{f} + \hat{g}$ is the Lovasz extension of $f + g$ and $\lambda \hat{f}$ is the Lovasz extension of λf.

(c) A set function $f : 2^V \to \mathbb{R}$ is submodular if and only if its Lovasz extension is convex.

Next, we state a proof for that any set function can be expressed as the difference of two submodular functions.

Let $f : 2^V \to \mathbb{R}$ be a set function. Let g be a submodular function such that g has a Hessian with all eigenvalues bounded below by $\varepsilon > 0$. Then for sufficiently large constant $lambda > 0$, $\hat{f} + \lambda\hat{g}$ has a positive definite Hessian and hence convex. Thus, $f + \lambda g$ is submodular and we have

$$f = (f + \lambda g) - \lambda g.$$

Question: Is this proof correct? Why?

6. (Multidimensional IC Model) In this model, we have the following rules.

- Each feature is diffused in IC model.
- Each customer buys the product if and only if all features are accepted.
- A customer is capable to influence others if and only if the customer bought the product.

Construct a counterexample to show that with this model, the influence spread is neither submodular nor supermodular.

7. Prove or disprove (by counterexample) that every set function can be decomposed into the difference of two nonnegative submodular functions.

8. Prove or disprove (by counterexample) that every set function can be decomposed into the difference of two nonnegative monotone nondecreasing submodular functions.

9. ([248]) Show that every set function $f : 2^X \to R$ can be decomposed into the difference of two monotone nondecreasing supermodular functions g and h, i.e., $f = g - h$.

10. ([247]) Consider a set function $f : 2^X \to R$. f is *strictly increasing* if for any two subsets $A \subset B$, $f(A) < f(B)$. f is *strictly submodular* if for and two subsets $A \subset B$ and any element $x \in X \setminus B$, $\Delta_x f(A) > \Delta_x f(B)$. Show that every set function can be decomposed into a difference of two strictly increasing and strictly submodular functions.

11. A set function f on 2^X is *strictly supermodular* if $-f$ is strictly submodular. Show that every set function can be decomposed into a difference of two strictly increasing and strictly supermodular functions.
12. Give a counterexample to show that not every set function can be decomposed into the sum of a monotone nondecreasing submodular function and a monotone nondecreasing supermodular function.
13. Show that for any monotone nondecreasing submodular function $f : 2^X \to R$ any $Y \subseteq X$, there exist a pair of monotone nondecreasing modular functions $u, l : 2^X \to R$ such that $u(Y) = f(Y) = l(Y)$ and $u(S) \geq f(S) \geq l(S)$ for any $S \subseteq X$.
14. ([398]) Consider a graph $G = (V, E)$ and a target-node subset S of V. A node subset A is called a *positive-influence target-dominating set* for S if every node $u \in S$ has at least one half of neighbors in A. Show the following.

    (a) Finding the minimum positive-influence target-dominating set for any given target-node set S has a polynomial-time $(1 + \log\lceil \frac{3}{2}\Delta\rceil)$-approximation solution where Δ is the maximum node degree of G.
    (b) If G is a power-law graph and $|S| = \Omega(|V|)$, then finding the minimum positive-influence target-dominating set for any given target-node set S has a polynomial-time $O(1)$-approximation solution.

15. Show, by giving the definition of threshold function fv at each node v, that the positive influence is a model of general threshold.
16. Describe the general cascade model which is equivalent to the positive influence model and prove the correctness of you definition.
17. In the linear threshold model, let threshold θ_v selected uniformly at random from $[1/2,1]$ at each node v. Discuss whether the influence spread is submodular.
18. (Wan et al. [414]) By the DS decomposition theorem, every set function is the difference of two monotone nondecreasing submodular functions g and h, i.e., $g - h$. Therefore, the minimization of such a set function can be worked out through solving a sequence of problems $\min\{g - c \mid h \geq c\}$ for a sequence of discrete constants c. Note that each monotone nondecreasing sbmodular function can be represented by the sum of a polymatroid function and a constant. Thus, $\min\{g - c \mid h \geq c\}$ can be transformed into the following form.

Problem 5.9.6 (*Submodular Set Cover with Submodular Objective*) Let g and f be polymatroid functions over 2^X for a finite set X.

$$\min \quad g(A)$$
$$\text{subject to} \quad f(A) = f(X).$$

This problem has Algorithm 30.

Algorithm 30 Greedy Approximation for Problem 5.9.6.

Input: two polymatroid functions g and f.
Output: A subset S of X.
1: $S \leftarrow \emptyset$;
2: **while** $f(S) < f(X)$ **do**
3:     choose x to maximize $\frac{\Delta_x f(S)}{\Delta_x g(S)}$
4:     and set $S \leftarrow S \cup \{x\}$;
5: **end while**
6: **return** S.

Now, we give an alternative definition of curvature: The *curvature* of g is defined by

$$\chi(g) = \min_{A \subseteq X} \frac{\sum_{x \in A} g(\{x\})}{g(A)}.$$

(a) Please establish the relationship between this curvature and the curvature in Definition 5.9.4.
(b) Please prove that Greedy Algorithm 30 produces an approximation solution with performance ratio at most $\chi(g) H(\gamma)$ where

$$\gamma = \max_{x \in X} f(\{x\})$$

and $H(\gamma) = \sum_{i=1}^{\gamma} 1/i$.

19. (Wan et al. [414]) Problem 5.9.6 is also closely related to the generalized hitting set problem as follows.

Problem 5.9.7 (*Generalized Hitting Set*) Given m nonempty collections $C_1, C_2, ..., C_m$ of subsets of a finite set X, find the minimum subset A of X such that every C_i has a member $S \subseteq A$.

Let $C = \cup_{i=1}^{m} C_i$. For every subcollection $\mathcal{A} \subseteq C$, define

$$g(\mathcal{A}) = |\cup_{A \in \mathcal{A}} A|$$

and

$$f(\mathcal{A}) = |\{C_i \mid \mathcal{A} \cap C_i \neq \emptyset, 1 \leq i \leq m\}|.$$

Prove the following.

(a) Both g and f are polymatroid functions.
(b) The generalized hitting set problem is equivalent to the following

$$\min \quad g(\mathcal{A}) \tag{5.2}$$
$$\text{subject to} \quad f(\mathcal{A}) = f(C),$$
$$\mathcal{A} \subseteq C.$$

This equivalence means that $\mathcal{A}$ is a minimum solution of problem (5.2) if and only if $\cup_{A \in \mathcal{A}} A$ is the minimum solution of the generalized hitting set problem.

20. (Feldman and Izsak [109]) Consider the nonnegative monotone nondecreasing set function maximization

$$\max \quad f(S)$$
$$\text{subject to} \quad |S| \leq k,$$

where $f : 2^V \rightarrow \mathbb{R}^+$ is nonnegative and monotone nondecreasing. Show that Algorithm 31 gives a $(1 - e^{-1/(\mathcal{D}^+ + 1)})$-approximation.

Algorithm 31 Greedy Approximation.

Input: a nonnegative monotone nondecreasing set function $f : 2^V \rightarrow \mathbb{R}^+$ and a positive integer k.
Output: A subset S of V with $|S| \leq k$.

1: $S \leftarrow \emptyset$;
2: **for** $d = 1$ to $\mathcal{D}^+$, $v \in V$, and C with $|C| = k \mod (d + 1)$ **do**
3:    $S_0 \leftarrow C$;
4:    $i \leftarrow 0$;
5:    **while** S_i is not a base **do**
6:      $i \leftarrow i + 1$;
7:      choose $u_i \in V \setminus S_{i-1}$ and $D_i \subseteq \mathcal{D}^+(u_i)$ to maximize $f(D_i \cup \{u_i\} \cup S_{i-1})$ subject to $|D_i \cup \{u_i\} \cup S_{i-1}| \leq k$ and $|D_i| \leq d$;
8:      set $S_i \leftarrow D_i \cup \{u_i\} \cup S_{i-1}$;
9:    **end while**
10:   $S \leftarrow \operatorname{argmax}(f(S), f(S_i))$;
11: **end for**
12: **return** S

---

21. Consider the active friending problem. Show that computing the success probability $\Pr_m(s, S, R, t)$ is #P-hard.
22. Show that Every supermodular function $f : 2^X \rightarrow \mathbb{R}$ can be expressed as $f = p_f + m_f$ where p_f is a monotone nondecreasing supermodular function with $p_f(\emptyset) = 0$ and m_f is a modular function.
23. Show that for a modular function m and any set Y,

$$m(Y) = m(\emptyset) + \sum_{y \in Y} (m(\{y\}) - m(\emptyset)).$$

24. Define $\alpha(h) = \min_{A \subset B \subseteq X \setminus \{j\}} \{\Delta_j h(A) - \Delta_j h(B)\}$. Show that if a lower bound of $\alpha(h)$ can be computed in polynomial-time, then two submodular functions f and g can be found in polynomial-time such that $h = f - g$.

25. (Lovasz Extension) Let $f : 2^X \to \mathbb{R}$ where $X = \{1, ..., n\}$. For each $z \in \mathbb{R}^n$ where $n = |X|$. Suppose z's components in ordering $z_1 \geq \cdots \geq z_n$. Define $S_i = \{1, ..., i\}$. The Lovasz extension of f at z is defined by

$$\hat{f} = \sum_{i=1}^{n-1} (z_i - z_{i+1}) f(S_i) + z_n f(S_n).$$

Prove the following.

(a) If f is a submodular function with $f(\emptyset) = 0$, then for any $z \in [0, 1]^n$,

$$\hat{f} = \max_{x \in P_f} z^T x,$$

where

$$P_f = \{x \mid x(S) \leq f(S) \text{ for any } S \subset X, x(X) = f(X)\}.$$

(b) $\hat{f} + \hat{g}$ is the Lovasz extension of $f + g$.
(c) $\lambda \cdot \hat{f}$ is the Lovasz extension of λf.
(d) A set function f is submodular if and only if its Lovasz extension $\hat{f}$ is convex.

26. The following is a proof for the DS decomposition theorem that any set function can be decomposed into the difference of two submodular functions. Please figure out whether the proof contains a flaw: Let $f : 2^V \to \mathbb{R}$ be a set function. Let g be a submodular function such that $\hat{g}$ has a Hession with all eigenvalues bounded below by $\varepsilon > 0$. Then, for sufficiently large constant λ, $\hat{f} + \lambda \hat{g}$ has positive definite Hessian and hence convex. Thus, $f + \lambda g$ is submodular.

Historical Notes

In the study of viral marketing of multi-products, the recommendation attracts many interesting research efforts since the recommendation introduces the influence between products, i.e., the item inference. It was initiated in [195] to consider both social influence and item inference. The item inference is formulated in a hypergraph, called the social item graph; each recommendation is represented by a hyperedge. For example, hyperedge $(a, b, c; d)$ represents recommendation "a customer who purchases products a, b and c may also purchases d." Guo and Wu [169] consider the case that the social item graph is a graph, that is, all recommendations are in form "a customer who purchases a may purchases b." They found that in this case, the viral marketing can be formulated as a maximization of monotone nondecreasing submodular function subject to a knapsack constraint. Actually, consider the product of the social network and the social item graph. Each node has label (u, a) where u is a node (customer) in the social network and a is a product. All nodes with the same product a form a layer G_a for product a and each node in this layer has cost $c(a)$ which is the cost of product a. The item inference gives the relationship between layers. Given a budget B for advertisement, the viral marketing problem is to select nodes in the product graph, within a given budget, to maximize the total cost of influence spread or the total revenue of the influence spread. Guo and Wu [164] studied the cases that each hyperedge in the social item graph contains at most three elements. In this case, the objective function is neither submodular nor supermodular. Therefore, they employed the sandwich method. Actually, the nonsubmodularity of objective function stems from the group influence in the social item graph. The group influence would introduce nonsubmodularity. This fact is also noted by earlier publications [496, 499], in which group influence and social hypergraph have been studied, respectively. To deal with nonsubmodularity, they also employed the sandwich method. However, two different pairs of upper and lower bounds are utilized. Which pair is better? The performance ratio of the sandwich methods cannot tell. That is a disadvantage of the sandwich method.

There is another situation for group activity to link to nonsubmodularity. When group activity generates benefit for objective function, this benefit may bring nonsubmodularity. This happens in online game advertising [124], interactive-aware influence maximization [125], and activity maximization [430]. It is worth mentioning that Gao et al. [125] found a technique to construct the DS decomposition for objective function studied in their paper. Could this technique be used in group influence for finding DS decomposition? This question may bring valuable research.

The sandwich method was initiated in [271] and later appeared in many publications [45, 122, 162, 164, 289, 317, 393, 430, 449, 451, 486, 487, 496, 499]. Therefore, the sandwich method becomes an important approach to deal with nonsubmodular optimization problems. The theoretical foundation of the sandwich method is established based on DS decomposition of set functions. Although the existence of DS decomposition is proved constructively, the time-complexity of the construction is very high. It is a very important open problem whether there exists an efficient way to find a DS decomposition for a given set function. For a specific set function, the influence spread of a general network with positive influence diffusion model, its DS decomposition is also unknown for a long time. The major critique on the sandwich method is about its approximation performance ratio, which is data-dependent. Especially, it is hard to see the approximation quality of the pair of upper and lower bounds. Therefore, the sandwich method does not look well theoretically, however, applications can be found well.

Friending is an active activity in online social networks. There are several nonsubmodular optimization problems motivated from the study of this activity [157, 399, 458, 478]. One interesting effort is about application of a variation of DS decomposition [248].

Parameterized methods have good looking theoretically [18, 70, 375]. To deal with nonsubmodular optimization, one intends to measure how far the function differs from the submodularity. Motivated from this intention, several parameters are introduced and theoretical results for submodular optimization are extended to non-submodular optimization, usually with those parameters involving in performance analysis. For example, the supermodular degree $\mathcal{D}_f^+$ for set function f is introduced in [105]. With this parameter, a few results on greedy algorithm for nonsubmodular maximization with matroid constraints have been established [109]. They have found to have applications in the secretary problem [110] and committee member selection [197].

However, in general, the parameterized method often provide good-looking mathematical theorem with very limited applications in the real world. For example, in many nonsubmodular maximization problems, $\mathcal{D}_f^+ = \infty$.

Chapter 6
Local Optimum

"Sometimes the best solution is to rest, relax and recharge. It's hard to be your best on empty."
- Sam Glenn

Many optimization problems in social networks are nonsubmodular optimization. In the study of nonsubmodular optimization, it is unlikely to be able efficiently to compute global optimal solution. Therefore, we have to move our attention to local optimal solution. In this chapter, we introduce new techniques for design of algorithms to compute local optimal solutions.

For example, linear relaxation and convex relaxation are important techniques in design of approximation algorithms and nonconvex Relaxation is a new technique in the study of local optimal solution. At the end of this chapter, we introduce the background and research potential in the study of nonconvex relaxation.

6.1 Effective Local Search

6.1.1 Iterated Sandwich Method

There exists a much interesting result about sandwich. Let us introduce it as follows.

Recall that a set function m over 2^X is a modular function if for any two sets A and B, $m(A) + m(B) = m(A \cup B) + m(A \cap B)$. The following lemma indicates that the modular function is similar to a linear function.

W. Wu et al., *Computational Aspects of Social Networks*, Springer Optimization and Its Applications 234, https://doi.org/10.1007/978-3-032-14833-9_6

Lemma 6.1.1 *For any modular function* $m : 2^X \to R,$

$$m(A) = m(\emptyset) + \sum_{x \in A} (m(x) - m(\emptyset))$$

for any set $A \subseteq X.$

Proof This lemma can be proved by induction on $|A|$. For $|A| = 1$, it is trivial. For $|A| \geq 2$, suppose $y \in A$. Then

$$m(y) + m(A \setminus y) = m(A) + m(\emptyset).$$

Therefore,

$$
\begin{aligned}
m(A) &= m(A \setminus y) + (m(y) - m(\emptyset)) \\
&= m(\emptyset) + \sum_{x \in A \setminus y} (m(x) - m(\emptyset)) + (m(y) - m(\emptyset)) \\
&= m(\emptyset) + \sum_{x \in A} (m(x) - m(\emptyset)).
\end{aligned}
$$

$\square$

The following theorem states a surprising property of the set function.

Theorem 6.1.2 (Sandwich Theorem [440]) *For any set function* $f : 2^X \to R$ *and any set* $Y \subseteq X$, *there are two modular functions* $m_u : 2^X \to R$ *and* $m_l : 2^X \to R$ *such that* $m_u \geq f \geq m_l$ *and* $m_u(Y) = f(Y) = m_l(Y).$

Note that the modular function is similar to linear function. Theorem 6.1.2 contains two different modular functions passing through the same set and one is always smaller than or equal to the other. This phenomenon cannot occur for continuous linear functions. A continuous linear function with n variables can be expressed as an n-dimensional plane in the $(n + 1)$-dimensional space. A pair of different n-dimensional planes with a point in common cannot have a coordinate along which one is always smaller than or equal to the other (Fig. 6.1.) Therefore, this theorem states a special property of the set function.

To prove the sandwich theorem, we first establish three lemmas.

First, using the modular function, we can decompose the submodular function as follows.

Lemma 6.1.3 (Decomposition of Submodular Function [68]) *Every submodular function* g *can be expressed as* $g = p + m$ *where* p *is a polymatroid function (i.e., a monotone nondecreasing submodular function with* $p(\emptyset) = 0$*) and* m *is a modular function.*

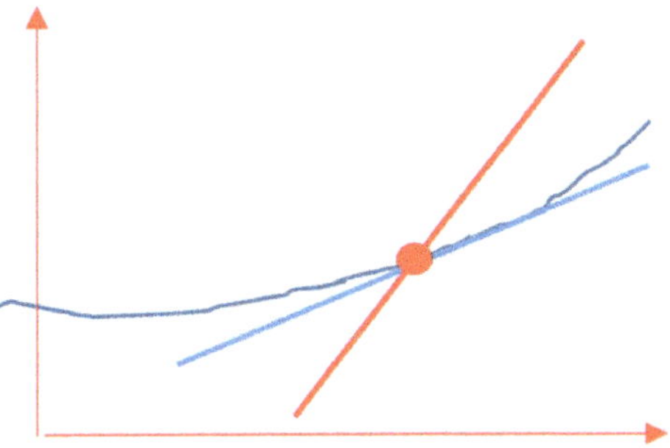

Fig. 6.1 It is impossible for two linear functions to cross and one is above the other one

Proof Define $m(A) = f(\emptyset) - \sum_{x \in A} \Delta_x f(X \setminus x)$ and $p = f - m$. Then, m is a modular function. Hence, p is a submodular function. Moreover, $p(\emptyset) = f(\emptyset) - m(\emptyset) = 0$ and for any set A and $x \in X \setminus A$, $\Delta_x p(A) = \Delta_x f(A) - \Delta_x f(X \setminus x) \geq 0$, i.e., p is monotone nondecreasing. Therefore, p is a polymatroid function. $\square$

Lemma 6.1.4 ([302]) *For any submodular function $f : 2^X \to R$ and any set $Y \subseteq X$, there exists a modular function $m_u : 2^X \to R$ such that $m_u \geq f$ and $m_u(Y) = f(Y)$.*

Proof Define

$$m_u(A) = f(Y) + \sum_{j \in A \setminus Y} \Delta_j f(\emptyset) - \sum_{j \in Y \setminus A} \Delta_j f(Y \setminus j).$$

Clearly, m_l is modular and $m_u(Y) = f(Y)$. Next, we show that $m_l \geq f$. Assume $A \setminus Y = \{j_1, \ldots, j_k\}$. Then

$$
\begin{aligned}
f(A) - f(A \cap Y) =& \Delta_{j_1} f(A \cap Y) + \Delta_{j_2} f((A \cap Y) \cup \{j_1\}) \\
& + \cdots + \Delta_{j_k} f((A \cap Y) \cup \{j_1, \ldots, j_{k-1}\}) \\
\leq & \sum_{j \in A \setminus Y} \Delta_j f(A \cap Y).
\end{aligned}
$$

Assume $Y \setminus A = \{i_1, \ldots, i_k\}$. Then

$$
\begin{aligned}
f(Y) - f(A \cap Y) =& \Delta_{i_1} f(A \cap Y) + \Delta_{i_2} f((A \cap Y) \cup \{i_1\}) \\
& + \cdots + \Delta_{i_k} f((A \cap Y) \cup \{i_1, \ldots, i_{k-1}\}) \\
\geq & \sum_{j \in A \setminus Y} \Delta_j f(Y \setminus j).
\end{aligned}
$$

Therefore,

$$
\begin{aligned}
f(A) \leq & f(Y) + \sum_{j \in A \setminus Y} \Delta_j f(A \cap Y) - \sum_{j \in A \setminus Y} \Delta_j f(Y \setminus j) \\
\leq & f(Y) + \sum_{j \in A \setminus Y} \Delta_j f(\emptyset) - \sum_{j \in A \setminus Y} \Delta_j f(Y \setminus j) \\
= & m_u(A).
\end{aligned}
$$

$\square$

Lemma 6.1.5 ([120]) *For any submodular function $f : 2^X \to R$ and any set $Y \subseteq X$, there exists a modular function $m_l : 2^X \to R$ such that $f \geq m_l$ and $f(Y) = m_l(Y)$.*

Proof Put all elements of X into an ordering $X = \{x_1, x_2, \ldots, x_n\}$ such that $Y = \{x_1, x_2, \ldots, x_{|Y|}\}$. Denote $S_i = \{x_1, x_2, \ldots, x_i\}$. Define $m_l(\emptyset) = f(\emptyset)$ and for $\emptyset \neq A \subseteq X$, define

$$
m_l(A) = f(\emptyset) + \sum_{x_i \in A} (f(S_i) - f(S_{i-1})).
$$

Clearly, m_l is modular and

$$
m_l(Y) = f(\emptyset) + \sum_{x_i \in Y} (f(S_i) - f(S_{i-1})) = f(Y).
$$

Moreover, for any set $A \subseteq X$ with $A \neq \emptyset$, suppose $A = \{x_{i_1}, x_{i_2}, \ldots, x_{i_k}\}$ and then we have

$$
\begin{aligned}
m_l(A) = & f(\emptyset) + (f(S_{i_1}) - f(S_{i_1-1})) + (f(S_{i_2}) - f(S_{i_2-1})) \\
& + \cdots + (f(S_{i_k}) - f(S_{i_k-1})) \\
\leq & f(\emptyset) + (f(\{x_{i_1}\}) - f(\emptyset)) + (f(\{x_{i_1}, x_{i_2}\}) - f(\{x_1\})) \\
& + \cdots + (f(A) - f(\{x_{i_1}, \ldots x_{i_{k-1}}\})) \\
= & f(A).
\end{aligned}
$$

$\square$

Now, we are ready to prove Theorem 6.1.2.

Proof (*Theorem* 6.1.2). By DS-Decomposition Theorem, there exist submodular functions g and h such that $f = g - h$. By Lemmas 6.1.4 and 6.1.5, there exist modular functions $m_{gu}, m_{gl}, m_{hu}, m_{hl}$ such that

$$
m_{gu} \geq g \geq m_{gl}, \, m_{gu}(Y) = g(Y) = m_{gl}(Y),
$$

and

$$
m_{hu} \geq g \geq m_{hl}, \, m_{hu}(Y) = h(Y) = m_{hl}(Y).
$$

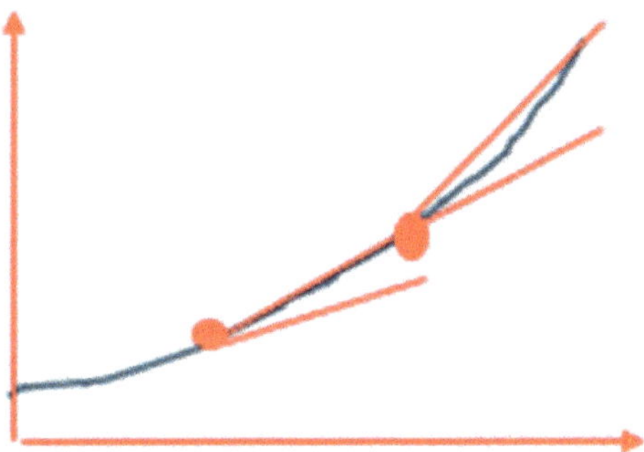

Fig. 6.2 Iterated sandwich method

Set $m_u = m_{gu} - m_{hl}$ and $m_l = m_{gl} - m_{hu}$. Then

$$m_u \geq f \geq m_l$$

and

$$m_u(Y) = g(Y) - h(Y) = f(Y) = g(Y) - h(Y) = m_l(Y).$$

$\square$

Motivated from the sandwich theorem, an iterated sandwich algorithm [440] can be designed immediately as follows (Fig. 6.2).

- Consider $\max_{X \in \Omega} f(X)$.
- Initially, select a set S from Ω.
- At each iteration, compute two modular function m_u and m_l such that $m_u(S) = m_l(S) = f(S)$ and $m_u(X) \geq f(X) \geq m_l(x)$ for all $X \in \Omega$.
- Find optimal/approximation solution S_u and S_l for $\max_{X \in \Omega} m_u(X)$ and $\max_{X \in \Omega} m_l(X)$, respectively.
- Set $S^+ = \mathrm{argmax}(f(S_u), f(S_l))$.
- If $f(S^+) > f(S)$, then $S \leftarrow S^+$ and go back to the beginning of the iteration.
- Return S.

Under certain condition, the iterated sandwich method returns a local optimal solution [125], i.e., a set at which its objective function value cannot be increased by adding or deleting an element. Such a condition would not be required by algorithms introduced in next two sections.

6.1.2 *Submodular-Supermodular Method*

Let f be a set function on 2^X. A is called a *local minimum* solution of f in 2^X if $f(A) \leq f(A \setminus \{x\})$ for any $x \in X \setminus A$ and $f(A) \leq f(A \cup \{x\})$ for any $x \in X$.

Similarly, A is called a *local maximum* solution of f in 2^X if $f(A) \geq f(A \setminus \{x\})$ for any $x \in X \setminus A$ and $f(A) \geq f(A \cup \{x\})$ for any $x \in X$.

In this and next sections, we introduce two algorithms for set function optimizations. They are designed based on DS-decomposition of set functions with a nice property that the algorithm will end at a local optimal solution. This property is established from the following observation.

Look at the proof of Lemma 6.1.5. It indicates the following fact.

Lemma 6.1.6 *For any submodular function $f : 2^X \to R$ and for any permutation σ of X, there exists a modular function m^σ such that $f(A) \geq m^\sigma(A)$ for any $A \in 2^X$ and*

$$f(S_i) = m^\sigma(S_i)$$

for any $i = 1, 2, \ldots, |X|$, where

$$S_i = \{\sigma(1), \sigma(2), \ldots, \sigma(i)\}$$

and $\sigma(i)$ denotes the ith element of X in the permutation σ.

Consider a set A. Let permutation σ satisfy $A = \{\sigma(1), \ldots, \sigma(|A|)\}$. Let $\theta = \max(|A|, |X| - |A|)$. Let $\Sigma(A)$ be a collection of θ such permutations σ of X such that $\sigma(|A|)$ goes over all elements of A and $\sigma(|A| + 1)$ goes over all elements of $X \setminus A$, that is, for any element $x \in A$, there exists $\sigma \in \Sigma_A$ such that $A \setminus \{x\} = \{\sigma(1), \ldots, \sigma|A| - 1\}$ and for any $x \in X \setminus A$, there exists $\sigma \in \Sigma_A$ such that $A \cup \{x\} = \{\sigma(1), \ldots, \sigma(|A| + 1)\}$.

Let m_{hl}^σ denote the modular function satisfying the following.

- $h(S) \geq m_{hl}^\sigma(S)$ for every $S \in 2^X$.
- $h(S_i) = m_{hl}^\sigma(S_i)$ for every $1 \leq i \leq |X|$.

As the first one, let us introduce the submodular-supermodular algorithm as shown in Algorithm 1 for the unconstrained minimization problem $\min_{S \in 2^X} f(S)$ (Fig. 6.3).

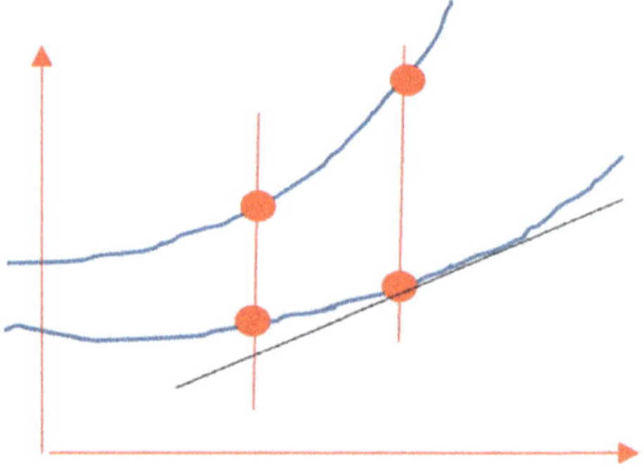

Fig. 6.3 Submodular-supermodular method

Algorithm 1 Submodular-Supermodular Algorithm for Minimization

---

Input: a set function $f : 2^X \to R$ and its DS decomposition $f = g - h$ where g and h are submodular functions.

Output: A subset A of X.

1: choose a set $A \subseteq X$;
2: **while** $f(A^+) < f(A)$ **do**
3:    $A \leftarrow A^+$;
4:    **for** $\sigma \in \Sigma(A)$ **do**
5:        compute a lower bound modular function m_{hl}^σ for h;
6:        compute a minimum solution A_σ^+ for $\min_{Y \in 2^X}[g(Y) - m_{hl}^\sigma(Y)]$;
7:    **end for**
8:    $\sigma \leftarrow \mathrm{argmin}_{\sigma \in \Sigma_A} f(A_\sigma^+)$;
9:    $A^+ \leftarrow A_\sigma^+$
10: **end while**
11: **return** A.

Theorem 6.1.7 *Algorithm 1 always ends at a local minimum solution.*

Proof First, note that for the unconstrained minimization of a submodular function, its optimal solution can be computed in polynomial-time [153, 322, 357]. Therefore, the line 6 can be done in polynomial-time.

When the algorithm stops, we must have $f(A^+) \geq f(A)$. Therefore,

$$f(A_\sigma^+) \geq f(A) \ \text{ for any } \sigma \in \Sigma(A).$$

Note that $f(A) = g(A) - m_{hl}^\sigma(A)$ for any $\sigma \in \Sigma(A)$. Thus, we have that for any $\sigma \in \Sigma(A)$,

$$g(A_\sigma^+) - m_{hl}^\sigma(A_\sigma^+) \geq f(A_\sigma^+) \geq f(A).$$

Note that for any $x \in A$, there exists $\sigma \in \Sigma(A)$ such that

$$f(A \setminus \{x\}) = g(A \setminus \{x\}) - m_{hl}^\sigma(A \setminus \{x\}) \geq g(A_\sigma^+) - m_{hl}^\sigma(A_\sigma^+).$$

Therefore,

$$f(A \setminus \{x\}) \geq f(A).$$

Similarly, we can show that for any $x \in X \setminus A$,

$$f(A \cup \{x\}) \geq f(A).$$

Thus, A is a local minimum. $\qquad\square$

In submodular-supermodular method, it employs a subproblem

$$\min_{Y \in 2^X}[g(Y) - m_{hl}^\sigma(Y)].$$

Note that $g(Y) - m_{hl}^{\sigma}(Y)$ is submodular. Thus, this subproblem can be solved in polynomial-time since the unconstrained minimization of submodular function is polynomial-time solvable [153, 322, 357]. However, for constrained submodular minimization, it is hard to find optimal solution or approximation of good quality [376]. The algorithm in next section is able to avoid this difficulty.

6.1.3 Modular-Modular Methods

The submodular-supermodular method works only for unconstrained minimization since for constrained minimization or maximization, it is unlikely to be able to find its optimal solution for submodular functions. This disadvantage will be improved for the modular-modular method as shown in Algorithm 2 (Fig. 6.4). This algorithm is for $\max_{S \in \Omega} f(S)$.

Algorithm 2 Modular-Modular Algorithm for Constrained Maximization

---

Input: a set function $f : 2^X \to R$ with its DS decomposition $f = g - h$, where g and h are submodular functions, and constraint $\Omega \subseteq 2^X$.
Output: A subset $A \in \Omega$.
1: choose a set $A \in \Omega$;
2: **while** $f(A^+) > f(A)$ **do**
3:     $A \leftarrow A^+$;
4:     **for** $\sigma \in \Sigma(A)$ **do**
5:         compute two modular functions m_{gl}^{σ} and m_{hl}^{σ};
6:         compute a maximum solution A_{σ}^+ for $\max_{Y \in \Omega}[m_{gl}^{\sigma}(Y) - m_{hl}^{\sigma}(Y)]$;
7:         **if** $f(A_{\sigma}^+) \leq f(A)$ **then**
8:             $A_{\sigma}^+ \leftarrow \mathrm{argmax}\{f(\{\sigma(1), \ldots, \sigma i\}) \mid 1 \leq i \leq |X|, \{\sigma(1), \ldots, \sigma i\} \in \Omega\}$
9:         **end if**
10:     **end for**
11:     $\sigma \leftarrow \mathrm{argmax}_{\sigma \in \Sigma_A} f(A_{\sigma}^+)$;
12:     $A^+ \leftarrow A_{\sigma}^+$
13: **end while**
14: **return** A.

Theorem 6.1.8 *Algorithm 2 always ends at a local maximum solution in Ω.*

Proof Suppose the algorithm stops at A. Then, $f(A^+) \leq f(A)$. For contradiction, assume that there exists $x \in A$ such that $A \setminus \{x\} \in \Omega$ and $f(A \setminus \{x\}) > f(A)$. Note that there exists $\sigma \in \Sigma(A)$ such that $f(A \setminus \{x\}) = m_{gl}^{\sigma}(A \setminus \{x\}) - m_{hl}^{\sigma}(A \setminus \{x\})$. Thus, for this σ, $m_{gl}^{\sigma}(A_{\sigma}^+) - m_{hl}^{\sigma}(A_{\sigma}^+) > f(A)$. Hence, $f(A_{\sigma}^+) > f(A)$. Therefore, $f(A^+) > f(A)$, a contradiction.

Similarly, we can prove that for any $x \in X \setminus A$ and $A \cup \{x\} \in \Omega$, $f(A \cup \{x\}) \leq f(A)$. Therefore, A is a local maximum. $\qquad \square$

Fig. 6.4 Modular-modular method

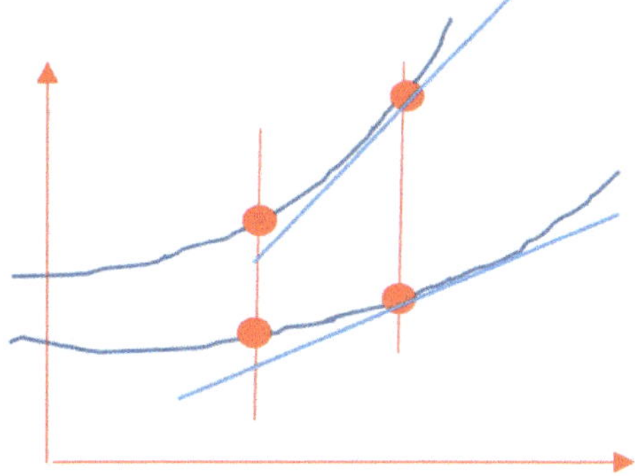

Modular-modular method has more applications since modular maximization/minimization with linear constraints can be solved in polynomial-time. Therefore, it can be employed to compute approximation solutions for many optimizations on social network, such as influence maximization, active friending, rumor blocking, and viral marketing.

6.2 PLS-Complete

To understand the hardness of computing a local optimum, we introduce the state-of-the-art in the study of computational complexity of local optimum, including related complexity class (e.g., FP, PLS, and FNP), reduction, and PLS-complete problems.

6.2.1 Class PLS

First, consider a few definitions.

Definition 6.2.1 (*Search Problem* [204]) Consider a relation $R \subseteq \Sigma^* \times \Sigma^*$. A search problem Q, for a given an input instance I, is to try to find a solution s such that $(I, s) \in R$. An algorithm is said to solve a search problem Q if for every input instance I, the algorithm can find either a solution s such that $(I, s) \in R$, or a conclusion that no such a solution exists.

Definition 6.2.2 (*Class FP* [291, 453]) There exists a polynomial-time algorithm which for any input I, either find a solution s such that $(I, s) \in R$, or determine that no such s exists.

Definition 6.2.3 ((*Class FNP* [291, 453]) A search problem Q with relation R belongs to class FNP if

- There exists a polynomial-time algorithm α to determine whether or not a given pair (I, s) is in R.
- If $(I, s) \in R$, then $|s|$ is polynomially bounded in $|I|$.

For each search problem Q, define

$$D(Q) = \{I \mid \text{There exists } s \text{ such that } (I, s) \in R\}.$$

Then, we have

Lemma 6.2.4 *A language $A \in NP$ if and only if there exists a search problem $Q \in FNP$ such that $A = D(Q)$.*

Proof Suppose $A \in NP$. Then, there exist a set $B \in P$ and a polynomial p such that

$$x \in A \iff (\exists y, |y| \leq p(|x|)) : \langle x, y \rangle \in B.$$

Set $R = B$ to define a search problem Q. Then, $Q \in FNP$ and $A = D(Q)$.

Conversely, let Q with relation R be a search problem in FNP. Then, $R \in P$ and there exists a polynomial p such that

$$I \in D(Q) \iff (\exists s, |s| \leq p(|I|)) : (I, s) \in R.$$

Thus, $D(Q) \in NP$. $\qquad\qquad\square$

Theorem 6.2.5 ([204]) $FP = FNP \iff P = NP$.

Proof ($\Rightarrow$) Let $A \in NP$. By Lemma 6.2.4, There exists a search problem $Q \in FNP$ such that $A = D(A)$. Since $FNP = FP$, $Q \in FP$. Therefore, $D(Q) in P$. It follows that $NP = P$.

($\Leftarrow$) Consider any $Q \in FNP$. Then, $D(Q) \in NP$. Since $NP = P$, we have $D(Q) \in P$, that is, there is a polynomial-time algorithm α to determine whether $D(Q) \in P$. It follows that for any input instance I, algorithm α should either generate a solution s such that $(I, s) \in R$ (for witness $I \in D(Q)$), or determine that no such solution s exists (for $I \notin D(Q)$). This means that $Q \in FP$. Therefore, $FNP = FP$. $\qquad\square$

Class PLS consists of local search problems which will be explained in the following definitions.

Definition 6.2.6 (*Combinatorial global optimization problem* [293]) A search problem Q is called a combinatorial global optimization problem if Q has a cost function $c : R \rightarrow \mathbb{R}$ where and for each input instance I, the problem is to search an optimal solution s^*, such that for maximization, $c(I, s^*) \geq c(I, s)$ for every $(I, s) \in R$, and for minimization, $c(I, s^*) \leq c(I, s)$ for every $(I, s) \in R$.

Definition 6.2.7 (*Local Search Problem*) A local search problem is a combinatorial global optimization problem with a local structure $N : (I, s) \rightarrow F$ where $F \subseteq F(I) = \{s \mid (I, s) \in R\}$. The target of the local search problem is to find a local optimum.

Definition 6.2.8 (*Local Optimum*) A local optimum is a solution that has no neighbor with better cost, i.e., for a local maximum s, $c(I, s) \geq c(I, s')$ for every $s' \in N(I, s)$, and for a local minimum s, $c(I, s) \leq c(I, s')$ for every $s' \in N(I, s)$.

Definition 6.2.9 (*Class PLS* [298]) The class PLS consists of all local search problems each having a polynomial p and three polynomial time computable functions $A, B, and C$ as follows.

- For every input I, $|s| \leq p(|I|)$ for s with $(I, s) \in R$.
- $A(I)$ determines whether or not $I \in D(L)$, and if yes, then returns a solution s.
- $B(I)$ computes cost $c(I, s)$.
- $C(I)$ determine whether or not a solution s is a local optimum, and if not, then returns a neighbor $s' \in N(I, s)$ with better cost.

Let LS be the class of all local search problems. The following theorem gives relations between PLS, FP and FNP.

Theorem 6.2.10 ([204]) $FP \cap LS \subseteq PLS \subseteq FNP \cap LS$.

Proof $(FP \cap LS \subseteq PLS)$ Consider a local search problem L. Suppose $L \in FP$, that is, there is a polynomial-time algorithm α which for a given input I, returns a local optimal solution s. Now, we can define three polynomial-time algorithms A, B, C as follows.

- $A(I)$ returns a local optimal solution s which is returned by algorithm α.
- $B(I, s)$ computes cost $c(I, s)$ as α computes.
- For any solution s', $c(I, s)$ returns s which is computed by α.

$(PLS \subseteq FNP \cap LS)$ Let $L \in PLS$ with three polynomial-time algorithms A, B, and C. By the definition, we would have the following.

- For any input I, if $I \in D(L)$, then $A(I)$ returns a solution s. Since A is a polynomial-time algorithm, we have $|s| \leq p(|I|)$ for some polynomial p.
- $C(I, s)$ would determine whether or not s is a local optimal solution.

Therefore, $L \in FNP$. $\qquad\qquad\qquad\qquad\qquad\qquad\qquad\qquad\qquad\qquad\qquad\square$

Lemma 6.2.11 *Every problem in PLS is nondeterministic polynomial-time computable.*

Proof For any problem in PLS, consider the following nondeterministic algorithm.

- For input I, determine whether or not $I \in D(L)$ by a polynomial-time algorithm $A(I)$. If no, then return "no solution" for this input I. If yes, continue.
- Guess a local optimal solution s with $|s| \leq p(|I|)$ where p is a polynomial in the definition of PLS on problem L.
- Using algorithm C, check if s is a local optimal solution. If yes, output S; otherwise, output nothing.

This algorithm runs in polynomial-time and for any input I, there exists a possible guess, which returns a correct output. $\square$

Theorem 6.2.12   ([204]) *If a problem in PLS is NP-hard, then $NP = co - NP$.*

Proof Let $L \in PLS$ be NP-hard. This means that for any $A \in NP$, there exists a polynomial-time deterministic Turing machine M with oracle L, which can decide A. Now, we replace the oracle by a nondeterministic polynomial-time algorithm for L. Then M would become a nondeterministic polynomial-time Turing machine to decide A. Thus, $\bar{A} \in NP$, that is, $A \in co - NP$. Therefore, $NP = co - NP$. $\square$

6.2.2   *PLS-Reduction*

Remind that for any local search problem L, $D(L)$ denotes the definition domain of L and for any input I of L,

$$F_L(I) = \{s \mid (I, s) \in R\}.$$

Let us define PLS-reduction.

Definition 6.2.13 (*PLS-Reduction* [204]) Consider two local search problems L_1 and L_2. L_1 is PLS-reducible to L_2 if there are two polynomial-time functions $f : D(L_1) \to D(L_2)$ and $g : D(L_1) \times F_{L_2}(f(I_1)) \to F_{L_1}(I_1)$ such that

- For any input instance I_1 of L_1, $f(I_1)$ is an input instance of L_2.
- If s_2 is a solution for $f(I_1)$ of L_2, then $g(I_1, s_2)$ is a solution for I_1 of L_1.
- If s_2 is a local optimal solution for $f(I_1)$ of L_2, then $g(I_1, s_2)$ is a local optimal solution for I_1 of L_1.

If L_1 is PLS-reducible to L_2, then we write $L_1 \leq_{pls} L_2$.

PLS-reduction has the following properties.

Lemma 6.2.14 *If $L_1 \leq_{pls} L_2$ and $L_2 \leq_{pls} L_3$, then $L_1 \leq_{pls} L_3$.*

Proof Suppose that $L_1 \leq_{pls}$ is witnessed by functions f_1 and g_1, and $L_2 \leq_{pls} L_2$ is witnessed by functions f_2 and g_2. Then $L_1 \leq_{pls} L_3$ can be witnessd by functions f_3 and g_3 defined as follows.

$$f_3(I_1) = f_2(f_1(I_1)), \forall I_1 \in D(L_1),$$
$$g_3(I_1, s_3) = g_1(I_1, g_2(I_1, s_3)).$$

$\square$

Lemma 6.2.15 *If $L_1 \leq_{pls} L_2$ and $L_2 \in FP$, then $L_1 \in FP$.*

Proof $L_2 \in FP$ means that for any input I_2 of L_2, there exists a polynomial-time algorithm α to generate a local optimal solution s_2 for I_2 of L_2. Then, for any input I_1 of L_1, let s_2 be a local optimal solution for $f(I_1)$, generated by algorithm α. Then, $g(I_1, s_2)$ is a local optimal solution for I_1 of L_1. Clearly, $g(I_1, s_2)$ can be computed in polynomial-time. $\square$

Definition 6.2.16 (*PLS-Complete* [204]) A local search problem L is *PLS*-complete if

- $L \in PLS$, and
- $L' \leq_{pls} L, \forall L' \in PLS$.

6.2.3 *Circuit/Flip and Max-Cut*

Next, we present the first *PLS*-complete problem.

Definition 6.2.17 (*Circuit/Flip*) The problem can be described as follows.

- Each instance is an acyclic circuit I with input variables $x_1, x_2, \ldots, x_n$ and outputs $y_1, y_2, \ldots, y_m$.
- Each solution is an assignment for input variables.
- The cost for an instance I and solution can be calculated as

$$c_I(x_1, x_2, \ldots, x_n) = \sum_{i=1}^{m} 2^{i-1} x_i.$$

- A solution τ is a neighbor of another solution τ' if τ can be obtained from τ' through a flip, where a flip is a changing of assignment on one input variable.

Max-Circuit/Flip is the problem of finding a solution to maximize the cost. Min-Circuit/Flip is the problem of finding a solution to minimize the cost.

Theorem 6.2.18 *Max-Circuit/Flip and Min-Circuit/Flip are equivalent.*

Proof For example I, construct I' by putting a negation gate on every output. Then, for the same assignment τ, the cost of (I', τ) is

$$c_{I'}(x_1, x_2, \ldots, x_n) = 2^m - 1 - c_I(x_1, x_2, \ldots, x_n).$$

$\square$

Theorem 6.2.19 *Min-Circuit/Flip is PLS-complete.*

Proof First, we show that Min-Circuit/Flip is in PLS. To do so, define a polynomial p and three polynomial-time functions A, B, C as follows.

- p can be an identity function since every solution has size n which is clearly bounded by the input size.
- $A(I)$ assigns 1 to every variable x_i.
- Given a variable assignment τ, $B(I, \tau)$ computes the cost of (I, τ), which depends on the evaluation of the circuit and clearly in polynomial time.
- Since each assignment τ has n neighbors, it is easy to select the one with the smallest cost within a polynomial time.

Next, we show that for any PLS problem L, $L \leq_{pls} Min - Circuit/Flip$. To do so, we will introduce the third PLS problem Q to show that $L \leq Q \leq Min - Circuit/Flip$.

Q has the same set of inputs as L. But the solution set, the cost function, and the neighborhood relationship are different. In the following, we will describe them one by one in detail.

First, let the set of solutions of Q consist of all binary strings of certain length. For any two solutions τ and θ, they are defined as neighborhood if and only if the Hamming distance of them is exactly equal to one, that is, they can reach each other through a flip. With this definition, it is easy to establish the reduction $Q \leq_{pls} Min - Circuit/Flip$.

- Since $Q \in PLS$, the cost function $c_Q(\tau)$ is polynomial-time computable. Hence, we can construct a polynomial-size circuit M to compute c_Q.
- Let f map each instance I of Q to a circuit M. Then each solution of I for Q becomes an assignment of input variables for M. Two solutions of I for Q have the Hamming distance one if and only if the corresponding two variable assignments can be obtained from each other through a flip.
- Let g map each assignment of the input variables of the circuit M to a binary string solution of Q. Then τ is a local minimal solution of $Min - Circuit/Flip$ if and only if $g(\tau)$ is a local minimal solution.

Next, we describe the set of solutions and the cost function of Q to help construct the reduction $L \leq_{pls} Q$. Before do so, let us make two assumptions for L.

- Since $L \in PLS$, there is a polynomial p such that for any input I and any solution s of I, $|s| \leq p(|I|)$. Assume that every solution of I for L is encoded into a binary string of length $p(|I|)$.

- Assume that every pair of solutions has a Hamming distance of at least two. In fact, if this is not satisfied, then we may double every digit in each solution. For example, instead of 101, use 110011.

The solution set for Q is defined to consist of all binary strings of length $2p(|I|) + 2$. To define the cost function, we first define a special subset of solutions, $\mathcal{R}$.

- For every solution u of I for L, $uu00$ belongs to $\mathcal{R}$.
- For each solution u of I for L, let w be the neighbor returned by $C_L(I, u)$. If $w \neq u$, then the Hamming distance of u and w is at least two. Suppose $v_1, \ldots, v_k$ is a sequence of binary strings such that in the sequence $(u, v_1, \ldots, v_k, w)$, every adjacent two strings have Hamming distance one, i.e., one can be obtained from the other one through one flip. Then every member of the following sequence belongs to $\mathcal{R}$: $uu00$, uv_100, ..., uv_k00, $uw00$, $uw10$, $uw11$, v_1w11, ..., v_kw11, $ww11$, $ww10$, $ww00$.
- Nothing else is $\mathcal{R}$.

Denote $p = p(|I|)$ and denote by $H(u, w)$ the Hamming distance between u and w. The cost of every solution in $\mathcal{R}$ is defined as follows.

$$c_Q(I, uu00) = (2p + 4)c_L(I, u),$$
$$c_Q(I, uv_100) = (2p + 4)c_L(I, w) + (p + 2) + H(v_1, w) + 2,$$
$$\cdots,$$
$$c_Q(I, uv_k00) = (2p + 4)c_L(I, w) + (p + 2) + H(v_k, w) + 2,$$
$$c_Q(I, uw00) = (2p + 4)c_L(I, w) + (p + 2) + 2,$$
$$c_Q(I, uw10) = (2p + 4)c_L(I, w) + (p + 2) + 1,$$
$$c_Q(I, uw11) = (2p + 4)c_L(I, w) + H(u, w) + 2,$$
$$c_Q(I, v_1w11) = (2p + 4)c_L(I, w) + H(v_1, w) + 2,$$
$$\cdots,$$
$$c_Q(I, v_kw11) = (2p + 4)c_L(I, w) + H(v_k, w) + 2,$$
$$c_Q(I, ww11) = (2p + 4)c_L(I, w) + 2,$$
$$c_Q(I, ww10) = (2p + 4)c_L(I, w) + 1,$$
$$c_Q(I, ww00) = (2p + 4)c_L(I, w).$$

Above definition also includes the case that $u = w$. Since $L \in PLS$, the cost function c_L is polynomial-time computable. Moreover, $|w| \leq p(|I|)$ for any solution w of I. Therefore, there exists a polynomial q such that $|c_l(I, w)| \leq q(|I|)$ for any solution w of I for L. Hence, for any solution w of I for L,

$$c_L(I, w) \leq 2^{q(|I|)}.$$

Let $a = A_L(I)$. For any solution s of I, not in $\mathcal{R}$, define

$$c_Q(I, s) = Z + H(s, aa00)$$

where $Z = (4p + 4)2^{q(|I|)}$.

With cost function defined as above, we claim the following.

Claim 6.2.20 *Every local optimal solution of I for Q must be in form $uu00$. Moreover, $uu00$ is a local optimal solution of I for Q if and only if u is a local optimal solution of I for L.*

Proof For $s \ni n\mathcal{R}$, there always exists a solution s' such that $H(s, s') = 1$ and $H(aa00, s') = H(aa00, s) - 1$. If $s' \ni n\mathcal{R}$, then clearly $c_Q(I, s') < c_Q(I, s)$. If $s' \in \mathcal{R}$, then we still have

$$c_Q(I, s') \leq Z + H(aa00, s') < c_Q(I, s).$$

Therefore, s cannot be a local optimal solution.

If s is a solution in $\mathcal{R}$, but not in form $uu00$, then s must lie on a path from $uu00$ to $ww00$ where $w = C_L(I, u)$ and $u \neq w$. Note that the cost on this path is decreased from $uu00$ to $ww00$. Therefore, s cannot be a local optimal solution.

Moreover, cost definition on the path also implies that $uu00$ is a local optimal solution of I for Q if and only if u is a local optimal solution of I for L. $\qquad\square$

Finally, we construct the reduction $L \leq_{pls} Q$. Let f be the identity from inputs of L to inputs of Q. Mapping g is defined by

$$g(I, s) = \begin{cases} u & \text{if } s = uu00, \\ A_L(I) & \text{otherwise.} \end{cases}$$

Clearly, $g(I, s)$ is a solution of I for L and if s is a local optimal solution, then $g(I, s)$ must be a local optimal solution. $\qquad\square$

Corollary 6.2.21 *Max-Circuit/Flip is PLS-complete.*

Definition 6.2.22 Given a graph $G = (V, E)$ with positive edge weight $w : E \to \mathbb{N}_+$, find a partition (A, B) of vertex set V to maximize the total weight of edges between A and B.

Theorem 6.2.23 ([355]) *Max-Cut/Flip is PLS-complete.*

6.2.4 The Standard Algorithm

It is unlikely for a PLS-complete problem to have a polynomial-time algorithm for computing a local optimal solution, since if this happens, then every PLS problem would have a polynomial-time algorithm for computing a local optimal solution. However, this hardness has not been successfully connected to NP-hardness, that is, the following is still open.

Open Problem 3 *Is it NP-hard for a PLS-complete problem to have a polynomial-time algorithm for computing a local optimal solution?*

It may be interesting to note that this open problem cannot be extended to some special type of algorithms. For example, in this section, we will study a type of algorithms, for which it is NP-hard for some PLS-complete problem to compute a local optimal solution in polynomial-time.

Definition 6.2.24 (*The Standard Algorithm*) For input I, this algorithm starts with a solution $s = A(I)$. In each iteration, it finds, if possible, a neighbor $s' = C(I, s)$ with better cost, until a local optimal solution is found. Its pseudo-code is as follows.

```
input I;
s ← A(I);
s' ← C(I, s);
while  s ≠ s' do
        s ← s';
        s' ← C(I, s);
end-while
return s.
```

For linear programming, the feasible domain is a polyhedron. All vertices are considered solutions. Two solutions are neighbors to each other if they are connected by an edge. In this setting, the standard algorithm is the simplex method. Under a non-degenerate condition, any local optimal solution for linear programming must be globally optimal. In the worst case, the simplex method runs in exponential time, that is, the following holds.

Theorem 6.2.25 *For linear programming, the standard algorithm runs in exponential time.*

Note that this type of result cannot be extended from one problem to another problem through a PLS-reduction because the standard algorithm may not become the standard algorithm through a PLS-reduction.

6.3 Approximate of Local Optimum

In this section, we study approximate local optimum. Let us start with explanation on question why we study approximation of local optimal solution.

So far, we already introduced three types of methods for nonsubmodular optimization. Let us make some comments on them.

- **Parameterized Method**: With parameters, we can obtain some theorems on approximation performance. However, those theorems are useless in specific problems.
- **Sandwich Method**: This method can be used in a wide range of problems. However, established performance ratio looks ugly and does not help in evaluation of approximation performance.
- **Stop at Local Optimum**: For this class of algorithms [196, 300], a serious problem is about the number of iterations. In [440], it indicates that for some optimization problems, computing a local optimal solution is PLS-complete [355], that is, it is unlikely to have a polynomial-time algorithm to compute a local optimal solution. Therefore, for this type of algorithms, it is unlikely to have a polynomial bound for the number of iterations since the algorithms work generally for a large family of problems, containing those optimization problems, for which computing a local optimal solution is PLS-complete.

The weakness of the first two types of methods illustrates that the performance ratio (current measure of approximation performance) does not fit in the study of nonsubmodular optimization, that is, based line of comparison should not be the global optimum. In other words, for nonsubmodular optimization, not only global optimum is not reachable, but also its good-quality approximation. Therefore, we have to move our attention to local optimum. The weakness of third type of algorithms illustrates that it is unlikely to have efficient way to reach a local optimal solution. Thus, our attention should be paid on approximate local optimum.

Definition 6.3.1 (α-*Approximate Local Solution*) A set S is said to be a α-**approximate local maximal optimal solution**, or (for short) **local α-approximation**, for $\max_{Y \in \Omega} f(Y)$ if for any element $s in S$ with $S \setminus \{s\} \in \Omega$, $f(S \setminus \{s\}) \leq \alpha \cdot f(S)$, and for any element $t \notin S$ with $S \cup \{t\} \in \Omega$, $f(S \cup \{t\}) \leq \alpha \cdot f(S)$. A set S is said to be a α-**approximate local minimal solution**, or (for short) **local α-approximation**, for $\min_{Y \in \Omega} f(Y)$ if for any element $s in S$ with $S \setminus \{s\} \in \Omega$, $\alpha \cdot f(S \setminus \{s\}) \geq f(S)$, and for any element $t \notin S$ with $S \cup \{t\} \in \Omega$, $\alpha \cdot f(S \cup \{t\}) \geq f(S)$.

Now, we may modify submodular-supermodular method into an efficient one to obtain approximate local solution.

Algorithm 3 Modified Submodular-Supermodular Algorithm for Minimization.

Input: a set function $f : 2^X \to R$ and its DS decomposition $f = g - h$ where g and h are submodular functions, and a positive number ε.

Output: A subset A of X.

1: choose a set $A \subseteq X$;
2: **while** $f(A^+) \cdot (1 + \varepsilon) < f(A)$ **do**
3:    $A \leftarrow A^+$;
4:    **for** $\sigma \in \Sigma_A$ **do**
5:       compute a lower bound modular function m_{hl}^σ for h;
6:       compute a minimum solution A_σ^+ for $\min_{Y \in 2^X}[g(Y) - m_{hl}^\sigma(Y)]$;
7:    **end for**
8:    $\sigma \leftarrow \mathrm{argmin}_{\sigma \in \Sigma_A} f(A_\sigma^+)$;
9:    $A^+ \leftarrow A_\sigma^+$
10: **end while**
11: **return** A.

Theorem 6.3.2 *Let f be a nonnegative function. Then Algorithm 3 ends at a local $(1 + \varepsilon)$-approximation within $O(\frac{1}{\varepsilon} \ln \zeta)$ iterations where ζ is the ratio of the maximum value and the minimum value of f.*

Proof In each iteration, we have that for any $x \in A$, $f(A^+) \leq f(A \setminus \{x\})$ and for any $x \in X \setminus A$, $f(A^+) \leq f(A \cup \{x\})$. When the algorithm stops, $f(A) \leq (1 + \varepsilon) \cdot f(A^+)$. Hence, for any $x \in A$, $f(A) \leq (1 + \varepsilon) \cdot f(A \setminus \{x\})$ and for any $x \in X \setminus A$, $f(A) \leq (1 + \varepsilon) \cdot f(A \cup \{x\})$.

Before algorithm stops, in each iteration, we have $f(A^+) \cdot (1 + \varepsilon) < f(A)$. Let k satisfy $opt \cdot (1 + \varepsilon)^k = f(A_0)$ for initial set A_0 and optimal value opt. Then the algorithm must stop within $\lfloor k \rfloor$ iterations. Note that $e < (1 + \varepsilon)^{1 + 1/\varepsilon}$ an $1 + 1/\varepsilon < 2/\varepsilon$ for $\varepsilon < 1$. Thus, $e^{k\varepsilon/2} < f(A)/opt$. Hence, $k\varepsilon < 2\ln(f(A_0)/opt) \leq 2\ln\zeta$. $\square$

It may be worth mentioning the similarity in developments of the discrete optimization theory and the continuous optimization theory. In discrete optimization, as the study moves from the submodular optimization to the nonsubmodular optimization, one's attention movers from approximation of global optimal solution to approxiomation of local optimal solution. In continuous optimization, at earlier stage, efforts were made on convex function minimization without/with constraints. At this stage, designed algorithms are asked to produce a sequence of points convergent to a global optimal solution. Later, when one's eyes move to nonconvex optimizations, designed algorithms are asked to have global convergence property.

Definition 6.3.3 (*Global Convergence* [82–84]) An algorithm for continuous optimization usually generates a sequence of infinitely many points. This sequence is said to be global convergent if every convergent subsequence converges to a Kuhn-Tucker point.

Here, every Khun-Tuker point is a candidate for local optimal solution. There are many interesting facts about global convergence. Let us mention one with easily-understanding background.

Fig. 6.5 A sequence has
more than one cluster points

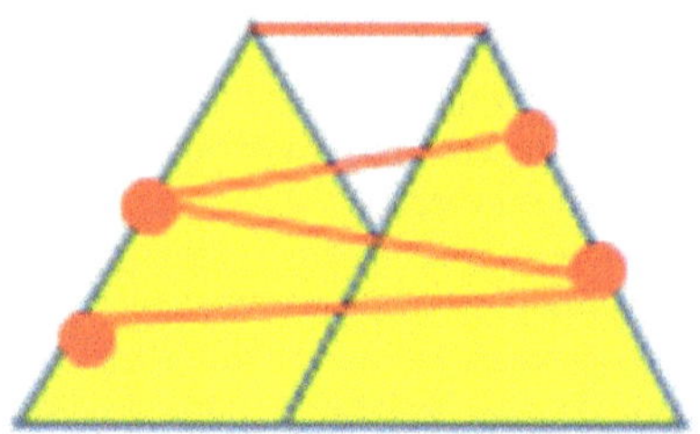

Consider a sequence of points, $\{x_k\}$ generated by an algorithm for maximization of a continuous function f. Assume $f(x_k) \leq f(x_k)$ for all k. Thus, the sequence looks like to climb a mountain (Fig. 6.5). Usually, a zigzag road is used for climbing a mountain. Such a road has longer length, but smaller slope. The average slope is equal to the mountain height divide the road length. The road length is longer, the road slope is smaller.

A point x^* is called a cluster point if there is a subsequence of $\{x_k\}$ convergent to x^*. Suppose that $\{x_k\}$ has two cluster points x^* and x^{**}. Then, sequence $\{x_k\}$ will be going in a zigzag road toward x^* and x^{**}. The road length $\|x_k - x_{k+1}\| + \|x_{k+1} - x_{k+2}\| + \cdots$ tends to infinity. Thus, the slope $\nabla f(x_k)$ approaches to $\vec{0}$, that is, $\nabla f(x^*) = \nabla f(x^{**}) = \vec{0}$. This is the intuition background of a slope lemma which plays an important role in the theory of global convergence.

Motivated from the beautiful theory of global convergence in continuous optimization, a new approximation performance measure is proposed in [81].

In [355], it has been indicated that the number of iterations is unlikely to have a polynomial bound. Therefore, those algorithms are unlikely to have a polynomial running time.

Based on above observations, a new approximation performance measure is proposed [81]. An algorithm is called a GL-approximation (global approximation of local optimality) for an optimization problem if it produces a solution S such that for some parameter α,

$$f(S) \leq \alpha \cdot \text{local-opt} \quad \text{for minimization}$$
$$f(S) \geq \alpha \cdot \text{local-opt} \quad \text{for maximization}$$

where f is the objective function, `local-opt` is the objective function value of a local optimal solution, and α is called a *GL-approximation performance ratio*.

The GL-approximation provides a new tool to study algorithms for nonsubmodular optimizations, which will introduce a lot of exploratory and innovative research opportunities. For example, we may study nonconvex-relaxation.

Sometimes, an algorithm may not be able to stop at a local optimal solution and instead, it stops at a local approximately-optimal solution. For example, consider the following problem.

$$\max \quad f(A)$$
$$\text{subject to} \quad |A| \leq k$$
$$A \in 2^X$$

where f is a set function over 2^X for a finite set X. Algorithm 4 is the submodular-supermodular algorithm for this problem.

Algorithm 4 Submodular-Supermodular Algorithm for Maximization.

Input: a set function $f : 2^X \to R$ and its DS decomposition $f = g - h$ where g and h are submodular functions.
Output: A subset A of X.
1: choose a set $A \subseteq X$;
2: **while** $f(A^+) > f(A)$ **do**
3:     $A \leftarrow A^+$;
4:     **for** $\sigma \in \Sigma_A$ **do**
5:         compute a lower bound modular function m_{hl}^{σ} for h;
6:         compute a $(1 - e^{-1}$-approximation solution A_σ^+ for $\max_{Y:|Y|\leq k}[g(Y) - m_{hl}^{\sigma}(Y)]$;
7:     **end for**
8:     $\sigma \leftarrow \mathrm{argmax}_{\sigma \in \Sigma_A} f(A_\sigma^+)$;
9:     $A^+ \leftarrow A_\sigma^+$
10: **end while**
11: **return** A.

Theorem 6.3.4 *Algorithm 4 stops at a solution A satisfying condition that for any $x \in A$, $f(A) \geq (1 - e^{-1}) \cdot f(A \setminus \{x\})$ and for any $x \in X \setminus A$, $f(A) \geq (1 - e^{-1}) \cdot f(A \cup \{x\})$.*

Proof In each iteration, A^+ satisfies that for any $x \in A$, $f(A^+) \geq (1 - e^{-1}) \cdot f(A \setminus \{x\})$ and for any $x \in X \setminus A$, $f(A^+) \geq (1 - e^{-1}) \cdot f(A \cup \{x\})$. When the algorithm stops, we have $f(A) \geq f(A^+)$. Therefore, A has the property. $\square$

Since the algorithm always stops at a local approximately-optimal solution, we may call it as a *global-optimal approximation*, or G-L approximation for a brief name. There is a more important reason for us to pay attention on G-L approximation, that is, the submodular-supermodular is unlikely to run in polynomial-time. Actually, computing a local minimum solution is PLS-complete (PLS stands for Polynomial Local Search), which is unlikely to run in polynomial-time. In order to obtain a polynomial-time algorithm, we have to modify the submodular-supermodular algorithm while accepting the G-L approximation solution.

Let us call A as a G-L ρ-approximation for minimization of set function $f : 2^X \to R$ if for any $x \in A$, $f(A) \leq (1 + \varepsilon) \cdot f(A \setminus \{x\})$ and for any $x \in X \setminus A$, $f(A) \leq (1 + \varepsilon) \cdot f(A \cup \{x\})$. Now, we modify the submodular-supermodular algorithm for minimization as shown in Algorithm 5, and obtain the following.

Algorithm 5 Submodular-Supermodular Algorithm for Minimization.

Input: a set function $f : 2^X \to R$ and its DS decomposition $f = g - h$ where g and h are submodular functions.
Output: A subset A of X.

1: choose a set $A \subseteq X$;
2: **while** $f(A^+) \cdot (1 + \varepsilon) < f(A)$ **do**
3:    $A \leftarrow A^+$;
4:    **for** $\sigma \in \Sigma_A$ **do**
5:       compute a lower bound modular function m_{hl}^σ for h;
6:       compute a minimum solution A_σ^+ for $\min_{Y \in 2^X}[g(Y) - m_{hl}^\sigma(Y)]$;
7:    **end for**
8:    $\sigma \leftarrow \operatorname{argmin}_{\sigma \in \Sigma_A} f(A_\sigma^+)$;
9:    $A^+ \leftarrow A_\sigma^+$
10: **end while**
11: **return** A.

Theorem 6.3.5 *Let f be a nonnegative function. Then Algorithm 5 ends at G-L $(1 + \varepsilon)$-approximation within $O(\frac{1}{\varepsilon} \ln \zeta)$ iterations where ζ is the ratio of the maximum value and the minimum value of f.*

Proof In each iteration, we have that for any $x \in A$, $f(A^+) \le f(A \setminus \{x\})$ and for any $x \in X \setminus A$, $f(A^+) \le f(A \cup \{x\})$. When the algorithm stops, $f(A) \le (1 + \varepsilon) \cdot f(A^+)$. Hence, for any $x \in A$, $f(A) \le (1 + \varepsilon) \cdot f(A \setminus \{x\})$ and for any $x \in X \setminus A$, $f(A) \le (1 + \varepsilon) \cdot f(A \cup \{x\})$.

Before algorithm stops, in each iteration, we have $f(A^+) \cdot (1 + \varepsilon) < f(A)$. Let k satisfy $opt \cdot (1 + \varepsilon)^k = f(A)$ for initial set A and optimal value opt. Then the algorithm must stop within $\lfloor k \rfloor$ iterations. Note that $e < (1 + \varepsilon)^{1+1/\varepsilon}$ an $1 + 1/\varepsilon < 2/\varepsilon$ for $\varepsilon < 1$. Thus, $e^{k\varepsilon/2} < f(A)/opt$. Hence, $k\varepsilon < 2\ln(f(A)/opt) \le 2\ln\zeta$. $\square$

6.4   Nonconvex Relaxation

LP-relaxation, semi-definite-relaxation, and convex-relaxation are important techniques for design of traditional approximation algorithms [80]. For GL-approximation, relaxation technique may be extendable to nonconvex functions. In fact, every set function $f : 2^X \to R$ can be extended to a multi-linear function $F : [0, 1]^{|X|} \to R$ as follows:

$$F(x) = \sum_{S \subseteq X} f(S) \prod_{i \in S} x_i \prod_{i \notin S}(1 - x_i).$$

This relaxation enable us to design GL-approximation by taking advantage of nice properties of multi-linear optimizations.

Relaxation is a popular technique for design of approximation algorithms. Many traditional approximation algorithms are designed with LP relaxation, semi-definite relaxation, and convex relaxation. Now, nonconvex relaxation may become a new development for global approximation of local optimality, which will also introduce some interesting new research directions.

Let us consider a specific problem to make explanation. Consider the following problem [88]:

Problem 6.4.1 (*Maximum Group Set Coverage with Composed Targets*) Consider m groups $\mathcal{G}_1, \mathcal{G}_2, \ldots, \mathcal{G}_m$ of subsets of a finite set X. All elements in X are partitioned into subsets $X_1, X_2, \ldots X_r$, each called a *composed target*. Suppose that for every composed target X_t, each subset in any group $\mathcal{G}_i$ covers at most one elements in X_t. The problem is to select one subset from each group to cover the maximum total number of composed targets, where a composed target is said to be covered if all elements in the composed target are covered.

This problem can be formulated as the following nonsubmodular maximization with linear constraints.

$$\max \ \sum_{t=1}^{\tau} \prod_{j \in X_t} y_j \qquad\qquad (INLP)$$

$$\text{s.t.} \ \ y_j \le \sum_{i=1}^{m} \sum_{S:j\in S\in\mathcal{G}_i} x_{iS} \ \ \forall j = 1, \ldots, n,$$

$$\sum_{S:S\in\mathcal{G}_i} x_{iS} \le 1 \ \ \forall i = 1, \ldots, m,$$

$$y_j \in \{0, 1\} \ \ \forall j = 1, \ldots, n,$$

$$x_{iS} \in \{0, 1\} \ \ \forall S \in \mathcal{G}_i \text{ and } i = 1, 2, \ldots, m.$$

Its relaxation is a linear constrained nonlinear programming as follows.

$$\max \ \sum_{t=1}^{\tau} \prod_{j \in X_t} y_j \qquad\qquad (NLP)$$

$$\text{s.t.} \ \ y_j \le \sum_{i=1}^{m} \sum_{S:j\in S\in\mathcal{G}_i} x_{iS} \ \ \forall j = 1, \ldots, n,$$

$$\sum_{S:S\in\mathcal{G}_i} x_{iS} \le 1 \ \ \forall i = 1, \ldots, m,$$

$$0 \le y_j \le 1 \ \ \forall j = 1, \ldots, n,$$

$$0 \le x_{iS} \le 1 \ \ \forall S \in \mathcal{G}_i \text{ and } i = 1, 2, \ldots, m.$$

Let (y_j^*, x_{iS}^*) be a $(1 - \varepsilon)$-approximate for a local optimal solution (KKT-point) of (NLP). Let (x_{iS}, y_j) be a solution obtained from the following randomized rounding.

Randomized Rounding: For each group $\mathcal{G}_i$, select one subset S (i.e., set $x_{iS} = 1$ and $x_{iS'} = 0$ for $S' \neq S$) with probability x_{iS}^* and no subset of group $\mathcal{G}_i$ is selected with probability $1 - \sum_{S \in \mathcal{G}_i} x_{iS}^*$. Set $y_j = 1$ if element j appears in a selected subset, and $y_j = 0$ otherwise.

Du et al. [88] showed the following.

Theorem 6.4.2 *Let* (y_j, x_{iS}) *be an approximate solution obtained by the above randomized rounding. Then*

$$E\left[\sum_{t=1}^{\tau} \prod_{j \in X_t} y_j \right] \geq (1 - e^{-1})^{\alpha}(1 - \varepsilon)lopt$$

where lopt is the objective value of a local optimal solution of (NLP) and $\alpha = \max_{1 \leq t \leq \tau} |X_t|$.

There are two research directions raised by this result.

First, what is complexity of computing approximation of KKT-points? Vavasis [408] gave an algorithm for computing an ε-KKT of a box-constrained quadratic programming with $O(n^3(M/\varepsilon)^2)$ arithmetic operations. Ye [469] presented an algorithm for computing an ε-KKT of a linearly constrained quadratic programming with $O((n^3/\varepsilon)\log(1/\varepsilon) + n \log n)$ iterations. Can those works be extended to other objective functions, especially multi-linear functions? It is an interesting problem. In fact, this is a research direction with a largely unexplored field. Research outcomes will become foundation for nonconvex relaxation.

Second, what is the relationship between local optimal solutions of (INLP) and (NLP)? That is, what is the relationship between local optimal solutions of an integer programming and local optimal solutions of its nonconvex relaxation? So far, there is no research outcome in this research direction although it seems a fundamental problem related to nonconvex relaxation.

6.5 Bipartite Influence Model

A bipartite influence model is proposed by Alon et al. [5] for marketing in social networks. The model is represented by a bipartite graph in which one node set S consists of marketing channels, such as social medias, TV, newspapers, and the other node set T consists of potential customers. Nodes representing marketing channels are called *source nodes* while nodes representing customers are called *target nodes*.

An edge between a source node s and a target node u represents the influence of s to u. s is allowed to try many times to influence u. Each time is called a trial. In the ith trial, the success probability is p_s^i. Here, we assume that the success probability for very target u is the same if the source is the same. It is nature to assume that if $i < j$, then $p_s^i \geq p_s^j$, that is an advertisement has more marketing effect at an earlier trial than later. Every source node s has a capacity $c(s) \in \mathbb{N}_+$.

A marketing strategy $\vec{m}$ is a vector with index set S. For each $s \in S$, $m(s)$ is the number of trials performed by s. For each target node t, let $N(t)$ be the set of source nodes adjacent to t. Then, the probability that t is activated is

$$p_t(\vec{m}) = 1 - \prod_{s \in N(t)} \prod_{i=1}^{m(s)} (1 - p_s^i).$$

Thus, the expected number of influenced target nodes is

$$\sigma(\vec{m}) = \sum_{t \in T} p_t(\vec{m}).$$

For each source node s, let δ_s be the unit cost for each trial. Then, the total cost for marketing strategy $\vec{m}$ is

$$\delta(\vec{m}) = \sum_{s \in S} m(s)\delta_s.$$

A marketing problem considered by Gu et al. [155] is as follows.

Problem 6.5.1 (*Profit Maximization*) Given a social network $G = (S, T; E)$, the marketing capacity vector $\vec{c}$, the influence probabilities $p_s^1, p_s^2, \ldots, p_s^{c(s)}$ for each node $s \in S$, and the unit marketing cost δ_s, find a marketing strategy $\vec{m}$ to maximize

$$f(\vec{m}) = \delta(\vec{(m})) - \delta(\vec{m}).$$

That is, the problem can be described as

$$\max \quad f(\vec{m})$$
$$\text{subject to} \quad \vec{0} \leq \vec{m} \leq \vec{c}$$

where $\vec{0} = (0, \ldots, 0)$.

Note that $\vec{m}$ is a point in the integer lattice. Therefore, this is a maximization problem on integer lattice.

When the concept of submodular set function is extended to integer lattice, there are two choices. Note that the submodular set function has two equivalent definitions. The first one is that $f : 2^E \to \mathbb{R}_+$ is submodular if for any two sets A and B in 2^E,

$$f(A) + f(B) \geq f(A \cup B) + f(A \cap B).$$

Its extension is that a function $g : \mathbb{Z}_+^E \to \mathbb{R}_+$ is submodular if for any $\vec{x}, \vec{y} \in \mathbb{Z}_+^E$,

$$g(\vec{x}) + g(\vec{y}) \geq g(\vec{x} \vee \vec{y}) + g(\vec{x} \wedge \vec{y})$$

where $(\vec{x} \vee \vec{y})(i) = \max\{\vec{x}(i), \vec{y}(i)\}$ and $(\vec{x} \wedge \vec{y})(i) = \min\{\vec{x}(i), \vec{y}(i)\}$.

The second definition of submodularity of set functions is that a set function $f : 2^E \to \mathbb{R}_+$ is submodular if it has the diminishing return property, i.e., for any two sets A and B in 2^E with $A \subseteq B$ and any element $x \in E$,

$$f(A \cup \{x\}) - f(A) \geq f(B \cup \{x\}) - f(B).$$

Its extension is that a function $g : \mathbb{Z}_+^E \to \mathbb{R}_+$ is DR-submodular if for any two points $\vec{x}, \vec{y} \in \mathbb{Z}_+^E$ with $\vec{x} \leq \vec{y}$ and any element $e \in E$,

$$g(\vec{x} + \chi_e) - g(\vec{x}) \geq g(\vec{y} + \chi_e) - g(\vec{y})$$

where χ_e is a unit vector such that $\chi_e(e) = 1$ and $\chi_e(x) = 0$ for $x \neq e$.

It is very important to note that two extension of submodularity to function on integer lattice are not equivalent. The concept of DR-submodularity seems to be more useful. For example, in the profit maximization problem, the objective function is DR-submodular. The proof is included in the following lemmas.

Lemma 6.5.2 (Gu et al. [155]) $\sigma(\vec{m})$ *is DR-submodular.*

Proof Since $\sigma(\vec{m})$ is a linear combination of $p_t(\vec{m})$'s, it is sufficient to prove that $p_t(\vec{m})$ is DR-submodular for any target t. Consider any $\vec{x}, \vec{y}$ in $\mathbb{Z}_+^S$ with $\vec{x} \leq \vec{y}$, and $e \in S$. Note that

$$p_t(\vec{x}) = 1 - \prod_{s \in N(t)} \prod_{i=1}^{\vec{x}(s)} (1 - p_s^i)$$

$$p_t(\vec{x} + \chi_e) = 1 - \left\{ \prod_{s \in N(t)} \prod_{i=1}^{\vec{x}(s)} (1 - p_s^i) \right\} \cdot (1 - p_e^{\vec{x}(e)+1}).$$

Thus,

$$p_t(\vec{x} + \chi_e) - p_t(\vec{x}) = \left\{ \prod_{s \in N(t)} \prod_{i=1}^{\vec{x}(s)} (1 - p_s^i) \right\} \cdot p_e^{\vec{x}(e)+1}.$$

Similarly,

$$p_t(\vec{y} + \chi_e) - p_t(\vec{y}) = \left\{ \prod_{s \in N(t)} \prod_{i=1}^{\vec{y}(s)} (1 - p_s^i) \right\} \cdot p_e^{\vec{y}(e)+1}.$$

Since $\vec{x} \leq \vec{y}$, we have

$$\prod_{s \in N(t)} \prod_{i=1}^{\vec{x}(s)} (1 - p_s^i) \geq \prod_{s \in N(t)} \prod_{i=1}^{\vec{y}(s)} (1 - p_s^i).$$

Moreover, since $\vec{x} \le \vec{y}$, we have

$$p_e^{\vec{x}+1} \ge p_e^{\vec{y}+1}.$$

Therefore,

$$p_t(\vec{x} + \chi_e) - p_t(\vec{x}) \ge p_t(\vec{y} + \chi_e) - p_t(\vec{y}).$$

$\square$

Lemma 6.5.3 (Gu et al. [155]) $\sigma(\vec{m}) - \delta(\vec{m})$ *is DR-submodular.*

Proof Note that $\delta(\vec{m})$ is linear and hence modular. Hence $-\delta(\vec{m})$ is DR-submodular. The sum of two DR-submodular functions is still DR-submodular. Therefore, the lemma holds. $\square$

6.6 DC Discrete Functions

Let us first introduce the subgradient of submodular functions.

For a submodular set function $f : 2^X \to R$, the subgradient of h at set A is defined by

$$\partial h(A) = \{c \in R^X \mid h(Y) \ge h(A) + \langle c, Y - A \rangle\},$$

i.e., it consists of all linear functions $c : X \to R$ satisfying $h(Y) \ge h(A) + c(Y) - c(A)$ where $c(Y) = \sum_{y \in Y} c(y)$. Each linear function c can also be seen as a vector in R^X, i.e., a vector c with components labeled by elements in X. The characteristic vector of each subset Y of X is a vector in $\{0, 1\}^X$ such that the component with label $x \in X$ is equal to 1 if and only if $x \in Y$. Here, for simplicity of notation, we use the same notation Y to represent the set Y and its characteristic vector.

Clearly, $\partial h(A)$ is a convex region. The following is an important property of $\partial h(A)$.

Lemma 6.6.1 (Fujishige [120]) *A point $c \in R^X$ is an extreme point of $\partial h(A)$ if and only if there is a permutation σ for elements in X, i.e., $X = \{\sigma(1), \sigma(2), \ldots, \sigma(|X|)\}$, such that $A = \{\sigma(1), \sigma(2), \ldots, \sigma(|A|)\}$ and $c(\{\sigma(i)\}) = h(S_i) - h(S_{i-1})$ for $1 \le i \le |X|$ where $S_0 = \emptyset$ and $S_i = \{\sigma(1), \sigma(2), \ldots, \sigma(i)\}$.*

Let f be a set function on 2^X. Suppose its DS decomposition $f = g - h$ is already known, where g and h are monotone nondecreasing and submodular functions on 2^X.

Theorem 6.6.2 *Let f be a set function on 2^X and $f = g - h$ its DS decomposition. Then, we have the following.*

1. *If A is a minimum solution for $\min_{Y \subseteq X} f(Y)$, then $\partial h(A) \subseteq \partial g(A)$.*

2. *If $\partial h(A) \subseteq \partial g(A)$, then $f(A) \leq f(Y)$ for $Y \in \mathcal{U}$, where*

$$\mathcal{U} = \{Y \mid \partial h(Y) \cap \partial g(A) \neq \emptyset\}.$$

3. *If $\partial h(A) \subseteq \partial g(A)$, then A is a local minimum for $\min_{Y \subseteq X} f(Y)$.*

Proof 1. Since A is a minimum solution for $\min_{Y \subseteq X} f(Y)$, we have $f(A) \leq f(Y)$ and hence $g(Y) - g(A) \geq h(Y) - h(A)$ for any $Y \subseteq X$. Therefore, for any $c \in \partial h(A)$, $g(Y) - g(A) \geq h(Y) - h(A) \geq c(Y) - c(A)$. This means that $\partial h(A) \subseteq \partial g(A)$.

2. Choose $c \in \partial h(Y) \cap \partial g(A)$. Then

$$h(A) \geq h(Y) + (c(A) - c(Y)) \text{ and } g(Y) \geq g(A) + (c(Y) - c(A)).$$

Hence $h(Y) - h(A) \leq c(Y) - c(A) \leq g(Y) - g(A)$. Therefore, $f(Y) \geq f(A)$.

3. For any $x \in A$, consider permutation $X = \{\sigma(1), \sigma(2), \ldots, \sigma(|X|)\}$ such that $A = \{\sigma(1), \sigma(2), \ldots, \sigma(|A|)\}$ and $\sigma(|A|) = x$. Define linear function c by $c(\{\sigma(i)\}) = h(S_i) - h(S_{i-1})$ for $1 \leq i \leq |X|$ where $S_0 = \emptyset$ and $S_i = \{\sigma(1), \sigma(2), \ldots, \sigma(i)\}$. Then $c \in \partial h(A \setminus \{x\}) \cap \partial h$. Since $\partial h(A) \subseteq \partial g(A)$, we have $c \in \partial h(A \setminus \{x\}) \cap \partial g(A)$. By 2, we have $f(A) \leq f(A \setminus \{x\})$.

Similarly, we can show that for any $x \in X \setminus A$, $f(A) \leq f(A \cup \{x\})$. $\square$

A continuous function is called a *DC function* if it can be represented as the difference of two convex functions. A discrete DC function is a DC function restricted in a discrete set in its definition domain. Every set function is a discrete DC function because every submodular set function can be extended to a convex function (Lovász extension) on domain $[0, 1]^n$. The definition of local minimum solution in Sect. 6.2 is consistent with the definition of local minimum solution of discrete DC functions [285].

There exist several algorithms for minimization of discrete DC functions [284, 285, 437], which terminate at local minimum solutions satisfying condition $\partial h(A) \subseteq \partial g(A)$.

Exercises

1. Consider a nonnegative submodular function $f : 2^X \to \mathbb{R}_{\geq 0}$ and unconstrained maximization $\max\{f(S) \mid S \subseteq X\}$. S is called a local maximal solution if it cannot be improved by the following two operations.

   - **Deletion**: If there exists $e \in S$ such that $f(S \setminus \{e\}) > f(S)$, then $S \leftarrow S \setminus \{e\}$.
   - **Addition**: If there exists $d \in X \setminus S$ such that $f(S \cup \{d\}) > f(S)$, then $S \leftarrow S \cup \{d\}$.

   Show that for any local maximal solution of type one, S, if $T \subset S$ or $T \supset S$, then $f(T) \leq f(S)$.

2. Consider a nonnegative submodular function $f : 2^X \to \mathbb{R}_{\geq 0}$ and unconstrained maximization $\max\{f(S) \mid S \subseteq X\}$. Let S be a local maximal solution under two operations, deletion and addition. Show that

$$\max(f(S), f(X \setminus S)) \geq \frac{1}{3} \cdot opt,$$

where opt is the maximum value of f. (Hint: Use the result in Problem 1.)
3. Let f be a nonnegative submodular function on all subsets of a finite set X. Consider a maximization problem

$$\max\{f(S) \mid S \in \mathcal{I}\}$$

where $(X, \mathcal{I})$ is a matroid. A subset S of X is called a local maximal solution of type two if it cannot be improved by the following three operations:

- **Deletion**: If there exists $e \in S$ such that $f(S \setminus \{e\}) > f(S)$, then $S \leftarrow S \setminus \{e\}$.
- **Addition**: If there exists $d \in X \setminus S$ such that $f(S \cup \{d\}) > f(S)$ and $S \cup \{d\} \in \mathcal{I}$, then $S \leftarrow S \cup \{d\}$.
- **Swap**: If there exist $d \in X \setminus S$ and $e \in S$ such that $f((S \setminus \{e\}) \cup \{d\}) > f(S)$ and $(S \setminus \{e\}) \cup \{d\} \in \mathcal{I}$, then $S \leftarrow S \setminus \{e\}) \cup \{d\}$.

Show that for any local maximal solution S and any solution C,

$$2f(S) \geq f(S \cup C) + f(S \cap C).$$

4. Consider a nonnegative submodular function f on all subsets of a finite set X. Let S_1 be a local maximal solution for

$$\max\{f(S) \mid S \in \mathcal{I}\}$$

where $(X, \mathcal{I})$ is a matroid. Let S_2 be a local maximal solution for

$$\max\{f(S) \mid S \in \mathcal{I} \cap 2^{X \setminus S_1}}.$$

Show that

$$\max(f(S_1), f(S_2)) \geq \frac{1}{4} \cdot opt$$

where opt is the maximum value of f in $\mathcal{I}$. (Hint: Use the result in Problem 3.)
5. Consider a nonnegative submodular function $f : 2^X \to \mathbb{R}_{\geq 0}$. Let S be a $(1 + \alpha)$-approximate local maximal solution. Denote $n = |X|$. Please the following.

(a) If $T \subset S$ or $T \supset S$, then $f(T) \leq (1 + n\alpha)f(S)$.
(b) $\max(f(S), f(X \setminus S)) \geq \frac{1}{3 + 2n\alpha} \cdot opt$ where opt is the maximum value of f.
(c) For $\alpha = \varepsilon/n^2$, $(1 + \alpha)$-approximate local maximal solution can be computed with $O(\frac{1}{\varepsilon} \cdot n^3 \log n)$ oracle calls on f's values.

6. Please modify the modular-modular algorithm into approximation algorithm with polynomial number of iterations.

7. Let $f(X) = \sqrt{|X| + 1}$ for $X \in 2^V$. Please find two modular functions m_u and m_l such that $m_l(\emptyset) = f(\emptyset) = m_u(\emptyset)$ and $m_l(X) \leq f(X) \leq m_u(X)$ for any $X \in 2^V$.

8. (Transition Graph) Consider a PLS problem L and an input I. The *transition graph* of L on I is a directed graph with all solutions of I as vertices and there exists an edge from s to s' if and only if s' is a neighbor of s, with better cost. In this graph, each sink is a local optimal solution. The height of a vertex s is the length of the shortest path from s to a sink. The height of the transition graph is the maximum height of the vertex s for s over all vertices.
Show the following:

   (a) Reference [355] There exists a PLS problem which has transition graph of exponential height for some inputs. (Hint: Consider a PLS problem with minimization. Given a natural number I, every number s, $1 \leq s \leq 2^I$, is considered as a solution. The cost function is $c(I, s) = s$. The neighbor of s for $1 < s \leq 2^I$ is $s - 1$.)

   (b) If a PLS problem has a transition graph of exponential height for some input, then the standard local search algorithm runs in exponential time in the worst case, independent from how to select the best neighbor.

9. (Tight PLS-Reduction [293]) A PLS-reduction (f, g) from L_1 to L_2 is said to be *tight* if for any input instance I_1 of L_1, a subset $\mathcal{R}$ of solutions of $f(I_1)$ for L_2 can be selected to satisfy the following.

   - $\mathcal{R}$ contains all local optimal solutions of $f(I_1)$.
   - For every solution p of I_1, s solution q in $\mathcal{R}$ can be constructed in polynomial-time such that $g(I_1, q) = p$.
   - Reference [355] If the transition graph $T_{f(I_1)}$ of $f(I_1)$ contains a directed path from q to q' for $q, q' \in \mathcal{R}$ such that all internal path vertices are not in $\mathcal{R}$, then for solutions $p = g(I_1, q)$ and $p' = (I_1, q')$, we have $p = p'$ or there is an edge from p to p' in the transition graph T_{I_1} of I_1.

   Show the following property of the tight PLS-reduction: Suppose that a PLS problem L_1 is tight PLS-reducible to another PLS problem L_2. If the standard algorithm for L_1 needs exponential time in the worst case, then the standard algorithm for L_2 also needs exponential time in the worst case.

10. Reference [453] Show that the Min-Circuit/Flip problem is PLS-complete under tight PLS-reduction.

11. Reference [25] Consider the positive-NAE-Max-3SAT problem. Sow the following.

   - The local optimum of the Kernighan-Lin neighborhood structure is always a local optimum of the Flip neighborhood structure.
   - However, a local optimum of the Flip neighborhood structure is not necessarily a local optimum of the Kernighan-Lin neighborhood structure.

Historical Notes

There are many nonsubmodular optimization problems raised from the study of computational social networks. For those problems, the sandwich method is used frequently. However, the theoretical guaranteed performance ratio is really ugly and useless in evaluation of the performance. There exists another approach, called the parameterized method, which can give good-looking theoretical results. Those results have very few applications. In fact, for each specific problem, the estimation on parameters result in non-significant out comes.

The third type of algorithms are designed based on DS decomposition. Several algorithms for nonsubmodular optimization have been designed in the literature, such as submodular-supermodular algorithm in [300], modular-modular algorithm [196], and iterated sandwich algorithm in [125, 440]. They all have a nice property that the algorithm will end at a local optimal solution where by locally optimal, it means that the objective function value cannot be improved by adding or deleting an element.

The trouble on them is about the running time. Actually, each iteration in those algorithms are proved to run in polynomial-time. However, there is no analysis on the number of iterations. In [355], it has been indicated that the number of iterations is unlikely to have a polynomial bound. Therefore, those algorithms are unlikely to have a polynomial running time. This is the motivation of proposing the concept of global approximation of local optimality [81, 441]. Nonconvex relaxation is technique with great potential in the study of this type of approximation solutions.

Chapter 7
Rumor

There are three fundamental problems related to rumor, rumor discovery, rumor source detection, and rumor blocking. How to discover a rumor, i.e., how to know whether a news is rumor or not. A popular way is to see whether the news comes from a reliable source or not. This means that the rumor discovery is actually reduced to the source detection. Therefore, essentially, rumor discovery and rumor source detection are overlapping seriously.

In the literature, research efforts on rumor blocking appear more frequently and are more successful. Therefore, we start this chapter with rumor blocking.

7.1 Node-Cuts

There are three ways to block a rumor, cutting network at nodes, cutting network at Edges, and placing protectors into network. The protector is used for sending positive information to clarify the rumor. In this section, we first introduce the approach of cutting at nodes.

Consider a social network represented by a directed graph $G = (V, E)$ with the IC model. Suppose that there exists a rumor which is spreading in G from one or more rumor sources. Due to budget limit, there are only k monitors available for screening out the rumor and carrying out node-cutting. This induces the following problem.

Problem 7.1.1 (*Rumor Blocking with Node-Cuts* [122]) Given a set of sources which spread a rumor, how to allocate k node-cuts to maximize the expectation of number of nodes blocked from the rumor.

© The Author(s), under exclusive license to Springer Nature Switzerland AG 2026
W. Wu et al., *Computational Aspects of Social Networks*, Springer Optimization and Its Applications 234, https://doi.org/10.1007/978-3-032-14833-9_7

Let S denote the set of rumor sources. Let $I_G(S)$ denote the set of nodes influenced by S. Define

$$\sigma^G(S) = \mathbb{E}[I_G(S)],$$

the expected number of nodes in $I_G(S)$. Then the expectation of number of blocked nodes is

$$\tau(C) = \sigma^G(S) - \sigma^{G\setminus C}(G) \tag{7.1}$$

where C is the set of node-cuts. The problem of the rumor blocking with node-cuts can be expressed as

$$\max_{C:|C|\leq k} \tau(C).$$

First, we show that this problem is NP-hard.

Theorem 7.1.2 (Gai et al. [122]) *The rumor blocking with node-cuts is NP-hard.*

Proof Let us construct a polynomial-time Turing reduction from the well-known NP-hard knapsack problem, to the rumor blocking with node-cuts.

The knapsack problem can be expressed as follows:

$$\max \quad c_1x_1 + c_2x_2 + \cdots + c_nx_n$$
$$\text{subject to} \quad b_1x_1 + b_2x_2 + \cdots + b_nx_n \leq B$$

where $c_1, c_2, \ldots, c_n, b_1, b_2, \ldots, b_n, B$ are positive integers and $b_i \leq B$ for every $i = 1, 2, \ldots, n$. For this instance, construct an instance of the rumor blocking with node-cuts as follows: First, create a rumor source node r. Then, for each i, construct a clique C_i consisting of $c_i + b_i$ nodes u_{ij} for $1 \leq j \leq c_i + b_i$ and add b_i arcs (r, u_{ij}) for $j = 1, \ldots, b_i$. All C_i are disjoint (Fig. 7.1). In every arc (x, y), set $p_{xy} = 1$. Now, if we intend to block the rumor influence to C_i, then we must allocate at least b_i monitors at nodes u_{ij} for $1 \leq j \leq b_i$. Clearly, the knapsack problem has a feasible

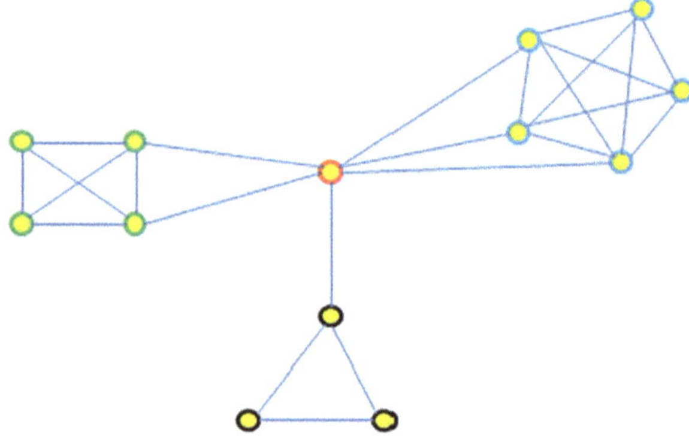

Fig. 7.1 An instance of rumor blocking with node-cuts

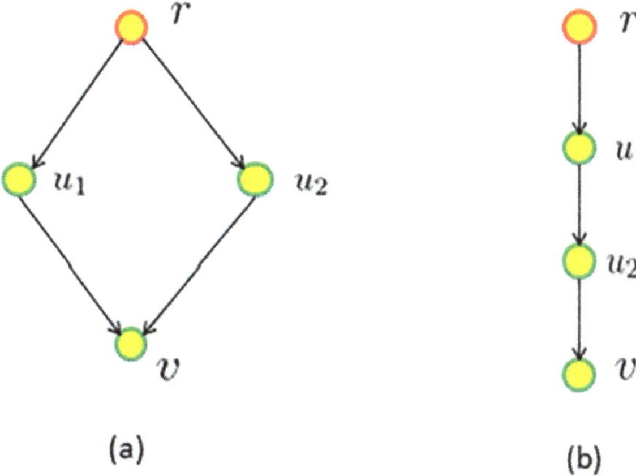

Fig. 7.2 Two counterexamples

solution with objective function value at least c if and only if B node-cuts can be allocated into the constructed social network to protect at least $B + c$ nodes from the rumor influence. $\qquad\square$

Proposition 7.1.3 (Gai et al. [122]) *The function $\tau(\cdot)$ defined in (7.1) is neither submodular nor supermodular.*

Proof First, we show that τ is not submodular. Consider a social network as shown in Fig. 7.2a. It has four nodes r, u_1, u_2, and v and two paths (r, u_1, v) and (r, u_2, v). For every arc (u, v), assign $p_{uv} = 1$. r is the unique rumor source. Then, $\Delta_{u_1}\tau(\emptyset) = 1$ and $\Delta_{u_1}\tau(\{u_2\}) = 2$. Therefore, $\Delta_{u_1}\tau(\emptyset) < \Delta_{p_1}\tau(\{p_2\})$, contradicting the definition of submodularity.

Now, we show that function $\tau(\cdot)$ defined in (7.1) is not supermodular. Consider a social network as shown in Fig. 7.2b. It has four nodes r, u_1, u_2, and v and a paths (r, u_1, u_2, v). For every arc (u, v), assign $p_{uv} = 1$. r is the unique rumor source. Then $\Delta_{u_1}\tau(\emptyset) = 3$ and $\Delta_{u_1}\tau(\{u_2\}) = 1$. Hence, $\Delta_{u_1}(\emptyset) > \Delta_{u_1}\tau(\{u_2\})$, contradicting the definition of supermodularity. $\qquad\square$

Using sandwich method, we may give a data-dependent approximation to the problem of rumor blocking with node-cuts as follows.

Define

$$\alpha_1(C) = \sum_{c \in C} \tau(\{c\})$$

and

$$\alpha_2(C) = \sum_{c, c' \in C : c \neq c'} \tau(\{c, c'\}).$$

Clearly, α_1 is modular, that is, for any two subsets $A \subset B$ and any element $x \notin B$,

$$\Delta_x \alpha_1(A) = \Delta_x \alpha_1(B).$$

However, we have

Lemma 7.1.4 $\alpha_1 - \alpha_2$ *is submodular.*

Proof Since α_1 is modular, it is sufficient to prove that α_2 is supermodular. By definition of α_2, for any two subsets $A \subset B$ and any element $x \notin B$,

$$\Delta_x \alpha_2(A) = \sum_{y \in A} \tau(\{x, y\}) \leq \Delta_x \alpha_2(B) = \sum_{y \in B} \tau(\{x, y\}).$$

Therefore, α_2 is supermodular. $\qquad\qquad\qquad\qquad\qquad\qquad\qquad\qquad\square$

By the inclusive-exclusive formula, we have

Lemma 7.1.5 (Gai et al. [122]) *For any set C,*

$$\alpha_1(C) \geq \tau(C) \geq \alpha_1(C) - \alpha_2(C).$$

Since α_1 is modular, $\max\{\alpha_1(C) \mid |C| \leq k\}$ can be solved in polynomial time. Define $\beta = \alpha_1 - \alpha_2$. Then by Lemma 7.1.4, β is a nonnegative submodular function. There exist several $1/e$-approximations in the literature for nonnegative submodular maximization with size-constraint. Next, we introduce one of them, a random greedy algorithm.

Algorithm 37 Random $1/e$-Approximation for Nonnegaive Submodular Maximization.

Input: a nonnegative submodular function f on 2^X and an integer $k > 0$.
Output: A subset A_k of X with $|A_k| \leq k$.
1: $A_0 \leftarrow \emptyset$.
2: **for** $i \leftarrow 1$ to k **do**
3:     Let $M_i \subseteq X \setminus A_{i-1}$ be a subset of size at most k maximizing $\sum_{u \in M_i} \Delta_u f(A_{i-1})$.
4:     Choose each element u_i with probability $1/k$ (mutually-exclusive, uniform random) from M_i and not choose any element with probability $1 - |M_i|/k$.
5:     **if** u_i is selected **then**
6:         $A_i \leftarrow A_{i-1} \cup \{u_i\}$
7:     **else**
8:         $A_i \leftarrow A_{i-1}$
9:     **end if**
10: **end for**
11: **return** A_k.

To analyze Algorithm 37, we first show two lemmas.

Lemma 7.1.6 *Let f be a submodular function on subsets of a finite set X. Let $A(p)$ be a random subset of A such that each element appears with probability at most p. Then*

$$\mathbb{E}[f(A(p))] \geq (1 - p)f(\emptyset).$$

Proof Let $A = \{u_1, u_2, \ldots, u_h\}$ and $p_i = \Pr[u_i \in A(p)]$. Suppose $p_1 \geq p_2 \geq \cdots \geq p_h$.

Let x_i be an indicator for event that $u_i \in A(p)$, i.e.:

$$x_i = \begin{cases} 1 & \text{if } u_i \in A(p), \\ 0 & \text{otherwise.} \end{cases}$$

Denote $A_i = \{u_1, u_2, \ldots, u_i\}$. Then, we have

$$\mathbb{E}[f(A(p))] = \mathbb{E}\left[f(\emptyset) + \sum_{i=1}^{h} x_i \cdot \Delta_{u_i} f(A(p) \cap A_{i-1})\right]$$

$$\geq \mathbb{E}\left[f(\emptyset) + \sum_{i=1}^{h} x_i \cdot \Delta_{u_i} f(A_{i-1})\right]$$

$$= f(\emptyset) + \sum_{i=1}^{h} \mathbb{E}[x_i] \cdot \Delta_{u_i} f(A_{i-1})$$

$$= f(\emptyset) + \sum_{i=1}^{h} p_i \cdot \Delta_{u_i} f(A_{i-1})$$

$$= (1 - p_1) \cdot f(\emptyset) + \sum_{i=1}^{h-1} (p_i - p_{i+1}) f(A_i) + p_h \cdot f(A)$$

$$\geq (1 - p) \cdot f(\emptyset).$$

$\square$

Lemma 7.1.7 *For $0 \leq i \leq k$,*

$$\mathbb{E}[f(OPT \cup A_i)] \geq \left(1 - \frac{1}{k}\right)^i \cdot opt$$

where OPT is an optimal solution and $opt = f(OPT)$.

Proof In each iteration i, every element of $X \setminus A_{i-1}$ stays outside of A_i with probability at least $1 - 1/k$. Thus, each element is selected to A_i with probability at most $1 - (1 - 1/k)^i$. Define $g(S) = f(S \cup OPT)$. Since f is submodular, so is g. By Lemma 7.1.6,

$$\mathbb{E}[f(OPT \cup A_i)] = \mathbb{E}[g(A_i)] \geq (1 - 1/k)^i \cdot g(\emptyset) = (1 - 1/k)^i \cdot opt.$$

$\square$

Now, we are ready to show the following.

Theorem 7.1.8 *Algorithm 37 has approximation performance ratio $1/e$ for non-negative submodular maximization.*

Proof Let us fix all random process until A_i is obtained for $1 \le i \le k$. Let OPT be an optimal solution and denote $opt = f(OPT)$. Then, we have

$$
\begin{aligned}
\mathbb{E}[\Delta_{u_i} f(A_{i-1})] &= \frac{1}{k} \cdot \sum_{u \in M_i} \Delta_u f(A_{i-1}) \\
&\ge \frac{1}{k} \cdot \sum_{u \in OPT \setminus A_{i-1}} \Delta_u f(A_{i-1}) \\
&\ge \frac{1}{k} \cdot (f(OPT \cup A_{i-1}) - f(A_{i-1})).
\end{aligned}
$$

The first inequality is due to greedy choice of A_i. The second inequality is true because f is submodular.

Release randomness of A_i and A_{i-1} and take expectation. By Lemma 7.1.7, we obtain

$$
\begin{aligned}
\mathbb{E}[\Delta_{u_i} f(A_{i-1})] &\ge \frac{1}{k} \cdot (\mathbb{E}[f(OPT \cup A_{i-1})] - \mathbb{E}[f(A_{i-1})]) \\
&\ge \frac{1}{k} \cdot ((1 - \frac{1}{k})^{i-1} \cdot opt - \mathbb{E}[f(A_{i-1})]).
\end{aligned}
$$

Thus,

$$
\begin{aligned}
\mathbb{E}[f(A_i)] &\ge \frac{1}{k} \cdot (1 - \frac{1}{k})^{i-1} \cdot opt + (1 - \frac{1}{k}) \cdot \mathbb{E}[f(A_{i-1})] \\
&\ge \frac{1}{k} \cdot (1 - \frac{1}{k})^{i-1} \cdot opt.
\end{aligned}
$$

Finally, set $i = k$ in above inequality.

$$
\mathbb{E}[f(A_k)] \ge (1 - \frac{1}{k})^{k-1} \cdot opt \ge e^{-1} \cdot opt.
$$

$\square$

Now, we can describe a data-dependent approximation algorithm as follows.

Data-Dependent Approximation

Compute an optimal solution C_{α_1} for $\max\{\alpha_1(C) \mid |C| \le k\}$.

Compute $1/e$-approximation C_β for $\max\{\beta(C) \mid |C| \le k\}$.

Compute a feasible solution C_τ for $\max\{\tau(C) \mid |C| \le k\}$.

Choose $C_{data} = \operatorname{argmax}(\tau(C_{\alpha_1}), \tau(C_\beta), \tau(C_\tau))$.

By Theorem 5.1.3, this solution C_{data} has the following performance.

Theorem 7.1.9

$$\tau(C_{data}) \geq \max\left(\frac{\tau(C_{\alpha_1})}{opt_{\alpha_1}}, \frac{1}{e} \cdot \frac{opt_{\beta}}{opt_{\tau}}\right) \cdot opt_{\tau}$$

where opt_{τ} (opt_{α_1}, and opt_{β}) is the objective function value of an optimal solution for problem $\max\{\tau(C) \mid |C| \leq k\}$ (problem $\max\{\alpha_1(C) \mid |C| \leq k\}$, and problem $\max\{\beta(C) \mid |C| \leq k\}$, respectively).

Note that τ is monotone nondecreasing, i.e., for $A \subset B, \tau(A) \leq \tau(B)$. Therefore, C_{τ} can be obtained by the following greedy algorithm:

Greedy Algorithm
    $C_0 \leftarrow \emptyset$;
    for $i = 1$ **to** k **do**
        $x = \text{argmax}_{x \in V \setminus C_{i-1}}(\tau(C_{i-1} \cup \{x\}) - \tau(C_{i-1}))$ and
        $C_i \leftarrow C_{i-1} \cup \{x\}$;
    end-for
    return $C_{\tau} = C_k$.

7.2 Edge-Cuts

The following problem will be studied in this section.

Problem 7.2.1 (*Rumor Blocking with Edge-Cuts* [451]) Consider a directed acyclic social network $G = (V, E)$ with the IC model. Given a rumor seed set S, a candidate edge set E', and a positive integer K, the problem is to find a set $\mathcal{E}$ of K edges from E' to minimize $f(\mathcal{E}) = \sum_{v \in V} \theta_{E \setminus \mathcal{E}}(v)$ where $\theta_{E \setminus \mathcal{E}}(v)$ is the probability that v gets activated by S in network $(V, E \setminus \mathcal{E})$.

This is a nonsubmodular optimization problem.

Theorem 7.2.2 (Yan et al. [451]) $f(\mathcal{E})$ *is neither supermodular nor submodular under IC model.*

Proof Consider a social network with the deterministic model, a special case of IC model, as shown in Fig. 7.3. Set the seed set $S = \{a\}$. Then,

$$f(\{(a, b)\}) - f(\emptyset) = 4 - 4 = 0$$
$$f(\{(a, c), (a, b)\}) - f(\{(a, c)\}) = 2 - 4 = -2.$$

Therefore, $f(\mathcal{E})$ is not supermodular.

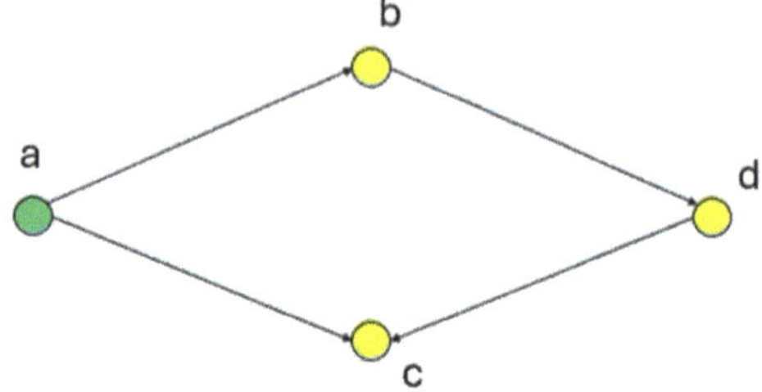

Fig. 7.3 A counterexample

On the other hand,

$$f(\{(a, b), (b, d), (a, b)\}) - f(\{(a, b).(b, d)\} = 1 - 2 = -1$$
$$> \ f(\{(a, c), (a, b)\}) - f(\{(a, c)\}) = -2.$$

Thus, $f(\mathcal{E})$ is not submodular. $\qquad\square$

A sandwich algorithm has been designed to solve this problem. Note that

$$f(\mathcal{E}) = \sum_{v \in V} \theta_{E \setminus \mathcal{E}}(v) = \sum_{v \in V}[\theta_E(v) - \Delta_{\mathcal{E}}\theta_{E \setminus \mathcal{E}}(v) = -g(\mathcal{E}) + \sum_{v \in V} theta_E(v)$$

where

$$g(\mathcal{E}) = \sum_{v \in V} \Delta_{\mathcal{E}}\theta_{E \setminus \mathcal{E}}(v).$$

Let $\mathcal{E} = \{e_1, \ldots, e_K\}$ and $\mathcal{E}^k = \{e_1, \ldots, e_k\}$. Then, $g(\mathcal{E})$ can be rewritten as

$$g(\mathcal{E}) = \sum_{k=1}^{K} \Delta_{e_k} g(\mathcal{E}^k).$$

Moreover,

$$\Delta_{e_k} g(\mathcal{E}^k) = \sum_{v \in V} \Delta_{e_k}\theta_{E \setminus \mathcal{E}^k}(v)$$

$$= \Delta_{e_k}\theta_{E \setminus \mathcal{E}^k}(t_k) + \sum_{v_1 \in N^+_{E \setminus \mathcal{E}^k}(t_k)} (\Delta_{e_k}\theta_{E \setminus \mathcal{E}^k}(v_1)$$

$$+ \sum_{v_2 \in N^+_{E \setminus \mathcal{E}^k}(v_1)} (\Delta_{e_k}\theta_{E \setminus \mathcal{E}^k}(v_2) + (\cdots))).$$

To study submodular lower and upper bound of $f(\mathcal{E})$, It is sufficient to give upper and lower bounds of $g(\mathcal{E})$.

Before do so, let us give two observations on $g(\mathcal{E})$.

- $\Delta_{e_k}\theta_{E\setminus\mathcal{E}^k}(v)$ is monotone increase. It causes $g(\mathcal{E})$ to increase.
- As $\mathcal{E}$ increases, the number of terms in the summation is decreased. It causes $g(\mathcal{E})$ to decrease.

Now, let us define two functions which are upper bound and lower bound of $g(\mathcal{E})$, respectively.

$$\bar{g}(\mathcal{E}) = \sum_{k=1}^{K} \Delta_{e_k}\bar{g}(\mathcal{E}^k)$$

where

$$\Delta_{e_k}\bar{g}(\mathcal{E}^k) = \Delta_{e_k}\theta_{E\setminus\mathcal{E}^k}(t_k) + \sum_{v_1 \in N_E^+(t_k)} (\Delta_{e_k}\theta_{E\setminus\mathcal{E}^k}(v_1) \\ + \sum_{v_2 \in N_E^+(v_1)} (\Delta_{e_k}\theta_{E\setminus\mathcal{E}^k}(v_2) + (\cdots))).$$

$$\underline{g}(\mathcal{E}) = \sum_{k=1}^{K} \Delta_{e_k}\underline{g}(\mathcal{E}^k)$$

where

$$\Delta_{e_k}\underline{g}(\mathcal{E}^k) = \Delta_{e_k}\theta_{E\setminus\mathcal{E}^k}(t_k) + \sum_{v_1 \in N_{E\setminus E'}^+(t_k)} (\Delta_{e_k}\theta_{E\setminus\mathcal{E}^k}(v_1) \\ + \sum_{v_2 \in N_{E\setminus E'}^+(v_1)} (\Delta_{e_k}\theta_{E\setminus\mathcal{E}^k}(v_2) + (\cdots))).$$

Clearly,

$$\bar{g}(\mathcal{E}) \geq g(\mathcal{E}) \geq \underline{g}(\mathcal{E}).$$

Moreover, those bounds are submodular.

Theorem 7.2.3 (Yan et al. [451]) *$\bar{g}(\mathcal{E})$ and $\underline{g}(\mathcal{E})$ are submodular.*

Proof Consider $A \subset B \subseteq E \setminus \mathcal{E}^k$ and $e = (s, t) \in (E \setminus \mathcal{E}^k) \setminus B$. Then, we have

$$\bar{g}(A \cup \{e\}) - \bar{g}(A) = \Delta_e \theta_{(E \setminus \mathcal{E}^k) \setminus A}(t) + \sum_{v_1 \in N_E^+(t)} (\Delta_e \theta_{(E \setminus \mathcal{E}^k) \setminus A}(v_1)$$

$$+ \sum_{v_2 \in N_E^+(v_1)} (\Delta_e \theta_{(E \setminus \mathcal{E}^k) \setminus A}(v_2) + (\cdots)))$$

$$\geq \Delta_e \theta_{(E \setminus \mathcal{E}^k) \setminus B}(t) + \sum_{v_1 \in N_E^+(t)} (\Delta_e \theta_{(E \setminus \mathcal{E}^k) \setminus B}(v_1)$$

$$+ \sum_{v_2 \in N_E^+(v_1)} (\Delta_e \theta_{(E \setminus \mathcal{E}^k) \setminus B}(v_2) + (\cdots)))$$

$$= \bar{g}(B \cup \{e\}) - \bar{g}(B).$$

A similar argument can apply to $\underline{g}$. $\qquad\square$

Now, let us study how to calculate the marginal decrement of each node in the network.

Lemma 7.2.4 (Yan et al. [451]) *If a directed edge $e = (s, t)$ is removed from the network, then the marginal decrement of each node v can be calculated as follows:*

- *First, compute the marginal decrement of t by*

$$\Delta_e \theta_{E \setminus \{e\}}(t) = \frac{p_e \cdot \theta_E(s)(1 - \theta_E(t))}{1 - p_e \cdot \theta_E(s)}.$$

- *Then, for every node v reached by one edge (t, v) from t, compute the marginal decrement of v by*

$$\Delta_e \theta_{E \setminus \{e\}}(v) = \Delta_e \theta_{E \setminus \{e\}}(t) \cdot p_{tv} \cdot \frac{1 - \theta_E(v)}{1 - p_{tv} \cdot \theta_E(t)}$$

where p_{tv} is the influence probability of t to v.

Proof Clearly, we have

$$\theta_{E \setminus \{e\}}(t) + \Delta_e \theta_{E \setminus \{e\}}(t) = \theta_E(t) \tag{7.2}$$

and

$$\theta_{E \setminus \{e\}}(t) + (1 - \theta_{E \setminus \{e\}}(t)) \cdot p_{st} \cdot \theta_E(s) = \theta_E(t). \tag{7.3}$$

By (7.3),

$$\theta_{E \setminus \{e\}}(t) = \frac{\theta_E(t) - p_{st} \cdot \theta_E(s)}{1 - p_{st} \cdot \theta_E(s)}.$$

Moreover, by (7.2),

$$\Delta_e \theta_{E \setminus \{e\}}(t) = \theta_E(t) - \theta_{E \setminus \{e\}}(t)$$
$$= \theta_E(t) - \frac{\theta_E(t) - p_{st} \cdot \theta_E(s)}{1 - p_{st} \cdot \theta_E(s)}$$
$$= \frac{p_{st} \cdot \theta_E(s) \cdot (1 - \theta_E(t))}{1 - p_{st} \cdot \theta_E(s)}.$$

The second part of the lemma may be proved similarly. $\square$

In [451], it claimed that this lemma can be used for computing the marginal decrement of each node recursively. However, the second part of the lemma is not powerful enough to do so. Therefore, this work is incomplete.

7.3 Protector

Protectors are information sources which are used for clarifying misinformation or rumor. There are many research efforts made on protectors. In this section, we select a few of them to present here.

7.3.1 With Community Structure

An earlier and simpler work is the rumor blocking with knowing community structure and the location of a rumor source. Suppose that a community partition is known for considered social network. (Please note that in a community partition, all communities are disjoint.) A rumor source is found in a community, denoted by community C. Since it is not easy to block the rumor spreads inside the community, a naive idea is to block the rumor at the boundary of this community. An edge is called a *bridge* to C if it contains one endpoint in C and the other endpoint outside of C. The one not in C is called a *bridge-end* for C. By blocking the rumor at the boundary of community C, we mean to block the rumor at all bridge-ends. In [103], it is considered to block rumor by placing protectors. Each *protector* is a source of anti-rumor message which can clarify the rumor. Thus, the spread of such a message will give immunization at bridge-ends. The following are assumed.

- The cascade priority is uniform, that is, if a node receives the rumor and the anti-rumor message at the same time, then the anti-rumor message has the priority to influence the node.

Under this assumptions and with deterministic information diffusion model, finding the minimum number of protector locations to protect all bridge-ends can be

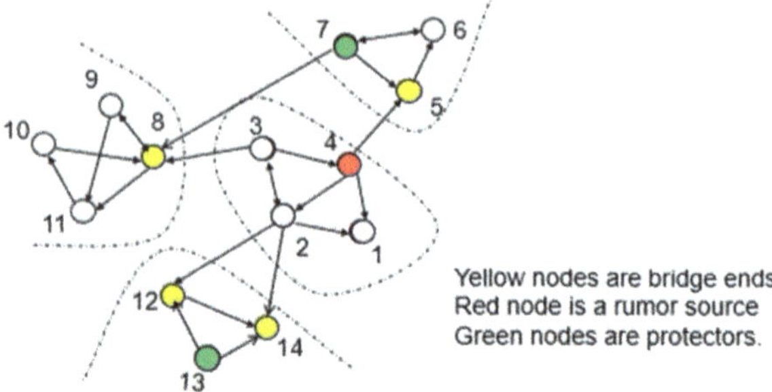

Fig. 7.4 Rumor blocking with community structure

formulated as a minimum set cover problem. To see this, for each node u not lying in community C, define B_u to be the subset of bridge-ends each of which can be protected by a protector allocated at u. Then, the problem is to select the minimum number of subsets B_u for covering all bridge-ends (Fig. 7.4). This fact implies the following.

Lemma 7.3.1 *Let A be a set of protectors. Consider deterministic information diffusion model. Let $\zeta(A)$ denote the number of bridge-ends protected by A. Then, $\zeta(A)$ is a monotone nondecreasing and submodular set function.*

As a consequence, we have the following.

Lemma 7.3.2 *Consider the IC model. Let $\zeta(A)$ denote the expected number of bridge-ends protected by protector set A. Then $\zeta(A)$ is a monotone nondecreasing and submodular set function.*

Proof It follows from the following equation:

$$\zeta(A) = \sum_{X \subseteq G} \Pr[X] \zeta^X(A)$$

where $\zeta^X(A)$ is the number of bridge-ends protected by A in subgraph X. □

With probabilistic information diffusion model, there are two ways to set optimization formulations.

- Given a positive integer k, find locations for k protectors to maximize the expected number of bridge-ends which can be protected by the k protectors.
- Given a number $r \in (0, 1)$, find the minimum number of protectors to cover at least r percentage of bridge-ends.

By Lemma 7.3.2, under the IC model, the first is a submodular maximization problem and the second is its inverse, a minimum submodular cover problem. Their approximation solutions have been well studied in the literature and also in previous chapters. Therefore, we will not discuss this further.

Next, we would like to consider removal of the assumption that community structure is already known. Actually, the community partition is not so easy to know since the community detection is a quite difficult research problem. In this case, we should first do community partition. In Chap. 1, we already found that in the literature there are many research efforts made on community partition, which produce many approaches based on different principals. What approach should we select in current situation? Here, for the purpose of studying an optimization problem about social influence, it may be better to do influence-based community partition so that in each community, members have bigger influence to each other.

Let us recall the influence-based community partition in [276], which is introduced in Sect. 1.5. Consider a social network with a directed graph structure $G = (V, E)$ and information diffusion model m. Let $(C_1, C_2, \ldots, C_k)$ be a partition of V. Denote by $G[C_i]$ the subgraph of G, induced by node subset C_i. For any node $u \in C_i$, let $\sigma_m^i(u)$ be the expected number of nodes in $C_i \setminus \{u\}$, influenced by u. Then, the influence between members inside C_i is measured by $\sigma_m[C_i] = \sum_{u \in C_i} \sigma_m^i(u)$. In Sect. 1.5, we formulate the influence-based community detection as a maximization problem as follows:

$$\max \quad \sigma_m[C_1] + \sigma_m[C_2] + \cdots + \sigma_m[C_k]$$
$$\text{subject to} \quad (C_1, C_2, \ldots, C_k) \text{ is a partition of } V \text{ without empty component.}$$

Let x_{ij} be a 0–1 variable which indicates whether a node $u_i \in V$ belongs to C_j or not. Then we can turn this maximization problem into a 0–1 integer programming with nonlinear objective function and linear constraints, which corresponds to a set function maximization problem. For $k = 2$ and m is the linear threshold model, it is showed that the objective function is supermodular and the problem can be solved in polynomial time. In general, the problem is NP-hard and especially, when m is the IC model, the objective function is neither submodular nor supermodular. Therefore, it is a case that approximation for nonsubmodular optimization would play an important role.

Note that we are only concerned about the community contaminated by the rumor. Therefore, we may simply to do community detection in the following way: In each iteration, partition the contaminated community into two subcommunities by the influence-based community detection method mentioned above (set $k = 2$). This process stops until two subcommunities are both contaminated or protector locations do not change after partition.

To make this algorithm efficient, we need to note that computing the spread of immunization at bridge-ends is #P-hard in the LT model and the IC model, similar to the computation of the influence spread [50, 53]. When Kempe, Kleinburg and Tardos [211] studied the influence maximization, they suggested to use Monte Carlo method to compute the influence, which requires 10000 samplings for one value. This heavy computation is reduced significantly in [22] with the reverse sampling technique. This technique is further improved in [310, 385, 386]. In the influence maximization problem, cascades initiated at seeds are all same type of positive ones.

However, in rumor blocking, we have to consider two types of cascades and that they are competing with each other. This issue would carry over additional difficulties in reverse sampling (see in next subsection). Now, the community detection must be introduced with new issues, which motivate new techniques.

Since the objective function is the expected number of immunized bridge-ends, we have opportunity to reduce the network size in the remaining part. There are several possible approaches to do so. One is to use metric approximation [98]. The network will be reduced into a tree-like structure so that the objective function can be easily computed. Any approximation solution obtained on this simpler network would have approximation performance ratio multiplying $O(\log n)$ for the problem in the original network.

The second one is to use a union of "shortest path" trees; each is rooted at a protector and consists of all edges in most influential paths (shortest paths with weight $\log p_{uv}$ for edge (u, v)) from the protector to bridge-ends. On such a simpler network, the misinformation clarification can be solved by a dynamic programming. However, the theoretical performance is hard to be analyzed, i.e., the approximation performance ratio is hard to be established in this approach.

The third possible approach is to give an upper bound for network size, say total number of nodes not exceed a constant number α. Under this constraint, compute a best or nearly best subnetwork to represent the original network. This can be formulated into a nonsubmodular maximization problem with an objective function chosen from our certain consideration, e.g., maximize the spread of influence on bridge-ends or maximize the spread of immunization on bridge-ends. (Note: When we do this maximization, the protector locations are assumed to be fixed and our purpose is to reduce the computation cost in the entire network.) Since the objective function in this problem can be DS-decomposed easily (detail will be given below), we may compute a solution or reasonably good approximation solution in a faster way. This is why this approach is considerable.

Now, let us find special techniques with which we can compute, in polynomial time, the DS decomposition for special classes of set functions appearing in rumor blocking.

Let x_i be the indicator for an element i selected into the set, i.e., $x_i = 1$ means that element i is selected and $x_i = 0$ means that element i is not selected. Consider a set function f which is a multivariable polynomial function. Let g be the sum of all terms with positive coefficients in f and h the sum of all terms with negative coefficients in f. Then $f = g + h$ where g is supermodular and h is submodular. This is because every term $x_{i_1} x_{i_2} \cdots x_{i_k}$ is supermodular. Let c be a sufficiently large positive constant. Set $h' = c - g$ and $g' = c + h$. Then $f = g' - h'$ and both g' and h' are submodular.

This DS decomposition would induce algorithms for solving the nonsubmodular maximization problem proposed in the third approach to reduce the network size.

Let us discuss this approach further. In this approach, we proposed to solve a nonsubmodular maximization under a constraint on network size α, i.e., the number of network nodes does not exceed α. The objective function is the spread of immunization at bridge-ends under condition that all positive cascades pass through a subnetwork with at most α nodes. Let x_v be the indicator for a node v to be selected into the subnetwork. Then in the LT model, the objective function is the sum of terms each of which is a product of x_v, corresponding to a path from a protector to a bridge-end and the coefficient is the probability that the bridge-end accepts the positive cascade provided from the protector through the path. Therefore, this objective function is monotone nondecreasing and supermodular. In the IC model, by the inclusive-exclusive principal, the objective function can be represented as

$$\Sigma_1 - \Sigma_2 + \Sigma_3 - \cdots$$

where Σ_1 is the sum of terms each of which is the union of i paths from protectors to bridge-ends and the coefficient is the expected number of bridge-ends that get immunized through those paths. Each Σ_i is monotone nondecreasing and supermodular and hence DS decomposition can be easily obtained.

There is a computational trouble in above decomposition. Especially, when the inclusive-exclusive principal is employed, the number of terms in the representation is too much and hence too hard to compute. Therefore, only Σ_1 and Σ_2 are useful. They can be used for establishing upper and lower bounds in the sandwich method.

7.3.2 Randomized Algorithm: Reverse Sampling

Using reverse sampling to design randomized algorithm for rumor blocking, it is a new challenge since there are two cascades that collider in the network. In particular, what should play the role of RR-set in this case? It is important to answer this question.

Let S_r be the seed set of rumor (or the set of rumor sources) and S_p the set of protectors, i.e., the seed set of positive cascade. Let $f(S_p)$ denote the expected number of nodes that are not activated by rumor. Our problem is as follows.

Problem 7.3.3 Given a social network $G = (V, E)$ with the IC model, a rumor seed set S_r, and a positive integer k, find a set S_p of k protectors to maximize $f(S_p)$.

For any realization g and any node u, define

$$f_g(S_p, u) = \begin{cases} 0 \text{ if } u \text{ is activated by rumor in } g \text{ under } S_p, \\ 1 \text{ otherwise} \end{cases}$$

Then, $f(S_p)$ can be represented as follows:

$$f(S_p) = \sum_{g \in \mathcal{G}} \sum_{u \in V} \Pr[g] f_g(S_p, u).$$

Let $dist_g(u, v)$ denote the shortest distance between nodes u and v in g.

Assume that if a node receives the rumor cascade and positive cascade at the same step, then it will be activated by rumor. In the other words, the rumor cascade has priority.

With this priority assumption, we have the following.

Lemma 7.3.4 $f_g(S_p, u) = 0$ *if and only if* $dist_g(S_r, u) \leq dist_g(S_p, u)$ *and* $dist_g(S_r, u) \neq +\infty$ *where* $dist_g(S_r, u) = \min_{v \in S_r} dist_g(v, u)$.

Proof If $dist_g(S_r, u) = +\infty$, then clearly, u cannot be activated by rumor.

If $dist_g(S_r, u) \neq +\infty$, then it follows immediately from the priority assumption that u is activated by rumor in g if and only if $dist_g(S_r, u) \leq dist_g(S_p, u)$. □

Theorem 7.3.5 $f(S_p)$ *is monotone nondecreasing and submodular.*

Proof From above representation of $f(S_p)$, it is easy to see that it is sufficient to prove that $f_g(S_p)$ is monotone nondecreasing and submodular with respect to S_p.

Clearly, $f_g(S_p, u)$ is monotone nondecreasing. Next, we prove that $f_g(S_p, u)$ is submodular. To do so, consider $S \subseteq S' \subseteq V$ and $v \in V \setminus S'$.

If $f_g(S' \cup \{v\}, u) - f_g(S', U) = 0$, then

$$f_g(S' \cup \{v\}, u) - f_g(S', u) \leq f_g(S \cup \{v\}, u) - f_g(S, u)$$

since $f_g(S_p, u)$ is monotone nondecreasing.

If $f_g(S' \cup \{v\}, u) - f_g(S', U) = 1$, then we must have $f_g(S' \cup \{v\}, u) = 1$ and $f_g(S', u) = 0$. $f_g(S', u) = 0$ implies $f_g(S, u) = 0$ due to monotonicity. $f_g(S' \cup \{v\}, u) = 1$ implies that $dist_g(S_r, u) = +\infty$ or

$$dist_g(S_r, u) > dist_g(S', u). \tag{7.4}$$

In the case that $dist_g(S_r, u) = +\infty$, we also have $f_g(S \cup \{v\}, u) = 1$. In the case that (7.4) holds

$$dist_g(S_r, u) > dist_g(S', u) \geq dist_g(S, u).$$

Hence, $f_g(S \cup \{v\}, u) = 1$. Therefore,

$$f_g(S' \cup \{v\}, u) - f_g(S', u) \leq f_g(S \cup \{v\}, u) - f_g(S, u).$$

□

By Theorem 7.3.5, we may use a greedy algorithm to obtain a $(1 - e^{-1})$-approximation solution if we ignore the computational difficulty of function $f(S_p)$. In fact, we have

Theorem 7.3.6 *Computing $f(S_p)$ is #-hard.*

Proof The proof is left to the reader as an exercise. □

Theorem 7.3.6 explains why randomized is designed and hence reverse sampling technique is utilized.

Now, "RR-sets" is defined as follows.

Definition 7.3.7 (*Random R-tuple of node v* [401]) For a node v, Random R-tuple (RR-tuple) of v is a four-tuple

$$\mathcal{T}_v = (V^*, E_t, E_f, B)$$

generated by Algorithm 38, which starts with node v and search in reverse direction whether each in-neighbor of nodes in V^* can be added to V^* until meet node in S_r or no in-neighbor of V^* exists. For each RR-tuple of v, four components have the following meaning:

- V^* is a set of nodes that can protect v if it becomes a protector.
- E_t is the set of all alive edges.
- E_f is the set of all fail edges.
- B is a Boolean variable. If $B = 1$, then the algorithm ends in case that the search meets a rumor seed. If $B = 0$, then the algorithm ends in case that no in-neighbor of V^* exists.

Definition 7.3.8 (*RR-Tuple* [401]) An RR-tuple is generated by Algorithm 39.

For any RR-tuple $\mathcal{T}$, denote its four components by $\mathcal{T}(V^*)$, $\mathcal{T}(E_t)$, $\mathcal{T}(E_f)$, and $\mathcal{T}(B)$. Define

$$x(S, \mathcal{T}) = \begin{cases} 1 & \text{if } S \cap \mathcal{T}(V^*) \neq \emptyset \text{ or } \mathcal{T}(B) = 0 \\ 0 & \text{otherwise.} \end{cases}$$

Theorem 7.3.9 (Tong et al. [401]) *For any $S \subseteq V$,*

$$\mathbb{E}[x(S, \mathcal{T})] = f(S)/n.$$

This is the key theorem with a little long proof. Let us put its proof at the end of this subsection and see its consequence right now.

Based on Theorem 7.3.9, a randomized algorithm can be designed as follows:

- Step 1: Generate a set of RR-tuples, $\mathcal{R} = \{\mathcal{T}_1, \mathcal{T}_2, \ldots, \mathcal{T}_\ell\}$, by Algorithm 39.
- Step 2: Find S to maximize

$$F(S, \mathcal{R}) = \sum_{i=1}^{\ell} x(S, \mathcal{T}_i).$$

Algorithm 38 RR-Tuple of v [401].

Input: G, S_r, v.
Output: $\mathcal{T} = (V^*, E_t, E_f, B)$.
1: $V^* \leftarrow \emptyset, E_t \leftarrow \emptyset, E_f \leftarrow \emptyset, V_1 \leftarrow \{v\}$;
2: **while** $V_1 \neq \emptyset$ and $V_1 \cap S_r = \emptyset$ **do**
3:     $V^* \leftarrow V^* \cup V_1$;
4:     $V_2 \leftarrow V \setminus V^*$;
5:     $V_1 \leftarrow \emptyset$;
6:     **for** each edge $(u_1, u_2) \in E$, obtained through bread-first search, with $u_1 \in V_2$ and $u_2 \in V^*$
        do
7:         Choose a random number $rand$ uniformly from $[0, 1]$;
8:         **if** $rand \leq p_{u_1, u_2}$ **then**
9:             $V_1 \leftarrow V_1 \cup \{u_1\}$;
10:             $E_t \leftarrow E_t \cup \{(u_1, u_2)\}$;
11:         **else**
12:             $E_f \leftarrow E_f \cup \{(u_1, u_2)\}$
13:         **end if**
14:     **end for**
15:     **if** $V_1 = \emptyset$ **then**
16:         $B = 0$
17:     **end if**
18:     **if** $V_1 \cap S_r \neq \emptyset$ **then**
19:         $B = 1$;
20:         Delete from V^* all nodes u such that $dist(u, v) = dist(S_r, v)$ in alive edges;
21:     **end if**
22: **end while**
23: **return** $\mathcal{T} = (V^*, E_t, E_f, B)$.

Algorithm 39 RR-Tuple [401].

Input: G, S_r.
Output: $\mathcal{T}$.
1: Select a node $v \in V$ uniformly in random.
2: Generate a RR-tuple $\mathcal{T}$ of u by Algorithm 38.
3: **return** $\mathcal{T}$.

To do so, for each RR-tuple $\mathcal{T}_i$, set

$$T_i = \begin{cases} V & \text{if } \mathcal{T}_i(B) = 0 \\ \mathcal{T}_i(V^*) & \text{if } \mathcal{T}(B) = 1 \end{cases}$$

Find the minimum hitting set of k nodes for $\{T_1, T_2, \ldots, T_\ell\}$.

It is well known that the maximum hitting set problem has a $(1 - e^{-1})$-approximation, which is the best possible. Therefore, in Step 2, there is a $(1 - e^{-1})$-approximation solution. For Step 1, a lower bound of ℓ should be determined in order to obtain a result that with probability at least $1 - \delta$, the algorithm produces a $(1 - 1/e - \varepsilon)$-approximation solution. How to establish such a lower bound? There are several approaches which have been discussed in Chapter 3. Therefore, we skip this part and put the one, used in [401], into exercises.

Proof of Theorem 7.3.9 [401]

We start with a definition.

Definition 7.3.10 (*Valid Pair*) A pair of ordered edge-sets (E_1, E_2) is valid, if $E_1 \subseteq E$, $E_2 \subseteq E$, and $E_1 \cap E_2 = \emptyset$.

For any realization g, $(E(g), E \setminus E(g))$ is a valid pair and for any valid pair (E_1, E_2) with $E_1 \cup E_2$, there exists a realization g such that $E(g) = E_1$. This relationship gives a bijection between realizations and valid pairs with $E_1 \cup E_2 = E$. In general, a valid pair (E_1, E_2) may not have $E_1 \cup E_2 = E$. A realization g is said to be *compatible* to a valid pair (E_1, E_2) if $E_1 \subseteq E(g)$ and $E_2 \cap E(g) = \emptyset$. Let $C(E_1, E_2)$ be the set of all realizations compatible to (E_1, E_2). For a realization g compatible to a valid pair (E_1, E_2), define

$$\Pr[(E_1, E_2)] = \prod_{e \in E_1} p_e \prod_{e \in E_2} (1 - p_e)$$

and

$$\Pr[g \mid (E_1, E_2)] = \prod_{e \in E(g) \setminus E_1} p_e \prod_{e \notin E(g) \cup E_2} (1 - p_e).$$

It is easy to verify that

$$\sum_{g \in C(E_1, E_2)} \Pr[g \mid (E_1, E - 2)] = 1. \tag{7.5}$$

and

$$\Pr[g] = \Pr[(E_1, E_2)] \cdot \Pr[g \mid (E_1, E_2)]. \tag{7.6}$$

Let T_v denote an RR-tuple generated by Algorithm 38 for node v. Let $\mathcal{R}$ denote the set of all possible T_v. Note that $(T_v(E_t), T_v(E_f))$ is a valid pair. It is nature to define

$$\Pr[T_v] = \Pr[(T_v(E_t), T_v(E_f))].$$

Lemma 7.3.11 *For any node v, the sets $C(T_v(E_t), T_v(E_f))$ for $T_v \in \mathcal{R}_v$ form a partition of $\mathcal{G}$.*

Proof First, for each $g \in \mathcal{G}$, we can find a T_v such that g is compatible to $(T_v(E_t),$ $T_v(E_f))$ as follows: Start from node v to search in reverse direction. If meet an edge in g, then put it in E_t. If meet an edge not in g, then put it in E_f.

Next, we show that for $T_v, T_v' \in \mathcal{R}_v$ with $T_v \neq T_v'$,

$$C(T_v(E_t), T(E_f)) \cap C(T_v'(E_t), T_{\iota}'E_f)) = \emptyset.$$

Since $T_v \neq T_v'$, there must exist an edge e such that $e \notin T_v(E_t)$, $e \in T_v'(E_t)$, or vice versa. For such an edge e, we must have $e \in T_v(E_f)$ so that no realization g can be compatible to both T_v and T_v'. $\square$

Lemma 7.3.12 *Consider a seed set S of positive cascade, an RR-tuple T_v, and a realization g compatible to $(T_v(E_t), T_v(E_f))$. Then, v is not activated by rumor in g if and only if $x(S, T_v) = 1$.*

Proof Note that $x(S, T_v) = 0$ if and only if $S \cap T_v(V^*) = \emptyset$ and $T_v(B) = 1$. Moreover, v is activated by rumor in g if and only if $dist_g(S_r, v) \leq dist_g(S, v)$ and $dist_g(S_r, v) \neq \infty$. Therefore, it is sufficient to prove that $S \cap T_v(V^*) = \emptyset$ and $T_v(B) = 1$ if and only if $dist_g(S_r, v) \leq dist_g(S, v)$ and $dist_g(S_r, v) \neq \infty$.

First, since g is compatible to $(T_v(E_t), T_v(E_f))$, v can be reached by some rumor seed node in g if and only if $T_v(B) = 1$. Therefore, $T_v(B) = 1$ if and only if $dist_g(S_r, v) \neq \infty$.

Next, suppose that $T_v(B) = 1$ and a is one of node in $V_1 \cap S_r$ at line 18 of Algorithm 38. Since the algorithm runs with a breadth-first search, we have $dist_g(a, v) = dist_g(S_r, v)$ and $T_v(V^*)$ consists of all nodes u such that $dist_g(u, v) < dist_g(S_r, v)$. Therefore, $S \cap T_v(V^*) = \emptyset$ if and only if $dist_g(S_r, v) \leq dist_g(S, v)$. $\square$

Proof (*Theorem* 7.3.9) Note that

$$\mathbb{E}[x(S, T)] = \frac{\sum\limits_{v \in V} \sum\limits_{T_v \in \mathcal{R}_v} \Pr[T_v] \cdot x(S, T_v)}{n}$$

and

$$f(S) = \sum_{g \in \mathcal{G}} \sum_{u \in V} \Pr[g] \cdot f_g(S, v).$$

It is sufficient to prove

$$\sum_{T_v \in \mathcal{R}_v} \Pr[T_v] \cdot x(S, T_v) = \sum_{g \in \mathcal{G}} \Pr[g] \cdot f_g(S, v).$$

Now, we show this equation.

$$\sum_{T_v \in \mathcal{R}_v} \Pr[T_v] \cdot x(S, T_v)$$

$$= \sum_{T_v \in \mathcal{R}_v} \Pr[T_v] \sum_{g \in C(T_v(E_t), T_v(E_f))} \Pr[g \mid (T_v(E_t), T_v(E_f))] \cdot x(S, T_v)$$

(by (7.5)

$$= \sum_{T_v \in \mathcal{R}_v} \sum_{g \in C(T_v(E_t), T_v(E_f))} \Pr[g] \cdot x(S, T_v)$$

(since $\Pr[T_v] = \Pr[(T_v(E_t), T_v(E_f))]$)

$$= \sum_{T_v \in \mathcal{R}_v} \sum_{g \in C(T_v(E_t), T_v(E_f))} \Pr[g] \cdot f_g(S, v) \quad \text{(by Lemma 7.3.12)}$$

$$= \sum_{g \in \mathcal{G}} \Pr[g] \cdot f_g(S, v). \quad \text{(by Lemma 7.3.11)}$$

$\square$

7.3.3 Composed Rumor

An object gets contaminated, possibly caused by a single influence. The composed contamination has been studied extensively in the area of bio-informatics [130, 254]. However, on social influence, the composition is considered only recently. This is due to the problem formulation, often a nonsubmodular optimization, which is not easy to find a good solution. In [499], Zhu et al. consider composed influence by following observation: The influence from two nodes u and v to one node w results in sum of two weights in the LT model and is treated as two independent events in the IC model. However, this may not give accurate result in the real world. Therefore, they add a term $(u, v) \to w$ to make up the difference. This results in a hypergraph model and the problem formulation for the influence maximization is turned out to be a nonsubmodular maximization. We present an approximation solution by employing the sandwich method. Now, we want to study.

Problem 7.3.13 Clarify composed misinformation.

In this problem, we will consider a more complicated background in the real world. An object gets contaminated due to misinformation from possibly two or more factors. Each factor may diffuse individually, but not cause individual contamination. For example, the reputation of a product may have several good features, such as quality and price. The misinformation can be different. The misinformation from competitors may mislead the consumers on different features in order to low down the reputation of the product. Therefore, to formulate such behavior, we have to use a model of composed influence. Meanwhile, for protectors, such a model is also required to formulate their positive cascades. Let us first give an example of composed influence model with positive cascades, and then discuss composed misinformation (with both positive and negative cascades) afterward.

Consider a product with r features and a directed OSN $G = (V, E)$. In the following, we list important facts about this model, consisting of discrete steps:

- Each node represents a customer. An active node represents a customer who has purchased the product.
- Every edge (u, v) is assigned with k probabilities $p_{uv}^{(1)}, \ldots, p_{uv}^{(r)}$. Those probabilities represent the expertise of u from the view point of v. When u just becomes active, v will be influenced by u and accepts the feature ϕ with probability $p_{uv}^{(\phi)}$.
- When v receives influence from more than one active in-neighbors, v treats them as independent events.
- Each customer v has a threshold θ_v and a weight $w_v^{(\phi)}$ for each factor ϕ. Customer v decides to purchase the product if and only if the total weight of accepted factors reaches at least θ_v.
- Initially, a set of seeds are selected and activated. At each subsequent step, every node checks if the activation condition is satisfied. This process ends if no node becomes newly active at current step.

Let us first indicate that for this model, the influence spread may not be a submodular function with respect to the seed set. To do so, we want to employ a property of marginal value of submodular function that f is submodular if and only if for any two sets $A \subseteq B$ and any element $x \notin B$, $\Delta_x f(A) \geq \Delta_x f(B)$ where $\Delta_x f(A) = f(A \cup \{x\}) - f(A)$. Consider five features and a star with center v and five in-neighbors u_1, u_2, u_3, u_4, u_5 of v. For each edge (u_i, v), define $p_{u_i v}^{(i)} = 1$ and $p_{u_i v}^{(\phi)} = 0$ for $\phi \neq i$. Assume customer v has threshold $\theta_v = 0.5$ and weight $w_v^{(\phi)} = 0.2$ on every feature ϕ. Let $\sigma(A)$ denote the influence spread with respect to the seed set A. Clearly, $\Delta_{u_3}\sigma(\{u_1\}) = 0 < \Delta_{u_3}\sigma(\{u_1, u_2\}) = 1$, contradicting the property of increasing marginal value.

In the composed rumor blocking, the negative cascade is spread in the same way as the positive cascade. However, we need to consider competition between them. Firstly, a node may accept a positive cascade on a feature while it accepts a negative node on another feature. In this case, we count this node immunized, but not have ability to diffuse any positive or negative cascade. Secondly, when a positive cascade and a negative cascade arrive a node at the same time, which has the priority to influence the node? We have to consider the cascade priority.

In [393], Tong et al. consider two classes of cascades, a class of negative cascades and a class of positive cascades. Sources of negative cascades and a subset of protectors (i.e., sources of positive cascades) are fixed and the task is to allocate remaining subclass of protectors. One of the main results in the paper is about the cascade priority. They showed that the spread of immunization is monotone nondecreasing and submodular if one of following occurs:

(b1) All positive cascades have priority higher than all negative cascades.

(b2) All negative cascades have priority higher than all positive cascades.

(b3) The cascade priority is strongly uniform, i.e., at every node, the priority ordering is the same.

They also indicate that, in general, the spread of immunization is nonsubmodular, and then designed data-dependent approximation with the sandwich method. In the sandwich method, a submodular upper bound and a submodular lower bound are usually required for the nonsubmodular objective function [45, 430]. The submodular special cases are often used for establishing the upper and lower bounds. In [393], these two bounds are established by using cases (b1) and (b2).

For composed rumor, the spread of immunization may not be submodular in cases (b1), (b2), and (b3) since it is nonsubmodular even in case that no negative cascade exists. To find submodular special cases, we must consider not only cascade priority, but also feature priority. Actually, each customer has his/her own favorite feature, or own feature priority, which reflects to setting of threshold θ_v and weights $w_v^{(\phi)}$. Is there some simple conditions on them to make the spread of immunization submodular? It is possible. Let us mention a special case on single feature, which is quite likely that the spread of composed influence is submodular.

In this case, every customer put positive weight on each feature and chooses the threshold equal to the sum of weights on all features. This constraint is equivalent to say that a customer decides to purchase the product if and only if all features are accepted. Hence, we may treat the above diffusion model as a multi-dimensional IC model. That is, each feature diffuses in its own dimension with the IC model and consults with other dimensions only when we make decision to purchase the product. According to a method given in Chap. 2, the IC model can be transformed into an equivalent general threshold model with monotone submodular threshold function at each node, so that the multi-dimensional IC model can be transformed into the multi-dimensional general threshold model with monotone submodular threshold functions at each node.

For the one-feature influence, the general threshold model is a generalization of the LT model (see in Chapter 2). In the definition of the LT model, each node is assigned a monotone set function f_v (a.k.a., a *threshold function*) on the set of active in-neighbors. f_v is used for determining whether v is activated or not. At each step, every node evaluates the value of f_v. If the value of f_v reaches at least the threshold θ_v, then v is activated; otherwise, v keeps inactive. Kempe, Kleinburg and Tados [211] conjectured that if for every node, the threshold function is monotone submodular, then the influence spread is a monotone submodular function. This conjecture is proved by Mossel and Roch in [297]. This proof is possibly to extend to the multi-dimensional general threshold model because of the following: In the multi-dimensional general threshold model, each node v is assigned with r threshold functions $f_v^{(\phi)}$ for $\phi = 1, \ldots, r$. At beginning, every node v chooses r thresholds $\theta_v^{(\phi)}$ randomly and uniformly from [0,1]. At each step, every node v checks the value of threshold functions $f_v^{(\phi)}$ for all features ϕ. If for all ϕ, the value of $f_v^{(\phi)}$ on the set of active in-neighbors is at least $\theta_v^{(\phi)}$, then v is activated; otherwise, v keeps inactive. It is likely to show that if at every node, all threshold functions are monotone submodular, then the influence spread is a monotone submodular function.

When put the cascade priority and the feature priority together, it is definitely challenging to find the submodularity in some special cases. However, it would give us an interesting research opportunity.

7.4 Rumor Source Detection

Rumor source detection is a very attractive research topic. In the literature, there are two types of methods, centrality-based methods and monitor placement. Let us present some examples in the following subsections.

7.4.1 Centrality-Based Method

The first centrality-based method is proposed by Shah and Zaman [359]. They propose the maximum likelihood as rumor source estimator.

Definition 7.4.1 (*Maximum Likelihood Estimator* [359]) Consider a undirected social network $G = (V, E)$ with the deterministic SI (susceptible-infected) model, i.e., the diffusion is in deterministic model and the rumor-infection is susceptible (cannot be recovered). Let G_N be the connected subgraph with all rumor-infected nodes. Suppose only one rumor source v^* exists. Then the maximum likelihood estimator of v^* is defined by

$$\hat{v} \in \mathrm{argmax}_{v \in G_N} \Pr[G_N \mid v],$$

that is, the probability that G_N gets infected under condition that v is the rumor source.

In general, it is hard to compute this probability. However, if G is a regular tree, then an explicit expression is given in [359]. To do so, it needs the concept of permitted permutation.

Definition 7.4.2 (*Permitted Permutation*) Given a tree $G = (V, E)$ and a source node $v \in V$, a permutation $\sigma : V \to \{1, 2, \ldots, |V|\}$ is *permitted* if

- $\sigma(v) = 1$.
- For any $(u, u') \in E$, if $d(v, u) < d(v, u')$, then $\sigma(u) < \sigma(u')$ where $d(v, u)$ is the shortest path distance from v to u.

Assume G is a regular tree with countably infinite node set V so that boundary effect can be avoided. Then, there is the following result.

Theorem 7.4.3 (Shah and Zaman [359]) *Let $R(v, G_n)$ be the number of distinct permitted permutations of nodes in G_N that begin with node v. Then for regular tree G, the maximum likelihood estimator can be represented as*

$$\hat{v} \in argma_{v \in G_N} R(v, G_N).$$

Proof Let $\Omega(v, G_N)$ be the set of permitted permutations of nodes in G_N that begin with node v. For $\sigma \in \Omega(v, G_N)$, let $\sigma(V(G_N)) = \{v_1 = v, v_2, \ldots, v_N\}$. Denote by $G_k(\sigma)$ the subgraph induced by nodes $\{v_1, \ldots, v_k\}$. Then

$$\Pr[\sigma \mid v] = \prod_{k=2}^{N} \Pr[k\text{th infected node } = v_k \mid G_{k-1}(\sigma), v].$$

Given $G_{k-1}(\sigma)$ and source node v, the next infected node could be any of the neighbors of $G_{k-1}(\sigma)$. Suppose G is d-regular (without boundary). Then $G_{k-1}(\sigma)$ has totally $d + (k-2)(d_2)$ neighbors; all of them have the same chance to be infected. Therefore,

$$\Pr[k\text{th infected node } = v_k \mid G_{k-1}(\sigma), v] = \frac{1}{d + (k-2)(d-2)}.$$

Hence,

$$\Pr[\sigma \mid v] = \prod_{k=2}^{N} \frac{1}{d + (k-2)(d-2)} = p(d, N).$$

Therefore,

$$\begin{aligned}
\hat{v} &\in \operatorname{argmax}_{v \in G_N} \Pr[G_N \mid v] \\
&= \operatorname{argmax}_{v \in G_N} \sum_{\sigma \in \Omega(v, G_N)} \Pr[\sigma \mid v] \\
&= \operatorname{argmax}_{v \in G_N} R(v, G_N) \cdot p(d, N) \\
&= \operatorname{argmax}_{v \in G_N} R(v, G_N).
\end{aligned}$$

$\square$

The following are other possible estimators for single rumor source.

Definition 7.4.4 (*Infection Eccentricity*) Let d_{uv} denote the shortest distance between u and v in social network G. Let M be a set of monitors. Then, the eccentricity of node v is the longest distance from v to a monitor u in m, denoted by $\bar{e}_M(v)$, i.e.:

$$\bar{e}_M(v) = \max_{u \in M} d_{vu}.$$

Definition 7.4.5 (*Jordan Infection Center*) A node v^* is called the Jordan infection center if

$$\bar{e}_M(v^*) = \min_{v \in V} \bar{e}_M(v).$$

Overall, for centrality-based approaches, there do not exist many good theoretical results in social networks, other than trees.

7.4.2 Monitor Selection

In this subsection, we study the following problem.

Problem 7.4.6 (*k-Monitor Selection*) Let O be a complete snapshot of the social network G with the IC model for rumor propagation. Let $O_a \subseteq O$ be the set of rumor-infected nodes. Given a positive integer k, select k nodes from Q_a to form a monitor set such that every node v in O_a is within distance r from a monitor and r reaches the minimum.

This is exactly the classical metric k-center problem since the shortest path length between two nodes in Q_a gives a metric in Q_a.

Problem 7.4.7 (*Metric k-Center Selection*) Given n cities with a metric distance table between them, and a positive integer k, select k cities as centers such that the maximum distance from a city to a nearest center is minimized.

We can easily obtain the following results.

Theorem 7.4.8 *There exists a polynomial-time 2-approximation for the k-center selection problem.*

Proof Consider the following selection: First, pick any city x_1 as a center. In each subsequent iteration, suppose $x_1, \ldots, x_{i-1}$ are centers selected in previous iteration, and then select x_i such that x_i has the maximum distance from a city to a center in $\{x_1, \ldots, x_{i-1}\}$. The algorithm ends after x_k is selected.

Let V be the set of all cities. Let OPT be the optimal solution of the metric k-center selection problem. We show that

$$\max_{v \in V} \min_{1 \le i \le k} d(v, x_i) \le 2 \cdot opt,$$

where $opt = \max_{v \in V} \min_{y \in OPT} d(v, y)$. Let $OPT = \{y_1, \ldots, y_k\}$. Partition $V = (V_1, V_2, \ldots, V_k)$ such that $\max_{v \in V_j} d(v, y_j) \le opt$. The proof is divided into two cases.

Case 1. Every V_j contains some x_i. In this case, for every $v \in V_j$,

$$d(v, x_i) \le d(v, y_j) + d(y_j, x_i) \le 2 \cdot opt.$$

Case 2. There exists a V_j which contains two selected centers x_i and $x_{i'}$. Suppose $i < i'$. Then, by the selection rule, we have that for any $v \in V$,

$$\min_{1 \le i'' < i'} d(v, x_{i''}) \le d(x_i . x_{i'}) \le d(x_i, y_j) + d(y_j, x_{i'}) \le 2 \cdot opt.$$

$\square$

Theorem 7.4.9 *If $NP \ne P$, then there is no polynomial-time $(2 - \varepsilon)$-approximation for the k-center selection problem and for any $\varepsilon > 0$.*

Proof Consider the dominating set problem as follows.

Dominating Set Problem: Given a graph $G = (V, E)$ and a positive integer k, determine whether G has a dominating set of size at most k where a node set D is called a *dominating set* if every node in V is either in D or adjacent to a node in D.

Now, given an instance (G, k) of the dominating set problem, define a distance between any two nodes u and v by

$$d(u, v) = \begin{cases} 1 \text{ if } (u, v) \in E, \\ 2 \text{ otherwise.} \end{cases}$$

It is easy to verify that this distance gives a metric for nodes in V. Consider (V, d, k) as an instance of the metric k-center selection problem. Then, there exist k centers such that the maximum distance from a node to the nearest center is at most 1 if and only if G has a dominating set of size at most k. Suppose that the metric k-center selection problem has a polynomial-time $(2 - varepsilon)$-approximation algorithm A with $\varepsilon > 0$. Apply A on instance (V, d, k). Then, A returns a solution with objective function value < 2 if and only if G has a dominating set of size at most k. Thus, A solves the dominating set problem. However, the dominating set problem is NP-complete. Therefore, $NP = P$. $\square$

7.4.3 Double Resolving Set

Consider a social network with the IC model and the influence distance d_{uv}.

Definition 7.4.10 *(Double Resolving Set)* A node subset A is called a *double resolving set (DRS)* if for any two distinct nodes x and y, there exist u and v in A such that

$$r_u(x) - r_u(y) \ne r_v(x) - r_v(y),$$

where $r_u(x) = d_{ux}$. An example can be found in Fig. 7.5.

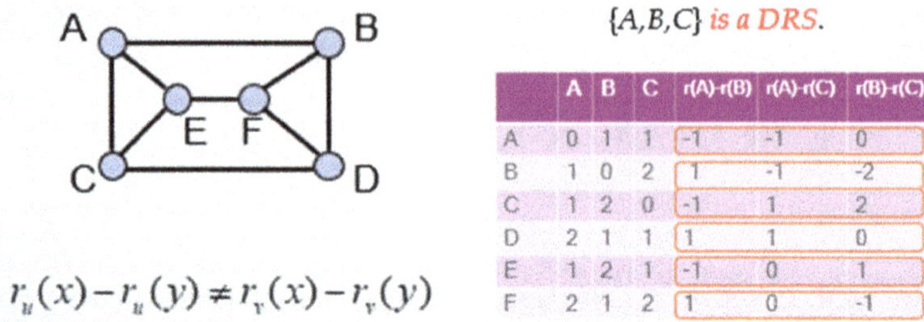

	A	B	C	r(A)-r(B)	r(A)-r(C)	r(B)-r(C)
A	0	1	1	-1	-1	0
B	1	0	2	1	-1	-2
C	1	2	0	-1	1	2
D	2	1	1	1	1	0
E	1	2	1	-1	0	1
F	2	1	2	1	0	-1

$$r_u(x) - r_u(y) \neq r_v(x) - r_v(y)$$

Fig. 7.5 An example of DRS

The DRS can be used for rumor source detection.

- For each node in a DRS, allocate a monitor.
- Every monitor reports the time that the rumor arrive.
- The differences between those recorded times would tell who is the rumor source.

Thus, a problem is raised.

Problem 7.4.11 (*Min DRS*) Give a graph G, find a minimum DRS of G. Here, the minimum means the minimum cardinality.

There is a greedy algorithm for this problem. To describe it, let us first define a potential function.

For each set A, define an equivalent relation $\equiv_A$ as follows: For two nodes u and v, $u \equiv_A v$ if and only if for any two nodes x and y in A, $r_u(x) - r_u(y) = r_v(x) - r_v(y)$. It is easy to verify that this is an equivalence relation. Suppose this relation divides V into equivalence classes $V_1, V_2, \ldots, V_k$. Then, define

$$f(A) = \log_2 \binom{|V|}{|V_1|, |V_2|, \ldots, |V_k|} = \log_2 \frac{|V|!}{|V_1|!\, |V_2|! \cdots |V_k|!}.$$

To show a property of function $f(A)$, let us first prove an inequality.

Lemma 7.4.12

$$\binom{n}{n_1, n_2, \ldots, n_k}$$
$$\geq \binom{m_1}{m_{11}, m_{12}, \ldots, m_{1k}} \binom{m_2}{m_{21}, m_{22}, \ldots, m_{2k}} \cdots \binom{m_h}{m_{h1}, m_{h2}, \ldots, m_{hk}}$$

where

$$n = m_1 + m_2 + \cdots + m_h$$
$$n_1 = m_{11} + m_{21} + \cdots + m_{h1}$$
$$n_2 = m_{12} + m_{22} + \cdots + m_{h2}$$
$$\cdots$$
$$m_h = m_{h1} + m_{h2} + \cdots + m_{hk},$$

and all numbers appearing in above are nonnegative integers.

Proof The left hand of inequality is the number of possibilities for distributing n balls to k buckets with cardinalities $n_1, n_2, \ldots, n_k$, respectively.

Suppose that among n balls, there are m_1 balls in color 1, m_2 balls in color 2, ..., m_h balls in color h. Then, the right hand of inequality is the number of possibilities for distributing n balls into k buckets such that n_1 balls are placed in bucket 1, n_2 balls are placed in bucket 2, ..., n_k balls are placed in bucket k, and, in addition, satisfying the condition that among n_1 balls in bucket 1, there are m_{11} balls in color 1, m_{21} balls in color 2, ..., m_{h1} balls in color h; among n_2 balls in bucket 2, there are m_{12} balls in color 1, m_{22} balls in color 2, ..., m_{h2} balls in color h; ...; among n_k balls in bucket k, there are m_{1k} balls in color 1, m_{2k} balls in color 2, ..., m_{hk} balls in color h.

Since additional condition may reduce the number of possibilities of distribution, the left hand should be not smaller than the right hand. $\square$

Lemma 7.4.13 (Chen et al. [56]) $f(A)$ *is monotone nondecreasing and submodular.*

Proof Consider $A \subset B$. Suppose that equivalence relation $\equiv_A$ partitions V into equivalence classes $V_1, V_2, \ldots, V_k$, and equivalence relation $\equiv_B$ further partitions V_1 into equivalence classes $V_{11}, \ldots, V_{1h_1}$, partitions V_2 into equivalence classes $V_{21}, \ldots, V_{2h_2}, \ldots$, and partitions V_k into equivalence classes $V_{k1} \ldots, V_{kh_k}$. Then,

$$f(B) - f(A) = \log_2 \frac{|V_1|!}{|V_{11}|! \cdots |V_{1h_1}|!} + \cdots + \log_2 \frac{|V_k|!}{|V_{k1}|! \cdots |V_{kh_k}|!}$$
$$= \log_2 \binom{|V_1|}{|V_{11}|, \ldots, |V_{1h_1}|} + \cdots + \log_2 \binom{|V_k|}{|V_{k1}|, \ldots, |V_{kh_k}|}$$
$$\geq 0.$$

Thus, $f(A)$ is monotone nondecreasing.

To show submodularity, suppose that equivalence $\equiv_{A \cup \{x\}}$ partitions V_1 into $U_{11}, U_{12}, \ldots, U_{1\ell}, \ldots$, partitions V_k into $U_{k1}, U_{k2}, \ldots, U_{k\ell}$. Moreover, suppose that equivalence $\equiv_{B \cup \{x\}}$ partitions V_{11} into $U_{111}, U_{112}, \ldots, U_{11\ell}, \ldots$, partitions V_{kh_k} into $U_{kh_k1}, U_{kh_k2}, \ldots, U_{kh_k\ell}$ such that

$$|U_{11}| = |U_{111}| + |U_{121}| + \cdots + |U_{1h_11}|$$
$$|U_{12}| = |U_{112}| + |U_{122}| + \cdots + |U_{1h_12}|$$
$$\cdots .$$

Thus,

$$f(B \cup \{x\}) - f(A \cup \{x\}) = \log_2 \left(\begin{array}{c} |U_{11}| \\ |U_{111}|, \ldots, |U_{1h_11}| \end{array} \right) \cdots \left(\begin{array}{c} |U_{1\ell}| \\ |U_{11\ell}|, \ldots, |U_{1h_1\ell}| \end{array} \right)$$
$$+ \cdots$$

By Lemma 7.4.12,

$$f(B) - f(A) \geq f(B \cup \{x\}) - f(A \cup \{x\}).$$

Hence,

$$f(A \cup \{x\}) - f(A) \geq f(B \cup \{x\}) - f(B).$$

Therefore, f is submodular. $\qquad\square$

Now, consider the greedy Algorithm 40 for the minimum submodular cover problem with non-integer potential function f.

Algorithm 40 Greedy Algorithm for Minimum Submodular Set Cover.

Input: A polymotroid function f over 2^X and a nonnegative cost function c on X.
Output: A subset A of X such that $f(A) = f(X)$.
1: $A \leftarrow \emptyset$;
2: **while** $f(A) < f(X)$ **do**
3:     choose $x \in X \setminus A$ to maximize $\Delta_x f(A)/c(x)$ and
4:     $A \leftarrow A \cup \{x\}$;
5: **end while**
6: **return** $A_G = A$.

Since f may not have integer value, we cannot use Theorem 1.6.4 to analyze this algorithm. The following result is proved in [80].

Theorem 7.4.14 (Du, Ko, Hu [80]) *Let $x_1, x_2, \ldots, x_k$ be the elements in A_g in the order of their selection into A_G. Denote $A_0 = \emptyset$ and $A_i = \{x_1, x_2, \ldots, x_i\}$, for $i = 1, \ldots, k$. Assume $\Delta_{x_i} f(A_{i-1})/c(x_i) \geq 1$ for all $i = 1, 2, \ldots, k$. Then*

$$c(A_G) \leq (1 + \ln \frac{f(A^*)}{c(A^*)}) \cdot c(A^*)$$

where A^* is an optimal solution, i.e.:

Proof Let $a_i = f(A^*) - f(A_i)$ for $1 \leq i \leq k$. Then $a_0 = f(A^*)$ and $\Delta_{x_i} f(A_{i-1}) = a_{i-1} - a_i$.

Suppose $A^* = \{y_1, y_2, \ldots, y_h\}$. Then by the greedy rule, for each $1 \leq j \leq k$, we have

$$\frac{a_{j-1} - a_j}{c(x_j)} \geq \max_{1 \leq i \leq h} \frac{\Delta_{y_i} f(A_{j-1})}{c(y_i)}$$

$$\geq \frac{\sum_{i=1}^{h} \Delta_{y_i} f(A_{j-1})}{c(A^*)}$$

$$\geq \frac{\Delta_{A^*} f(A_{j-1})}{c(A^*)}$$

$$= \frac{f(A^*) - f(A_{j-1})}{C(A^*)}$$

$$= \frac{a_{j-1}}{c(A^*)}.$$

Thus, for $1 \leq j \leq k$

$$\frac{a_{j-1} - a_j}{c(x_j)} \geq \frac{a_{j-1}}{c(A^*)} \tag{7.7}$$

and

$$a_j \leq a_{j-1} \cdot \left(1 - \frac{c(x_j)}{c(A^*)}\right). \tag{7.8}$$

Note that

$$a_0 = f(A^*) = f(A_G) = \sum_{i=1}^{k} \Delta_{x_i} f(A_{i-1}) \geq \sum_{i=1}^{k} c(x_i) = c(A_G) \geq c(A^*)$$

and $a_k = f(A^*) - f(A_G) = 0$. Moreover, since f is monotone nondecreasing, we have $a_i \leq a_{i-1}$ for $1 \leq i \leq k$. Therefore, there exists $0 \leq r \leq k$ such that $a_{r+1} < c(A^*) \leq a_r$. By (7.7),

$$\frac{a_r - a_{r+1}}{c(x_{r+1})} \geq \frac{a_r}{c(A^*)}.$$

Divide the numerator and the denominator of the left-hand side into two parts

$$a' = c(A^*) - a_{r+1}, a'' = a_r - c(A^*),$$
$$a_r - a_{r+1} = a' + a'',$$
$$c(x_{r+1}) = c' + c'',$$

such that

$$\frac{a'}{c'} = \frac{a''}{c''} = \frac{a_r - a_{r+1}}{c(x_{r+1})}.$$

Then

$$\frac{a''}{c''} = \frac{a_r - a_{r+1} - a'}{c''} \geq \frac{a_r}{c(A^*)}.$$

Therefore, by repeatedly using (7.8), we obtain

$$\begin{aligned}
c(A^*) &= a_{r+1} + a' \\
&\leq a_r \left(1 - \frac{c''}{c(A^*)}\right) \\
&\leq a_r \cdot \exp\left(-\frac{c''}{c(A^*)}\right) \\
&\leq \cdots \\
&\leq a_0 \cdot \exp\left(-\frac{c'' + c(x_1) + \cdots + c(x_r)}{c(A^*)}\right),
\end{aligned}$$

since $1 + x \leq e^x$. It follows that

$$c'' + c(x_1) + \cdots + c(x_r) \leq c(A^*) \cdot \ln \frac{a_0}{c(A^*)}.$$

Moreover,

$$\begin{aligned}
c(x_{r+2}) + \cdots + c(x_k) &\leq \Delta_{x_{r+2}} f(A_{r+1}) + \cdots + \Delta_{x_k} f(A_{k-1}) \\
&= f(A_G) - f(A_{r+1}) = a_{r+1}.
\end{aligned}$$

Also, $a'/c' \geq a_r/c(A^*) \geq 1$. Therefore, we have

$$\begin{aligned}
C(A_G) &\leq c(A^*) \cdot \ln \frac{a_0}{c(A^*)} + c' + a_{r+1} \\
&\leq c(A^*) \cdot \ln \frac{a_0}{c(A^*)} + a' + a_{r+1} \\
&= c(A^*)\left(1 + \ln \frac{f(A^*)}{c(A^*)}\right).
\end{aligned}$$

$\square$

Theorem 7.4.15 (Chen et al. [56]) *There exists a greedy algorithm for Problem 7.4.11 with performance ratio* $\ln n + \ln \log_2 n + 1$.

Proof In proof of Lemma 7.4.13, replace B by $A \cup \{x\}$. Note that if $h_i \geq 2$, then

$$\binom{|V_i|}{|V_{i1}|, \ldots, V_{ih_i}} \geq 2.$$

This implies that at each iteration of the greedy algorithm, for selected x, $\Delta_x f(A \cup \{x\}) \geq 1$. Thus, the condition in Theorem 7.4.14 is satisfied.

Furthermore, note that $f(A^*) \leq \log_2 n! \leq n \cdot \log_2 n$. This theorem follows immediately from Theorem 7.4.14.

$\square$

7.5 Effector Detection

There are several methods for solving the rumor discovery problem. However, only one has algorithmic aspect, which is to find whether the information comes from a reliable source. Effector detection can be used for such a purpose.

Consider a social network $G = (V, E)$ with the IC model. Let $V = \{u_1, u_2, \ldots, u_n\}$. An *activation state* $A = (a_1, a_2, \ldots, a_n)$ is a vector in $\{0, 1\}^n$ where

$$a_i = \begin{cases} 1 & \text{if node } i \text{ is active,} \\ 0 & \text{if node } i \text{ is inactive.} \end{cases}$$

For each seed set S, $I(S) = (\alpha_1, \alpha_2, \ldots, \alpha_n)$ is a random vector in $[0, 1]^n$ where α_i represents a probability that node i becomes active at the end of information diffusion process. For a realization g, $I_g(S) = (\beta_1, \beta_2, \ldots, \beta_n)$ is a vector in $\{0, 1\}^n$ where $\beta_i = 1$ represents the fact that node i becomes active at the end of information process under realization g, and otherwise $\beta_i = 0$. For any vector $x \in \{0, 1\}^n$, define $x^1 = \{i \mid x_i = 1\}$ and $x^0 = \{i \mid x_i = 0\}$. The effector detection is the problem defined as follows.

Problem 7.5.1 (*Effector Detection*) Given an activation state A, a positive integer k, and a number $\lambda \in (0, 1)$, find a solution S for the following:

$$\max \quad f(S) = \lambda \cdot \mathbb{E}[|A^1 \cap I_g^1(S)|] + (1 - \lambda) \cdot \mathbb{E}[|A^0 \cap I_g^0(S)|]$$
$$\text{s.t.} \quad |S| = k,$$
$$S \subseteq V.$$

Lemma 7.5.2 (Zhang et al. [488]) *For $\lambda \in (0, 1)$, $f(S)$ is neither submodular nor supermodular.*

Proof Consider a counterexample: Consider a social network $G = (V, E)$ with $V = \{v_1, v_2, v_3\}$, $E = \{(v_1, v_2), (v_2, v_3)\}$, and $p_{12} = p_{23} = 1$. Let $A = (0, 1, 0)$. If $\lambda = 2/3$, then

$$f(\{v_2\}) - f(\emptyset) = (\frac{2}{3} \cdot 1 + \frac{1}{3} \cdot 1) - \frac{1}{3} \cdot 2 = \frac{1}{3},$$
$$f(\{v_2, v_3\}) - f(\{v_3\}) = (\frac{2}{3} \cdot 1 + \frac{1}{3} \cdot 1) - \frac{1}{3} \cdot 1 = \frac{2}{3}.$$

Thus, $f(\{v_2\}) - f(\emptyset) < f(\{v_2, v_3\}) - f(\{v_3\})$. Hence, $f(S)$ is not submodular. Moreover,

$$f(\{v_1, v_2\}) - f(\{v_1\}) = 0 < f(\{v_2\}) - f(\emptyset).$$

Hence, $f(S)$ is not supermodular. $\square$

Lemma 7.5.3 $\mathbb{E}[|A^1 \cap I_g^1(S)|]$ *is submodular.*

Proof Consider subsets $S \subseteq T \subseteq V$ and a node $v \in V \setminus T$. Then

$$|A^1 \cap I_g^1(S \cup \{v\})| - |A^1 \cap I_g^1(S)| = |(A^1 \cap (I_g^1(S \cap \{v\}) - I_g^1(S))|$$
$$|A^1 \cap I_g^1(T \cup \{v\})| - |A^1 \cap I_g^1(T)| = |(A^1 \cap (I_g^1(T \cap \{v\}) - I_g^1(T))|.$$

Since $S \subseteq T$, we have

$$I_g^1(T \cap \{v\}) - I_g^1(T) \subseteq I_g^1(S \cap \{v\}) - I_g^1(S)).$$

Hence,

$$|A^1 \cap I_g^1(T \cup \{v\})| - |A^1 \cap I_g^1(T)| \leq |A^1 \cap I_g^1(S \cup \{v\})| - |A^1 \cap I_g^1(S)|.$$

Therefore, $\mathbb{E}[|A^1 \cap I_g^1(S)|]$ is submodular. $\qquad\square$

Theorem 7.5.4 (Zhang et al. [488]) *Set*

$$f_l(S) = \lambda \cdot \mathbb{E}[|A^1 \cap I_g^1(S)|],$$
$$f_u(S) = \lambda \cdot \mathbb{E}[|A^1 \cap I_g^1(S)|] + |A^0|.$$

Then, both $f_l(S)$ and $f_u(S)$ are submodular, and moreover

$$f_l(S) \leq f(S) \leq f_u(S).$$

Exercises

1. Let $s(u)$ denote the starting time of node u for spreading a rumor. If u is not a rumor source, then $s(u) = \infty$. Let $r_A(u)$ denote the time when node u receives the rumor and where A is the set of rumor sources. Then

$$r_A(u) = \min_{v \in A}[s(v) + dist(v, u)]$$

   where $dist(v, u)$ is the distance from node v to node u.
   Two node subsets A and B are *separable* if there exist x and y such that $r_A(x) - r_A(y) \neq r_B(x) - r_B(y)$. A node x in the set A is *detectable* if A and $A \setminus \{x\}$ are separable. A set A is said to be detectable if every node $x \in A$ is detectable. Show that a node set A is detectable if and only if for every $x \in A$, $s(x) < r_{A \setminus \{x\}}(x)$.
2. Suppose each node u reports an initial time $s(u)$ immediately when it receives the rumor. Show that for any node u, there is unique maximal node subset A such that u is detectable in A.

3. (Set-Resolving Set) Let $\mathcal{F}$ be a collection of detectable sets. A set K is called a set-resolving set (SRS) for $\mathcal{F}$ if for any two detectable A and B in $\mathcal{F}$, K has two elements which separate A and B. Show that for any collection of detectable sets, $\mathcal{F}$, there exists an SRS. (Hint: The node set V of the social network is an SRS.)

4. Show that any subset of a detectable set is detectable.

5. Consider a social network $G = (V, E)$ with the IC model. For a rumor source set R and a protector set S, let $\sigma(S \mid R)$ be the expected number of protected nodes, i.e., nodes that accepted the influence of S. Suppose that each node can select its own priority, that is, each node u can decide to accept the rumor or the positive cascade if they arrive u at the same time. Assume that all protectors send the same information and hence their cascades receive the same priority at each node. Also assume that all rumor sources send out the same rumor and hence receive the same priority at each node. Show that $\sigma(S \mid R)$ is submodular does not matter what is the selection of each node.

6. Zhang et al. [489] design a greedy approximation for the following problem: Given a social network G, a potential initial time function s, and a positive integer q, let

$$\mathcal{F} = \{A \mid A \text{ is a detectable set of } V \text{ and } |A| \leq q\}$$

be the collection of potential rumor source sets, find and SRS K for $\mathcal{F}$ with the smallest cardinality.

7. Tong et al. [393], in a social network with the IC model, consider two classes of cascades, a class of negative cascades and a class of positive cascades. Sources of negative cascades and a subset of protectors (i.e., sources of positive cascades) are fixed and the task is to allocate remaining subclass of protectors. One of the main results in the paper is about the cascade priority. Show that the spread of immunization is monotone nondecreasing and submodular if one of following occurs:
(b1) All positive cascades have priority higher than all negative cascades.
(b2) All negative cascades have priority higher than all positive cascades.
(b3) The cascade priority is strongly uniform, i.e., at every node, the priority ordering is the same.

8. Tong et al. [393] consider a social network $G = (V, E)$ with the IC model. For a rumor source set R and a protector set S, let $\sigma(S \mid R)$ be the expected number of protected nodes. It is possible that different protectors send different positive cascades and receive different priorities at each node. Meanwhile, different rumor sources may send different rumors and hence receive different priorities at a node. Show that, in general, $\sigma(S \mid R)$ is not submodular with respect to S.

9. Tong et al. [393] consider a social network $G = (V, E)$ with the IC model. For a rumor source set R and a protector set S, let $\bar{\sigma}(S \mid R)$ be the expected number of protected nodes under condition that all positive cascades have priority compared with rumor. Let $\underline{\sigma}(S \mid R)$ be the expected number of rumor-effected nodes under condition that all rumor cascades have priority compared with positive cascades.

Show that

$$\underline{\sigma}(S \mid R) \le \sigma(S, R) \le \bar{\sigma}(S \mid R).$$

10. Show that $\alpha_1 - \alpha_2$ is not monotone nondecreasing, where

$$\alpha_1(C) = \sum_{c \in C} \tau(\{c\})(c),$$

$$\alpha_2(C) = \sum_{c,c' \in C : c \ne c'} \tau(\{c, c'\}),$$

and τ is defined by (7.1).

11. Let S_r and S_p be the seed set of rumor and the set of protectors, respectively. Suppose that the positive cascade has higher priority. Show that u is activated by rumor in realization g under S_p if and only if $dist_g(S_r, u) < dist_g(S_p, u)$ and $dist_g(S_r, u) \ne +\infty$ where $dist_g(S_r, u) = \min_{v \in S_r} dist_g(v, u)$.

12. (*opt*-Estimation in [401]) Consider Problem 7.3.3 in Sect. 7.3. Let *opt* be the objective function value of optimal solution. Let opt_k^* be returned from Algorithm 41. Show that the following holds.

Algorithm 41 *opt*-Estimation.

Input: (k, δ, N).
Output: opt_k^*.
1: $\mathcal{R} \leftarrow \emptyset$;
2: $\lambda_3 = \frac{n(2+\delta)\cdot \ln(N \cdot \binom{n}{k}) \cdot \log n}{\delta^2}$;
3: **for** $i = 1$ to $\log(n - 1)$ **do**
4:      $x_i \leftarrow \frac{n}{2^i}, \ell_i \leftarrow \frac{\lambda_3}{x_i}$;
5:      **while** $|\mathcal{R}| \le \ell_3$ **do**
6:          Generate an RR-tuple and insert into $\mathcal{R}$;
7:          Select a node set S' by $(1 - 1/e)$-approximation algorithm for maximizing $F(S, \mathcal{R})$ subject to $|S| \le k$ in Subsection 7.3.3;
8:          **if** $\frac{n \cdot F(S', R)}{\ell_i} \ge (1 + \delta)x_i$ **then**
9:              $opt_k^* \leftarrow \frac{n \cdot F(S', R)}{\ell_i(1+\delta)}$
10:          **end if**
11:      **end while**
12: **end for**
13: **return** opt_k^*.

(a) With probability at least $1 - 2/N$, $opt_k^* \le opt$ where N is predetermined in Algorithm 41.

(b) With probability at least $1 - 1/N$,

$$opt_k^* \ge \frac{(1 - 1/e) \cdot opt}{2(1 + \delta)^2}.$$

(c) With probability at least $1 - 3/N$,

$$opt \geq opt_k^* \geq \frac{(1 - 1/e) \cdot opt}{2(1 + \delta)^2}. \tag{7.9}$$

13. (Randomized Algorithm in [401]) By Lemma 7.3.2, $\frac{n \cdot F(S, \mathcal{R})}{|\mathcal{R}|}$ is an unbiased estimate of $f(S)$ as $|\mathcal{R}|$ is sufficiently large. As described in Algorithm 42, given adjustable parameters $N > 0, \delta_1 > 0, \delta_2 > (1 - 1/e)\delta_1$ and $\delta_3 > 0$, let opt_k^* be obtained by Algorithm 41 with input (k, δ_3, N). Set

$$\ell_1 = \frac{2n \cdot \ln N}{\delta_1^2 \cdot opt_k^*}$$

$$\ell_2 = \frac{(2 + \delta_2 - (1 - 1/e)\delta_1)n \cdot \ln(N \binom{n}{k}))}{(\delta_2 - (1 - 1/e)\delta_1)^2 \cdot opt_k^*},$$

and $\ell^* = \max(\ell_1, \ell_2)$.

Algorithm 42 Randomized Algorithm for Rumor Blocking [401].

Input: $G, S_r, k, N, \delta_1, \delta_2, \delta_3$.
Output: S'.
1: Compute opt_k^* by Algorithm 41 on input (k, δ_3, N).
2: $\ell_1 = \frac{2n \cdot \ln N}{\delta_1^2 \cdot opt_k^*}$;
3: $\ell_2 = \frac{(2 + \delta_2 - (1 - 1/e)\delta_1)n \cdot \ln(N \binom{n}{k}))}{(\delta_2 - (1 - 1/e)\delta_1)^2 \cdot opt_k^*}$;
4: $\ell^* = \max(\ell_1, \ell_2)$;
5: Generate ℓ^* RR-tuples $\mathcal{R}_{\ell^*} = \{\mathcal{T}_1, ..., \mathcal{T}_{\ell^*}\}$ by Algorithm 39;
6: Compute a $(1 - 1/e)$-approximation S' for maximization of $F(S, \mathcal{R}_{\ell^*})$ subject to $|S| \leq k$;
7: **return** S'.

Let S_k be an optimal solution of maximizing $F(S, \mathcal{R}_{\ell^*})$ subject to $|S| \leq k$. Let S' be returned from Algorithm 42. Please prove the following.

(a) Suppose (7.9) holds. If $\ell^* > \ell_1$, then with probability at least $1 - 1/N$,

$$\frac{n}{\ell^*} \cdot F(S_k, \mathcal{R}_{\ell^*} \geq (1 - \delta_1) \cdot opt.$$

(b) Suppose (7.9) holds. If $\ell^* > \ell_2$, then with probability at least $1 - 1/N$,

$$F(S', \mathcal{R}_{\ell^*}) \leq (\delta_2 - (1 - 1/e)\delta_1) \cdot \frac{\ell^*}{n} \cdot opt.$$

(c) With probability at least $1 - 5/N$,

$$f(S') \geq (1 - 1/e - \delta_2) \cdot opt.$$

14. (Running Time Analysis in [401]) Let $TIME$ be the expected running time of Algorithm 39. Show the following.

   (a) Let opt_1 be the opt when $k = 1$. Then

   $$TIME \leq \frac{m}{n} \cdot opt_1.$$

   (b) Algorithm 42 runs in $O(\frac{km \log n}{\delta_2^2})$ time.

15. Please find a proof for Theorem 7.3.6.
16. [488] Show that $\mathbb{E}[|A^0 \cap I_g^0(S)|]$ is supermodular where notations are defined in Sect. 7.5.
17. (Causal-related rumors [467]) When users believe one rumor, the probability of being influenced by another causal-related rumor is greater. To describe this scenario, Yao et al. [467] propose the Causal Rumors Enhance Cascade (CREC) model. In this model, two rumors A^* and B^* are considered. After a node u gets influenced by A^*, the probability that u gets influenced by B^* is increased. Given two rumor seed sets R_1 and R_2 for A^* and B^*, respectively, and a positive integer k, find k protector locations to minimize the expected number of nodes influenced by rumor B^*. Please take the IC model as a base to give the detailed description of the CREC model and show the following:

   (a) This problem is NP-hard.
   (b) The objective function is not submodular.

Historical Notes

Budak et al. [33] presented the first research work on misinformation (or rumor). They worked with the multi-campaign independent cascade model. He et al. [188] and Fan et al. [103] followed up to study the problem with competitive linear threshold model and the OPOAO model, respectively. Later, this problem became more popular and several variations appear in the literature [245, 333, 483].

   Over all, there are three research issues about rumor:

- Rumor identification: Identify whether an information is rumor or not.
- Rumor blocking: Reduce the influence of rumor cascade.
- Rumor source detection: Find the location of rumor source.

For rumor identification, several methods are designed for special types of information. For example, Leskovec et al. [236] describe a method to track the spread of misinformation in the new cycle; Ratkiewicz et al. [345] present a machine learning framework to detect political misinformation; Qazvinian et al. [336] introduce a content-based method to work on misinformation in microblogs; and Takahashi et al. [378] design a system for rumor identification in Twitter.

A general and efficient way is to clarify whether the information source is reliable or not. Since at this stage, it is unknown whether the information is a rumor or not, finding the source is called as effector detection. Lappas et al. [229] are among the first who study the effector detection. They design a dynamic program solution for tree networks. Tong et al. [394] established several mathematical formulations for this problem and designed approximation solutions for some of them. Zhang et al. [488] proposed two formulations, a unconstrained optimization and a constrained optimization. They indicated that both problems are nonsubmodular and have to be solved by methods in Chaps. 5 and 6.

The rumor is an important research subject in study of social networks since its spread can make a lot of negative effects. For example, a rumor on earthquake will cause people's panic and a rumor on a political leader's health will cause a shaking of stock market. Therefore, there exist many publications in the literature, which proposed many methods to block the spread of rumor. For rumor blocking, there are three popular methods:

- Node-cut.
- Link-cut.
- Protector placement.

Kimura et al. [216] propose a method to cut a limited number of links for rumor blocking. Khali et al. [212] study an edge deletion problem for controlling the spread of flu. Tong et al. [403] and Yan et al. [451] also study rumor blocking with link deletions. Wang et al. [423] and Gai et al. [122] block rumor with node-cut.

There are many efforts that are made on protector placement. Fan et al. [103] assume to know a community structure and the location of rumor source. They place protectors to resist the rumor from the community, where the source is located, to spread to other communities. Zhu et al. [495, 498] assume that the location of rumor sources is completely unknown, Tong et al. [401] design a randomized approximation algorithm for protector placement using reverse influence sampling technique. Tong et al. [401] present a randomized $(1 - 1/e - \varepsilon)$-approximation by using reverse sampling. Guo et al. [161] study multiple feature rumor blocking with protector placement. Yao et al. [466] consider causal-related rumors.

For rumor source detection, there are two types methods. The first one is to use certain type of center as source estimator. The second one is to study the monitor placement. The single rumor source detection is first studied by Shah and Zaman [359]. They propose the maximum likelihood estimator (MLE) and establish a method to determine the MLE in regular tree under the epidemic SI model. Zhu and Ying [500, 501] propose a Jordan infection center as source estimator. Luo and Ying [277] worked under the SIS model. Chen et al. [57] and Luo et al. [278] study the multiple sources detection. Other centrality-based source detection can be found in [261, 446]. For monitor placement, one can find research works in [56, 202, 489].

Finally, it may be worth mentioning two surveys related to rumor or misinformation [203, 223], which may give you more information and references.

Chapter 8
Competitive Influence

"Competition is always a good thing. It forces us to do our best. A monopoly renders people complacent and satisfied with mediocrity."
— Nancy Pearcy

When many innovations face active opposition spreading by word of mouth, or multiple companies with comparable products attempt for sales with competing word-of-mouth cascades, competitive influence occurs. This chapter is contributed to the study of problems regarding this type of influence.

8.1 Community Expansion

In the situation that competitive influence occurs, from view of one company, how to expand its business. This is the background of the community expansion problem.

Consider the business environment as a social network. Each community consists of all persons adopting the products from the same company. Given a community, called the *target community (TC)*, the community expansion problem is to maximize expansion of TC by allocating seeds (discount or free samples) within certain budget [15].

In some sense, the community expansion can be seen as a generalization of the rumor blocking in the situation that multiple rumors appear.

Bi et al. [16, 17] gave a solution of the community expansion problem by using electric charge system.

First, the features in social networks are transformed into the characters in the physical charged system. In the physical charge system, every charge point i has an electric field which acts on other charged objects in the field to force them moving closer or further to i. Let i and j be two charge points. Let q_i and q_j be electric

W. Wu et al., *Computational Aspects of Social Networks*, Springer Optimization and Its Applications 234, https://doi.org/10.1007/978-3-032-14833-9_8

quantities of i and j, respectively. Then, the electric force between i and j can be computed by the following formula (Halliday et al. [178]):

$$\mathbf{F}_{ij} = h \cdot \frac{q_i q_j}{r_{ij}^2}$$

where h is the Coulomb constant, r_{ij} is the distance between i and j, and $\mathbf{F}_{ij}$ is a vector quantity, which has direction appearing attractive if the charges are of opposite sign and repulsive, otherwise. For a set of charges, the resultant force putting on a test point equals the vector sum of forces exerted by each individual charges.

Let us see an individual in a social network as a charged particle. Consider an individual who may not belong to any community, or belong to several communities in the social network, in this case, that communities are overlapping. The quantity of a particle q can be considered as the attractiveness to a community C when C wants to expand. Actually, in real world, people known by more people have much more influence toward others. Let us define the degree of the node as its charged quantity, i.e., $n_i = q_i$.

When n_i is larger, this means that the person i has more friends, i.e., the electric quantity of his/her corresponding electric particle is larger. Thus, this particle appears more attractive to a community and it has more opportunity to have impact on other particles. That is, in the marketing strategy, these attractive celebrities would obtain more attentions.

Recall that TC is a community that one wants to expand its size. TC can be treated as a large electric particle consisting of all the nodes in the TC. To determine the electrical quantity of TC, we should consider not only the number of nodes in TC but also the connection density inside the TC, i.e.:

$$Q_{TC} = N_{TC} \times Den_{TC}$$

where Q_{TC} is the electrical quantity of TC, N_{TC} is the number of nodes in TC, and Den_{TC} is the connection density inside TC. Actually, a community which has more members and tighter density connection looks like more powerful to nodes outside TC. The electrical polarity of an TC is set as positive while the electrical particle outside TC is set as negative. Therefore, TC has attractive force to particles outside TC.

The distance r_{ij} between two particles i and j is the shortest path length between them. But, the distance between a particle i and the community C, r_{iC} is not simply the shortest path length between i and node set C. It is defined as follows:

$$r_{iC} = \frac{dist(i, C)}{\log(m_i + 1) + 1}$$

where $dist(i, C)$ denotes the shortest path length from i to C and m_i is the number of i's neighbors lying in the community C. In [9], the experiments indicate that the main feature to affect an individual to join a community is the number of friends in a community. Moreover, the relation between the probability of joining a community and the number of friends in that community is under the "law of diminishing returns." Thus, $log(m + 1)$ can be used for indicating that individual who is more likely to join in C appears more closer to C. To make the function keep physical meaning when $m_i = 0$, one is added in two places.

Since particles which do not belong to TC are seen as negative, the attractive force from TC to a certain particle i can be computed by the following formula:

$$F_{TC \to i} = h \cdot \frac{Q_{TC} \cdot n_i}{r_{iTC}^2}.$$

Next, community expansion will be explained by the electric field theory.

Initial State. The snapshot of social network at a given time point can be seen as the initial state. As shown in Fig. 8.1, for a node i, it receives the electric force $F_{TC \to i}$ exerted by TC, and is balanced with forces exerted by other communities.

Promotion State. To attract new members for TC, TC should add additional force to make nodes outside TC move toward TC. This additional force can be generated by distributing promotion discount to new members. Suppose the promotion budget allows TC to select K nodes to receive discount. Let ζ denote the collection of all node subsets of K nodes outside TC. Then, node set S should be selected from ζ to maximize the following function:

$$f_1(S) = \sum_{i \in S} F_{TC \to i} = \sum_{i \in S} h \cdot \frac{q_i \cdot Q_{TC}}{r_{iTC}^2}.$$

Remark: This state is to increase TC by promotion.

Expansion State. In promotion state, some nodes move to TC. Their moving will affect their neighbors, i.e., produce the benefit. Suppose once a vertex outside TC

Fig. 8.1 Initial state

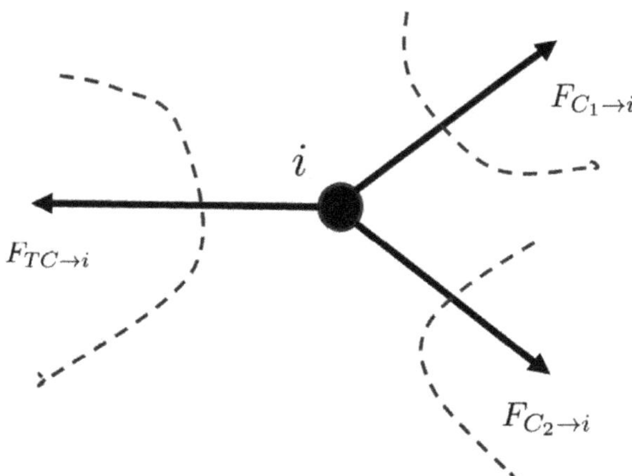

decides to join in TC, its polarity becomes positive and hence it will exert attractive force to its neighbors not in TC. Then, the benefit can be computed by the following function:

$$f_2(S) = \sum_{i \in S} \sum_{j \in W_j} (F^*_{TC \to j} - F_{TC \to j})$$

where W_i denote the set of i's neighbors not in TC and $F^*_{TC \to j}$ is the attractive force from TC to j.

Now, based on the charged system, the community expansion problem can be formulated as a multi-objective optimization problem as follows: Given a social graph $G = (V, E)$, a target community TC, and a positive integer K, find a set S of K nodes outside TC, to maximize $f(S)$ and $f_2(S)$.

To combine two objectives into one, a popular method is to consider the objective function

$$f(S) = \alpha f_1(S) + (1 - \alpha) f_2(S)$$

where α is a constant in $(0, 1)$, which can be adjusted based on which one should be considered more, $f_1(S)$ or $f_2(S)$.

8.2   Business Expansion in Competitive Environment

Consider IC model. In order to extend the community expansion problem to the social network with IC model, the IC model has to be extended to fit the competitive environment.

Definition 8.2.1 (*Competitive IC Model [14]*) IC model can be extended to multiple players as follows:

- Each of p players selects a color and a set S_i of at most k_i nodes.
- A node activated by players with the same color will take the color.
- A node activated by players with multi-color will take the color of one of the players uniformly at random.
- No node with a color can take another color.
- Process ends until no new activation occurs.

Let $\sigma(S_1, S_2, \ldots, S_p)$ denote the expected number of colored nodes if seed sets S_j are selected by players, respectively. Let $\sigma_i(S_1, S_2, \ldots, S_p)$ denote the expected number of nodes in color i.

Lemma 8.2.2 $\sigma(S_1, S_2, \ldots, S_p)$ *is a monotone nondecreasing and submodular function with respect to* $S = \cup_{j=1}^{p} S_j$.

Proof When consider function $\sigma(S_1, S_2, \ldots, S_p)$ with respect to S, the color does not make significant impact. Therefore, the monotone nondecreasing and submodular properties follow from Lemma 2.1.6, that is, the influence spread is monotone nondecreasing and submodular in social networks with the IC model. $\qquad\square$

Lemma 8.2.3 *If seed sets S_j for $j \neq i$ are fixed, then player i's payoff $\sigma_i(S_1, S_2, \ldots, S_p)$ is a monotone nondecreasing and submodular function with respect to S_i.*

Proof First, consider a special case of the IC model, the deterministic model. In the deterministic model, if a node u is in color i at the end of diffusion process, then the distance from u to S_i is no more than the distance from u to S_j for any $j \neq i$. Moreover, if the distance from u to S_i is shorter than the distance from u to S_j for any $j \neq i$, then u will be in color i at the end of process. If the distance from u to S_i is equal to or shorter than the distance from u to S_j for any $j \neq i$, and, in addition, there are q j's such that the distance from u to S_i is equal to the distance from u to S_j, then u will be in color i with probability $1/q$.

Consider seed set $S_i \subset S_i'$ for player i. From above necessary condition and sufficient conditions for a node to be in color i, it is easy to see that

$$\sigma_i(S_1, \ldots, S_{i-1}, S_i, S_{i+1}, \ldots, S_p) \leq \sigma_i(S_1, \ldots, S_{i-1}, S_i', S_{i+1}, \ldots, S_p).$$

Now, suppose v is a seed for player i such that $x \notin S_i'$. Then,

$$\sigma_i(S_1, \ldots, S_{i-1}, S_i \cup \{v\}, S_{i+1}, \ldots, S_p) - \sigma_i(S_1, \ldots, S_{i-1}, S_i, S_{i+1}, \ldots, S_p)$$
$$\geq \sigma_i(S_1, \ldots, S_{i-1}, S_i' \cup \{v\}, S_{i+1}, \ldots, S_p) - \sigma_i(S_1, \ldots, S_{i-1}, S_i', S_{i+1}, \ldots, S_p).$$

In fact, as shown in Fig. 8.2, marginal nodes in color i obtained by adding seed v to S_i' must also be marginal by adding seed v to S_i.

For general IC model, the lemma can be seen from the following formula:

$$\sigma_i(S_1, S_2, \ldots, S_p) = \sum_{H \in \mathcal{G}} P[H]\sigma_i^H(S_1, S_2, \ldots, S_p)$$

where H consists of all alive edges, realized with probability $P[H]$,

$$\mathrm{Pr}[H] = \left(\prod_{(u,v):\text{alive}} p_{uv} \right) \left(\prod_{(u,v):\text{blocked}} (1 - p_{uv}) \right).$$

Moreover, $\mathcal{G}$ is the set of all realizations, and $\sigma_i^H(S_1, S_2, \ldots, S_p)$ is the expected number of nodes in color i in subgraph H under deterministic model. $\qquad\square$

Similar to community expansion, we may study the following problem.

Fig. 8.2 Submodularity

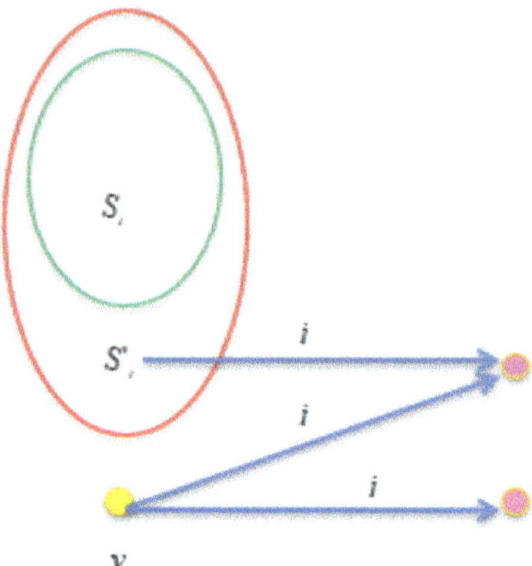

Problem 8.2.4 (*Business Expansion*) Let i be a target player. Consider the following process and raised the expansion problem.

- **Initial State**. The snapshot of information process at a given time point can be seen as the initial state.
- **Promotion State**. To give more promotion power for player i, add k seeds for player i at the initial state. Suppose any seed can activate a node to have color i. Then, there is no optimization in this state.
- **Expansion State**. After adding k seeds, player i obtains more power to expand his influence. How to allocate these k seeds to maximize such influence? This is the problem of the business expansion.

Actually, in the business expansion, it studies the problem: How to introduce a new product into an market where a competing product is also being introduced? There are two assumptions:

- A company can hide itself from a competitor until the moment of introduction.
- Every company has a fixed budget for targeting consumers and knows early adopters of its competitor.

Theorem 8.2.5 *The business expansion is a monotone nondecreasing submodular maximization problem.*

Proof Similar to the proof of Lemma 8.2.3, we may first consider the deterministic information diffusion model. Suppose that at the t's step of diffusion process, additional k seeds are given to the player i for increasing his power. Let $\hat{S}_j$ denote the set of nodes in color j at this step. Then, $\sigma_i(\hat{S}_1, \ldots, \hat{S}_{i-1}, \hat{S}_i + A, \hat{S}_{i+1}, \ldots, \hat{S}_p)$ is monotone nondecreasing and submodular with respect to $\hat{S}_i + A$, and hence with respect to A when $\hat{S}_j$ is fixed for all j.

By the same argument as in the end of the proof of Lemma 8.2.3, the conclusion can be extended from the deterministic model to the general IC model. $\square$

Based on Theorem 8.2.5, a reverse sampling algorithm can be designed to produce a good approximation solution for the business expansion problem.

8.3 Non-cooperative Game

The competitive IC model is a special case of non-cooperative game. In general, a non-cooperative game can be described as follows.

Consider a set of non-cooperative agents. They have action spaces which are subsets of an underlying ground set. Each agent seeks to maximize own private utility function. However, the actions of the agents induce some social utility function and their equilibrium reaches certain level of maximization of social utility.

To establish mathematical formulation, let $V_1, \ldots, V_k$ be disjoint ground sets of k agents. Each element in V_i is called an *act* of agent i. Each *action* a_i of agent i is a subset of V_i. The action space $\mathcal{A}_i$ is the set of all possible actions of agent i (Fig. 8.3).

A *pure strategy* is a specific action which the agent decides to carry out. A *mixed strategy* is the one in which the agent decides upon an action according to a probabilistic distribution.

Let $\mathcal{S}_i$ denote the set of mixed strategies of agent i. Denote

$$\mathcal{A} = \mathcal{A}_1 \times \cdots \mathcal{A}_k$$
$$\mathcal{S} = \mathcal{S}_1 \times \cdots \mathcal{S}_k$$
$$V = V_1 \cup \cdots \cup V_k.$$

Fig. 8.3 Ground set, acts, actions, and action space

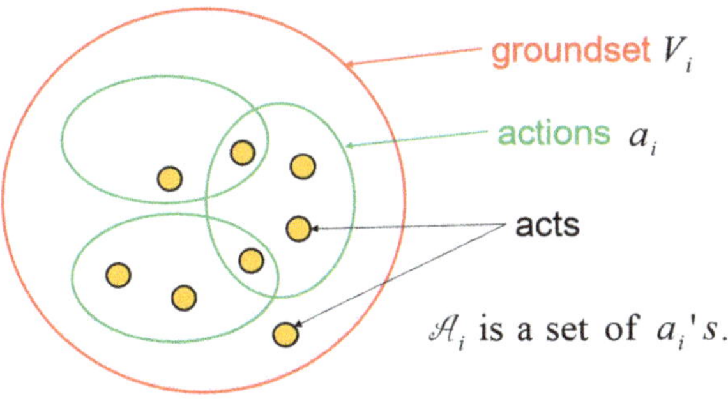

For $f : 2^V \to R_+$, define $\bar{f} : \mathcal{S} \to R$ by

$$\bar{f}(S) = \sum_{A \in \mathcal{A}} f(A)\Pr[A \mid S]$$

where for $A = (a_1, \ldots, a_k) \in \mathcal{A}$,

$$f(A) = f(a_1 \cup \cdots \cup a_k).$$

For $A = (a_1, \ldots, a_k) \in \mathcal{A}$, define

$$A \odot (i, a_i') = (a_1, \ldots, a_{i-1}, a_i', a_{i+1}, \ldots, a_k).$$

Similarly, for $S = (s_1, \ldots, s_k) \in \mathcal{S}$, define

$$S \odot (i, s_i') = (s_1, \ldots, s_{i-1}, s_i', s_{i+1}, \ldots, s_k).$$

Let $\gamma : 2^V \to R$ be the social utility function and $\alpha_i : 2^V \to R$ the private utility function of agent i, satisfying

$$\gamma(\emptyset) = \alpha_i(\emptyset) = 0$$

where $\emptyset = (\emptyset_1, \ldots, \emptyset_k)$ and $\emptyset_i$ means that agent i declines to take part in the game.

Definition 8.3.1 (*Nash Equilibrium*) $S \in \mathcal{S}$ is a Nash equilibrium if for any $s_i' \in \mathcal{S}_i$,

$$\bar{\alpha}_i \geq \bar{\alpha}_i(S \odot (i, s_i')).$$

A Nash equilibrium $S \in \mathcal{S}$ is pure if every strategy in S is pure. Otherwise, it is called a mixed-strategy Nash equilibrium.

In 1951, Nash established the following:

Theorem 8.3.2 (Nash, 1951) *Any finite k-person, non-cooperative game has at least one Nash equilibrium.*

The utility functions γ and α_i are required to satisfy certain condition in a utility system.

Definition 8.3.3 (*Utility System*) Consider submodular social utility function $\gamma : 2^V \to R$ and submodular private utility function $\alpha_i : 2^V \to R$. The system $(\gamma, \cup_i \alpha_i)$ is called a utility system if for any i,

$$\bar{\alpha}_i(S) \geq \bar{\gamma}(S) - \bar{\gamma}(S \odot (i, \emptyset_i))$$

where $\emptyset_i$ is the null strategy (action) for agent i.

There are several special types of utility systems.

Definition 8.3.4 (*Valid Utility System*) Consider submodular social utility function $\gamma : 2^V \to R$ and submodular private utility function $\alpha_i : 2^V \to R$. The utility system $(\gamma, \cup_i \alpha_i)$ is valid if

$$\sum_i \bar{\alpha}_i(S) \le \bar{\gamma}(S).$$

Definition 8.3.5 (*Basic Utility System*) Consider submodular social utility function $\gamma : 2^V \to R$ and submodular private utility function $\alpha_i : 2^V \to R$. The utility system $(\gamma, \cup_i \alpha_i)$ is basic if

$$\bar{\alpha}_i(S) = \bar{\gamma}(S) - \bar{\gamma}(S \odot (i, \emptyset_i)).$$

Theorem 8.3.6 *Every basic utility system* $(\gamma, \cup \alpha_i)$ *is valid.*

Proof Denote

$$\Delta_{s_i'} f(S) = f(S \odot (i, s_i')) - f(S).$$

Then,

$$
\begin{aligned}
\bar{\gamma}(S) =& \bar{\gamma}(s_1, \emptyset_2, \ldots, \emptyset_k) - \bar{\gamma}(\emptyset) \\
& + \bar{\gamma}(s_1, s_2, \emptyset_3, \ldots, \emptyset_k) - \bar{\gamma}(s_1, \emptyset_2, \ldots, \emptyset_k) \\
& + \cdots + \bar{\gamma}(s_1, \ldots, s_k) - \bar{\gamma}(s_1, \ldots, s_{k-1}, \emptyset_k) \\
=& \sum_{i=1}^{k} \Delta_{s_i} \bar{\gamma}(s_1, \ldots, s_{i-1}, \emptyset_i, \ldots, \emptyset_k) \\
=& \sum_{i=1}^{k} \Delta_{s_i} \bar{\gamma}((s_1, \ldots, s_i, \emptyset_{i+1}, \ldots, \emptyset_k) \odot (i, \emptyset_i)) \\
\ge& \sum_{i=1}^{k} \Delta_{s_i} \bar{\gamma}(S \odot (i, \emptyset_i)) \quad \text{(due to submodularity)} \\
=& \sum_{i=1}^{k} \bar{\alpha}_i(S),
\end{aligned}
$$

since the utility system is basic. $\square$

Theorem 8.3.7 *If* $(\gamma, \cup_i \alpha_i)$ *is a valid and finite utility system, there exists pure Nash equilibrium in the game.*

Proof This pure equilibrium can be obtained by the following algorithm.

$a \leftarrow (\emptyset_1, \ldots, \emptyset_k)$
while there exists $a_i' \in A_i$ such that
   $\alpha_i(a) < \alpha_i(a \odot (i, a_i'))$
do
    $a \leftarrow a \odot (i, a_i')$
end-while
return a.

Since this system is finite, the algorithm must terminate within finitely many steps. At the end, a pure Nash equilibrium is obtained. $\qquad\square$

Theorem 8.3.8 (Vetta [410]) *Suppose that social utility function γ is monotone nondecreasing submodular. Let a be a pure Nash equilibrium. Then*

$$\gamma(a) \geq \frac{1}{2} \cdot \gamma(\omega)$$

where ω is a full action such that $\gamma(\omega) = \max_{p'} \bar{\gamma}(p')$.

Proof Let $a = (a_1, \ldots, a_k)$ and $\omega = (\omega_1, \ldots, \omega_k)$. Denote $w_i = (\omega_1, \ldots, \omega_i, \emptyset, \ldots, \emptyset)$. Then,

$$\gamma(a \cup \omega) - \gamma(a) = \sum_{i=1}^{k}[\gamma(a \cup w_i) - \gamma(a \cup w_{i-1})].$$

Due to submodularity and monotonicity of γ,

$$
\begin{aligned}
&\gamma(a \cup w_i) - \gamma(a \cup w_{i-1}) \\
&\leq \gamma(a \cup (\emptyset, \ldots, \emptyset, \omega_i, \emptyset, \ldots, \emptyset)) - \gamma(a) \\
&\leq \gamma(a \odot (i, \emptyset) \cup (\emptyset, \ldots, \emptyset, \omega_i, \emptyset, \ldots, \emptyset)) - \gamma(a \odot (i, \emptyset)) \\
&= \gamma(a \odot (i, \omega_i)) - \gamma(a \odot (i, \emptyset)) \\
&\leq \alpha_i(a \odot (i, \omega_i)) \\
&\leq \alpha_i(a).
\end{aligned}
$$

Thus,

$$\gamma(\omega) - \gamma(a) \leq \gamma(a \cup \omega) - \gamma(a) \leq \sum_{i=1}^{k} \alpha_i(a) \leq \gamma(a).$$

Therefore,

$$\gamma(a) \geq \frac{1}{2} \cdot \gamma(\omega).$$

$\qquad\square$

Theorem 8.3.9 (Vetta [410]) *Consider monotone nondecreasing submodular social utility function γ. If each agent adopts $(1 - e^{-1})$-approximation response, then the approximate Nash equilibrium a would satisfy that for any pure strategy a' and agent i,*

$$\alpha_i(a) \geq (1 - e^{-1}) \cdot \alpha_i(a \odot (i, a_i')).$$

Moreover, we have

$$\gamma(a) \geq \frac{e - 1}{2e - 1} \cdot \gamma(\omega)$$

where ω is a full action such that $\gamma(\omega) = \max_{p'} \bar{\gamma}(p')$.

Proof Let $a = (a_1, \ldots, a_k)$ and $\omega = (\omega_1, \ldots, \omega_k)$. Denote $w_i = (\omega_1, \ldots, \omega_i, \emptyset, \ldots, \emptyset)$. Then,

$$\gamma(a \cup \omega) - \gamma(a) = \sum_{i=1}^{k} [\gamma(a \cup w_i) - \gamma(a \cup w_{i-1}).$$

Due to submodularity and monotonicity of γ,

$$\begin{aligned}
&\gamma(a \cup w_i) - \gamma(a \cup w_{i-1}) \\
&\leq \gamma(a \cup (\emptyset, \ldots, \emptyset, \omega_i, \emptyset, \ldots, \emptyset)) - \gamma(a) \\
&\leq \gamma(a \odot (i, \emptyset) \cup (\emptyset, \ldots, \emptyset, \omega_i, \emptyset, \ldots, \emptyset)) - \gamma(a \odot (i, \emptyset)) \\
&= \gamma(a \odot (i, \omega_i)) - \gamma(a \odot (i, \emptyset)) \\
&\leq \alpha_i(a \odot (i, \omega_i)) \\
&\leq (1 - e^{-1})^{-1} \alpha_i(a).
\end{aligned}$$

Thus,

$$\gamma(\omega) - \gamma(a) \leq \gamma(a \cup \omega) - \gamma(a) \leq (1 - e^{-1})^{-1} \sum_{i=1}^{k} \alpha_i(a) \leq (1 - e^{-1})^{-1} \gamma(a).$$

Therefore,

$$\gamma(a) \geq \frac{e - 1}{2e - 1} \cdot \gamma(\omega).$$

$\square$

Now, let us look at how to put competitive influence into non-cooperative game model.

- Each node is an act.
- Each action for player i is a subset of k_i nodes.

- Private utility function $\alpha_i = \sigma_i$.
- Social utility function $\gamma = \sigma$.

Lemma 8.3.10 *The above game formulated by influence under competitive IC model is a valid utility system. In fact,*

$$\sigma_i(s_1, \ldots, s_k) \geq \sigma(s_1, \ldots, s_k) - \sigma_i(s_1, \ldots, s_{i-1}, \emptyset, s_{i+1}, \ldots, s_k) \tag{8.1}$$

and

$$\sigma(s_1, \ldots, s_k) = \sum_{i=1}^{k} \sigma_i(s_1, \ldots, s_k). \tag{8.2}$$

Proof It is easy to see that (8.1) and (8.2) hold under deterministic information diffusion model. This can be extended to the extension of IC model since the extension of IC model can be represented by linear combination of deterministic models. $\qquad\square$

Next, consider the following problem.

Definition 8.3.11 (*Competitive Influence Maximization*) Consider a social network $G = (V, E)$ with competitive information diffusion model, find seed allocation strategy to maximize the social utility.

Since σ and σ_i are monotone nondecreasing submodular, the following holds.

Theorem 8.3.12 *In competitive influence, Nash equilibrium gives a 2-approximation for maximization of social utility σ under competitive IC model.*

8.4 Distributed Rumor Blocking

Another example of non-cooperative is given in this section.

Suppose there are several agents who want to block the rumor. However, the network is too large to make cooperation. What guaranteed performance can be reached?

First, introduce the following notations:

- C_r: the cascade of the rumor with fixed seed set a_r.
- C_i: the cascade of agent i selected seed set a_i.
- A_i: the action space of agent i, i.e.:

$$A_i = \{a_i \subseteq V \mid |a_i| = b_i\}.$$

- a: a full action, i.e.:

$$a = (a_1, \ldots, a_k) \in A_1 \times \cdots \times A_k.$$

- $\gamma(a)$: the expected number of non-rumored nodes under full action a.
- σ_i: the expected number of C_i-influenced nodes under the full action a.

Definition 8.4.1 (*Cascade Priority*) The rumor has the highest priority. All agents have equal priority.

Next, show some properties.

Lemma 8.4.2 *Social utility function γ is monotone nondecreasing submodular.*

Lemma 8.4.3 *Every private utility function σ_i is monotone nondecreasing submodular.*

Lemma 8.4.4 *For any full action a and agent i,*

$$\sigma_i(a) \geq \gamma(a) - \gamma(a \odot (i, \emptyset)).$$

Proof This inequality holds since, if cascade C_i does not exist, then those C_i-influenced nodes may be influenced by other cascades. Therefore, only a smaller portion is left for the difference. $\square$

Lemma 8.4.5 *For any full action a,*

$$\gamma(a) \geq \sigma_1(a) + \cdots + \sigma_k(a).$$

Proof As shown in Fig. 8.4, the set of non-rumored nodes contains all C_i-influenced nodes for every $i = 1, 2, \ldots, k$, and nodes which are not reached by any cascade. $\square$

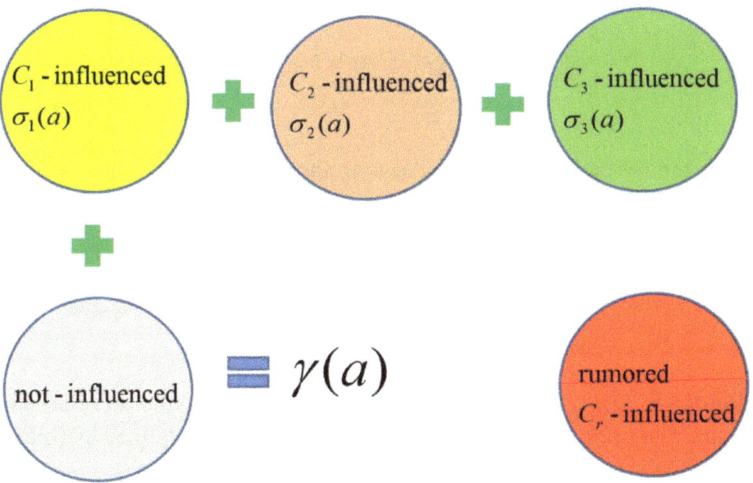

Fig. 8.4 The proof of Lemma 8.4.5

Above two lemmas indicate the following.

Theorem 8.4.6 $(\gamma, \cup_i \sigma_i)$ *is a valid utility system.*

proof. $\square$

Corollary 8.4.7 *Every pure Nash equilibrium a in system* $(\gamma, \cup_i \sigma_i)$ *gives a* $\frac{1}{2}$-*approximation solution for the maximization of social utility* γ.

8.5 Price of Anarchy

The meaning of anarchy has made a tight relationship with the non-cooperative game. In fact, to understand what is the price of anarchy, we have to learn a related non-cooperative game model. The price of anarchy will be defined after setting up this model.

Consider p players on a social network $G = (V, E)$. Each player i is a company. Its strategy is a multi-set M^i of nodes. Denote by $M = (M^1, \ldots, M^p)$ the strategy vector for all players and by α_v^i the number of times that node v appears in M^i. These strategies are required to satisfy the following constraints:

$$|M^i| \leq B^i$$

$$\alpha_v^i \leq K_v^i$$

where $K_v^i \leq B^i$ are their budgets.

To set up the game, all players allocate their budgets to the nodes of G and assign a weight $w_v \geq 0$ to every node v. w_v measures the importance of node v. The payoff function of player i is

$$\sigma^i(M) = \mathbb{E}[\sum_{v \in S_\ell^i} w_v]$$

where S_ℓ^i is the set of i-adopters at step ℓ of adoption process. The social utility is defined by

$$\gamma(S_0) = \sum_{i=1}^{p} \sigma^i(M)$$

where $S_0 = (S_0^1, \ldots, S_0^p)$.

Definition 8.5.1 (*The Price of Anarchy*) Let EQ_{pure} denote the set of all pure Nash equilibrium. The price of anarchy pure Nash equilibrium is defined as follows:

$$\text{Pure Price of Anarchy} = \max_{M \in \text{EQ}_{\text{pure}}} \frac{\gamma(OPT)}{\gamma(M)}$$

where OPT is a strategy to reach the maximum social utility. Let EQ_{mixed} denote the set of all mixed Nash equilibrium. The price of anarchy mixed Nash equilibrium is defined as follows:

$$\text{Mixed Price of Anarchy} = \max_{M \in \text{EQ}_{\text{mixed}}} \frac{\gamma(OPT)}{\gamma(M)}.$$

Consider *general adoption model* as follows.

- Each node has $p + 1$ states, inactive and i-active for $i = 1, \ldots, p$. By i-active, it means that the node is an adopter of product of player i. Diffusion process works in two stages.
- *Seeding Stage*: Based on strategy M, determine the initial state of each node v. If $Z_v = \sum_{i=1}^{p} \alpha_v^i = 0$, then the initial state of v is inactive. Otherwise, the initial state of v is i-active with probability α_v^i / Z_v.
- *Diffusion Stage*: An update schedule, i.e., a node sequence $\{v_1, \ldots, v_\ell\}$ of length ℓ, is given in advance. Each node may appear in the sequence multiple times. At step t, consider node v_t. Let S_{t-1}^i be the set of i-active nodes at the end of previous step. (S_0^i is the seed set of player i.) If v_t is i-active, then keep v_i in i-active state. In fact, no active node is allowed to change its state, i.e., the influence process is *progressive*. If v_t is inactive, then the state of v_t is changed to i-active with probability

$$h_v^i(S_{t-1})$$

where $S_{t-1} = (S_{t-1}^1, \ldots, S_{t-1}^p)$ and $h_v^i(\cdot)$ satisfying the following conditions.

$$0 \leq h_v^i(S) \leq 1 \quad \forall v \in V, i = 1, \ldots, p$$

$$\sum_{i=1}^{p} h_v^i(S) \leq 1 \quad \forall v \in V$$

where S is a p-dimensional vector of sets.
- The process ends after ℓ update steps.

Is $(\gamma, \cup \sigma_i)$ a valid utility system? That is, does the following hold?

- γ is submodular.
- For every i, $\sigma^i(S_0) \geq \gamma(S_0) - \gamma(S_0 \odot (i, \emptyset))$.
- $\sum_{i=1}^{p} \sigma^i(S_0) \leq \gamma(S_0)$.

The reader may find an answer in [186].

Exercises

1. An older statement of Theorem 8.3.7 is as follows: "If $(\gamma, \cup_i \alpha_i)$ is a basic valid utility system, there exists pure Nash equilibrium in the game." Now, we removed the condition that $(\gamma, \cup_i \alpha_i)$ is a basic utility system. Could you explain why we can do so?

2. Consider a rumor blocking environment is symmetric as follows: There are two teams. The duty of one team is to choose locations for allocating protectors. The duty of the other team is to select locations for staring the rumor. Therefore, they are playing a game. In rumor blocking, destructive competition is not allowed, that is, if a node is influenced by one cascade, then it cannot be influenced by the other one. This rule meets the requirement of the non-cooperative game. Would you suggest to study the rumor blocking problem by employing the non-operative game as a tool?

3. In the non-cooperative game model of distributed rumor blocking, suppose that each node has own choice for the priority of cascades. Is it possible to make private utility or social utility nonsubmodular?

4. In the non-cooperative game model of competitive influence, the private utility $\sigma_i(S_1, \ldots, S_k)$ is submodular with respect to S_i. Can $sigma_i$ be submodular with respect to S_j for $j \neq i$? If your answer is yes, then please give a proof. If your answer is no, then please provide a counterexample.

5. In the non-cooperative game model of competitive influence, suppose that each node has own choice for the priority of cascades. Is it possible to make private utility or social utility nonsubmodular?

6. (Expansion with Positive Influence Model) Consider the positive influence diffusion model as follows: Let $G = (V, E)$ be a directed graph. Each node has two states, active and inactive. Initially, every node is inactive. To start, activate a subset of nodes, called seeds. At each iteration, every inactive node v evaluates how many active in-neighbors exist. If the number of active in-neighbors is not less than the number of inactive in-neighbors, then the inactive node v would be activated, i.e., becomes active. This process repeats until no new inactive node becomes active. Can the positive influence be formulated into the non-cooperative game? Why?

7. (A Distance-Based Model [36]) There are two companies A and B selling the same type of products. Let I_A and I_B be the sets of adopters for products of A and B, respectively. Suppose u receives influence starting from I_A and I_B at the same time step. Then, in this model, u will adopt product of A with probability

$$\frac{n_A}{n_A + n_B}$$

and will adopt product of B with probability

$$\frac{n_B}{n_A + n_B}$$

where n_A (n_B) denotes the number of nodes in I_A (I_B) with the distance to u equal to the shortest distance from I_A (I_B) to u with alive edges. Show that in this model, the business expansion is monotone nondecreasing and submodular.

8. In above problem, suppose that for node u, n_A (n_b) is the number of neighbors who are adopters of A (B) and try to influence u. Please discuss that in this model, whether the business expansion is monotone nondecreasing and submodular.

9. In [14], it states: "In the influence maximization game, each of b players selects a set S_i of at most k_i nodes. A node selected by multiple players will take the color of one of the players uniformly at random. Then, with S_i being active for influence i, the process unfolds as described above until no new activations occur. Letting $T_1, \ldots, T_b$ be the active sets at that point, the goal of each player i is to maximize $\mathbb{E}[|T_i|]$. Player i is indifferent between strategies S_i and S_i' if their expected gain is the same. Simple examples show that in general, this game has no pure strategy Nash Equilibria; however, it does have mixed-strategy Nash Equilibria." Does this statement contradict Theorem 8.3.7? If not, please construct a simple example mentioned as above.

10. Consider a social network $G = (V, E)$ with IC model. There exist three cascades, protector 1, protector 2, and rumor, with seed sets A_1, A_2, and A_r, respectively. For any node v, let t_{rv} $(t_{1v}$ and t_{2u}, resp.) denote the time step which v is firstly influenced by A_r $(A_1$ and $A_2)$. Define

$$S_p = \{u \mid t_{pu} < t_{ru}\}$$
$$S_r = \{u \mid t_{pu} > t_{ru}\}$$
$$S_e = \{u \mid t_{pu} = t_{ru}\}.$$

Show that the expectation of the following function is monotone nondecreasing submodular with respect to A_p:

(a) $|S_p| + |S_e|$.
(b) $|S_p| + \frac{1}{2}|S_e|$.
(c) $|S_p|$.

11. Chen et al. [54] consider a social network $G = (V, E)$ with IC model. There exist two cascades, rumor and protector, with seed sets A_r and A_p, respectively. For any node v, let t_{rv} $(t_{pv}$ resp.) denote the time step which v is firstly influenced by A_r (A_p), and $t_{pu} = \min(t_{1u}, t_{2u})$. Define

$$S_p = \{u \mid t_{pu} < t_{ru}\}$$
$$S_r = \{u \mid t_{pu} > t_{ru}\}$$
$$S_e = \{u \mid t_{pu} = t_{ru}\}.$$

S_p can be partitioned into five subsets:

$$S_p = S_1 \cup S_2 \cup S_3 \cup S_4 \cup S_5$$

where

$$S_1 = \{u \mid t_{1u} < t_{ru} < t_{2u}\}$$
$$S_2 = \{u \mid t_{2u} < t_{ru} < t_{1u}\}$$
$$S_3 = \{u \mid t_{1u} < t_{ru}, t_{2u} < t_{ru}\}$$
$$S_4 = \{u \mid t_{1u} < t_{ru}, t_{2u} = t_{ru}\}$$
$$S_3 = \{u \mid t_{2u} < t_{ru}, t_{1u} = t_{ru}\}.$$

S_e can be partitioned into three subsets:

$$S_e = S_6 \cup S_7 \cup S_8$$

where

$$S_6 = \{u \mid t_{1u} = t_{ru} < t_{2u}\}$$
$$S_7 = \{u \mid t_{2u} = t_{ru} < t_{1u}\}$$
$$S_8 = \{u \mid t_{1u} = t_{ru} = t_{2u}\}.$$

Given $A = (A_1, A_2)$, the expectation of $|S_i|$ can be consider a function of A. Define

$$\sigma_1(A) = \mathbb{E}[|S_1| + \frac{1}{2} \cdot (|S_4| + |S_8| + |S_6|)]$$

$$\sigma_2(A) = \mathbb{E}[|S_2| + \frac{1}{2} \cdot (|S_5| + |S_8| + |S_7|)]$$

and

$$\mu(A) = \mathbb{E}[\sum_{i=1}^{5} |S_i| + |S_8| + \frac{1}{2}(|S_6| + |S_7|)].$$

Show that $\mu(A)$, $\sigma_1(A)$, and $\sigma_2(A)$ are monotone nondecreasing submodular, and $(\mu, \sigma_1 \cup \sigma_2)$ is a valid utility system.

12. (Private IC [400]) Under this model, an active node u can only attempt to activate one of its out-neighbors at a time step. This model represents the private communication, such as email. Given a set of seeds, show that executing the diffusion process on the network G with private IC model is equivalent to the following:

- Randomly generate a realization g of G as follows: For every edge (u, v), set (u, v) to be alive with probability p_{uv} and for each node u, select a permutation β_u of out-neighbors uniformly at random.
- Execute the deterministic diffusion process on g.

13. In the study of decentralized rumor blocking, the cascade priority is *homogeneous* if the priority of the cascades is same for each agent. Please construct a counterexample to show that under the homogeneous cascade priority, if rumor does not have the highest priority, then social utility $\gamma(A)$ is not a set function anymore under private IC model.

14. In the study of decentralized rumor blocking, the cascade priority is *heterogeneous* if the priority of the cascades is not homogeneous. Please construct a counterexample to show that under the heterogeneous cascade priority, the social utility $\gamma(A)$ is not monotone nondecreasing under private IC model.

15. Please construct a counterexample to show that under the homogeneous cascade priority, if an active node is allowed only to activate inactive neighbors, then the social utility γ is not monotone nondecreasing under private IC model.

16. Consider a social network $G = (V, E)$ with LT information diffusion model. Both companies A and B want to expand their business in this network. They sell the same type of products. Let I_A and I_B be the sets of initial adopters of products of companies A and B, respectively. Each customer u makes his adoption decision in two stages. At the first stage, u decides whether u adopts a product of this type (it doesn't matter from either A or B). In the second stage, u selects the product from which company. Let $N_A^-(u)$ ($N_B^-(u)$) denote the set of in-neighbors who are adopters of product from A (B). According to LT model, initially, every node u selects a threshold θ_u (randomly uniform) from $[0, 1]$. At each subsequent step, each inactive node u (i.e., customer has not adopted the product) compute

$$w(N_A^-(u) \cup N_B^-(u)) = \sum_{v \in N_A^-(u) \cup N_B^-(u)} w_{vu}$$

where w_{vu} is the label on the edge (v, u), given in the LT model. If $w(N_A^-(u) \cup N_B^-(u)) \geq \theta_u$, then u is activated in the first stage. Then, u will select product from A with probability

$$\frac{w(N_A^-(u))}{w(N_A^-(u) \cup N_B^-(u))}$$

and select product from B with probability

$$\frac{w(N_B^-(u))}{w(N_A^-(u) \cup N_B^-(u))}.$$

Let $\sigma_A(I_A, I_B)$ denote the expected number of adopters of A at the end of information process. Show that $\sigma_A(I_A, I_B)$ is neither monotone nondecreasing nor submodular with respect to I_A.

17. Consider a social network $G = (V, E)$ with LT information diffusion model. Both companies A and B want to expand their business in this network. They sell the same type of products. Let I_A and I_B be the sets of initial adopters of

products of companies A and B, respectively. Starting with I_A and I_B, companies A and B diffuse their information independently. When a node u satisfies conditions to become A-adopter and B-adopter at the same time step, u will choose A or B with probability $1/2$. Let $\sigma_A(I_A, I_B)$ denote the expected number of adopters of A at the end of information process. Show that $\sigma_A(I_A, I_B)$ is monotone nondecreasing, but not submodular with respect to I_A.

18. Consider a social network $G = (V, E)$ with LT information diffusion model. Both companies A and B want to expand their business in this network. They sell the same type of products. Let I_A and I_B be the sets of initial adopters of products of companies A and B, respectively. Starting with I_A and I_B, companies A and B diffuse their information independently. When a node u satisfies conditions to become A-adopter and B-adopter at the same time step, u will choose A or B with probability $1/2$. After diffusion process ends, each inactive node will select to become A-adopter or B-adopter with equal probability. Let $\sigma_A(I_A, I_B)$ denote the expected number of adopters of A at the end of information process. Show that $\sigma_A(I_A, I_B)$ is monotone nondecreasing, but not submodular with respect to I_A.

19. Borodin et al. [24] consider a social network $G = (V, E)$ with LT information diffusion model. Both companies A and B want to expand their business in this network. They sell the same type of products. Let I_A and I_B be the sets of initial adopters of products of companies A and B, respectively. Starting with I_A and I_B (not necessary at the same time), companies A and B diffuse their information completely independent until two processes end. At the end, let R_A be the set of A-adopters and R_B the set of B-adopters. At every node u, give two monotone nondecreasing functions $f_A^u(R_A, R_B)$ and $f_B^u(R_A, R_B)$ satisfying $f_A^u(R_A, R_B) + f_B^u(R_A, R_B) \leq 1$. Assign product of A (B) to node $u \notin I_A \cup I_B$ with probability $f_A^u(R_A, R_B)$ $(f_B^u(R_A, R_B))$. Let $\sigma_A(I_A, I_B)$ denote the expected number of adopters of A at the end of whole process. Show that $\sigma_A(I_A, I_B)$ is monotone nondecreasing submodular with respect to I_A.

20. (Relative Influence [457]) In viral marketing and political election, two opposite opinions (positive opinion and negative opinion) spread simultaneously. Consider a competitive IC model with the higher priority of negative opinion. Given a seed set of negative opinion, I_N, and a positive integer k, find a seed set of positive opinion, I_P with $|I_P| = k$, to maximize

$$\sigma_P(I_P, I_N) - \sigma_N(I_P, I_N)$$

where $\sigma_P(I_P, I_N)$ and σ_N are the influence spreads of positive and negative opinions, respectively. Show the following:

(a) This problem is NP-hard.
(b) The objective function is monotone nondecreasing and submodular.
(c) For any $\varepsilon > 0$, design a $(1 - 1/e - \varepsilon)$-approximation with reverse sampling.

21. (Timeliness IC Model [473]) Similar to the IC model, for every edge (u, v), there is a probability p_{uv}, which means v can be successfully activated by u

with probability p_{uv}. In this model, each node has three states, inactive, active, and decided. Suppose there are m influences (from m companies) diffusing in the social network, denoted by m colors $c_1, \ldots, c_m$. An active node does not have a certain color and it may spread multiple influences. However, each node v has a deadline D_v for making decision. Before time D_v, v needs to record the number b_i of successful activations of c_i, Then at time D_v, v will turn state to "decided to color c_i" with probability $b_i / \sum_{i=1}^{m}$. After time D_v, v will spread only decided color c^v. If v receives no successful activation before time D_v, then v keeps inactive forever. Initially, every seed node is decided to a color and only one color. The information process is terminated if all deadlines are passed over. Let $I_i(S_i, S)$ denote the expected number of nodes in color c_i at the end of process where S is the total number of seeds and S_i is the number of seeds in color c_i. Please show that $I_i(S_i, S)$ is monotone nondecreasing and submodular with respect to S_i.

22. (Individual Spread [473]) Consider the Timeliness IC model and a fixed total seed set S. For any $s \in S$, define its individual spread $I_{-S}(s)$ as the expected i-spread when s is the only seed in color c_i. Please prove

$$I_i(S_i, S) = \sum_{s \in S_i} I_{-S}(s).$$

23. (Fair Multi-Influence [473]) Consider the Timeliness IC model. Let μ_i be the budget of influence c_i. The Earning-to-Budget Ratio (EBR) is defined as

$$e_i = \frac{I_i(S_i, S)}{\mu_i}.$$

The fairness is defined as

$$f = \frac{\min_{1 \le i \le m} e_i}{\max_{1 \le i \le m} e_i}.$$

Please prove that allocating the seed s to the influence c_i that achieves current lowest EBR e_i is the greedy choice of maximizing fairness f.

24. (Competitive and Complementary Influence [191]) There are two cascades A and B diffusing in given social network. The seed set of cascade B is given and fixed. When a node v is already activated with cascade B, v is activated by cascade A with probability reduced (increased) by a ratio λ. Find locations of k seeds for cascade A to maximize the influence spread of cascade A. Show that this problem is monotone submodular maximization. ($\lambda < 1$ means that two cascades are competitive while $\lambda > 1$ means that two cascades are complementary.)

25. (Competition Coefficient [258]) There are two cascades A and B diffusing in given social network. The seed set of cascade B is given and fixed. When a node v is already activated with cascade B, v is activated by cascade A with probability reduced (increased) by a ratio λ and v's influence power along edge (v, u) is also reduced to $p_{v,u}/c(v, u)$ where $c(v, u)$ is called *competition coefficient*. Find

locations of k seeds for cascade A to maximize the influence spread of cascade A. Show that this problem is monotone submodular maximization.

26. (User-Driven Competitive Influence [267]) Consider a social network with the LT model. There are two cascades A and B with seed sets S_A and S_B, respectively. Every node is classified to have preference for A, or for B, or have no preference. At initial step of the LT model, each node v selected a threshold θ_v. At each subsequent step, every node evaluates the total weight w_A of A-active in-neighbors, and the total weight w_B of B-active in-neighbors:

- For node v with preference for A, if $w_A \geq \theta_v$, then v becomes A-active; otherwise, if $w_B \geq \theta_v + \theta_e$, then v becomes B-active. If $w_A < \theta_v$ and $w_B < \theta_v + \theta_e$, then v keeps inactive.
- For node v with preference for B, if $w_B \geq \theta_v$, then v becomes B-active; otherwise, if $w_A \geq \theta_v + \theta_e$, then v becomes A-active. If $w_B < \theta_v$ and $w_A < \theta_v + \theta_e$, then v keeps inactive.
- For node v with no preference, if $w_A \geq \max(w_B, \theta_v)$, then v becomes A-active; if $w_B \geq (w_A, \theta_v)$, then v becomes B-active. If $w_A = w_B \geq theta_v$, then v can choose A or B arbitrarily. If $\max(w_A, w_B) < \theta_v$, then v keeps inactive.

Given $S_B, \theta_e > 0$, and a positive integer k, find S_A with $|S_A| = k$ to maximize the influence spread of cascade A. Show that this problem is a monotone submodular maximization.

Historical Notes

Community expansion [15, 16] is closely related to business expansion in environment of competitive influence.

In the study of competitive influence, there are two optimization approaches. The first one is to look for a best strategy for a company while fixing all competitors' strategies. The second one is to model the competition as a simultaneous game and to maximize the social utility.

Along this first optimization approach, there are three formulations. In the first formulation, for products of each company, consider that all adopters form a community. Then, business expansion of a company becomes a community expansion problem. Actually, online social network is dynamic, so is its community structure. In the study of the dynamics, Aggarwal and Yu [2] named three transitions of a community; one of them is expansion. Bi et al. [16] motivated from this work to study community expansion by using electric charge system. They utilized electric charge systems in more works [15, 17]. Actually, there exist many efforts made on the study of community evolution [8, 20, 37, 235, 308].

The second formulation is the positive influence. When fixing seed allocations for all companies except one, find the best seed allocation to maximize the influence spread for this remaining company. This influence maximization problem has been known as the maximization of a set function which is neither submodular nor supermodular. No progress has been made on this problem. Nobody has found a successful approximation solution, including to employ the sandwich approach.

In the third formulation, consider social networks with competitive information diffusion models, such as competitive independent cascade model [14] and competitive threshold model [24, 28, 44, 188]. The optimization is to optimize the benefit of one company while fixing other companies' strategies, such as maximizing the influence spread of the one company [36] and minimizing the influence of competitors [33].

Actually, those competitive information diffusion models are game formulations. For example, the competitive independent cascade model defined by Bharathi et al. [14] is a typical instance of the non-cooperative game. The theoretical foundation of non-cooperative game is established by Vetta [410]. Furthermore, Tong et al. [400] and Chen et al. [54] employ the non-cooperative game to study distributed or decentralized rumor blocking. The price of anarchy [21, 151, 186, 349] is an interesting topic in the study of the game model.

In the study of competitive influence, there are several other interesting subjects, such as limited budget [329, 330], fair competitive influence [270], minimum cost seed set [452], and dominated competitive [426]. They involve various optimization problems, the Colonel Blotto game [319], and comparative influence [271].

Chapter 9
Adaptive Influence

Adaptability is being able to adjust to any situation at any given time.
— John Wooden

Recently a lot of researchers are interested in studying adaptive process of information, or adaptive construction of objects, e.g., adaptive influence in social networks. This chapter is contributed to this subject.

9.1 Feedback Model

In the influence maximization, a social network is given together with information diffusion model and a positive integer k is also given. The problem is to find locations for k seeds to maximize the influence spread. These k seeds will be deployed into those locations at the initial step of information process.

In the adaptive influence maximization, seeds may not be deployed into the social network all at the initial step. Instead, they will be deployed adaptively, i.e., they may be deployed one by one and locations can be selected after receiving feedback on the outcome of previous seed-deployments.

Suppose that the second seed is deployed at the a_1's step after the first seed is deployed, the third seed is deployed at the a_2's step after the second seed is deployed, and so on. The sequence $(a_1, a_2, ..., a_{k-1})$ is called the *feedback pattern*, which represents the feedback model. When $a_1 = a_2 = \cdots = a_{k-1} = 0$, it is the *nonadaptive model*. When $a_1 = a_2 = \cdots = a_{k-1} = 1$, it is called the *myopic model*. When $a_1 = a_2 = \cdots = a_{k-1} = \infty$, i.e., each seed is deployed after the influence process of previous seeds ends, it is called the *full feedback model*. When $a_1 = a_2 = \cdots = a_{k-1} = d$ and $1 \leq d < \infty$, it is called the *partial feedback model*.

© The Author(s), under exclusive license to Springer Nature Switzerland AG 2026
W. Wu et al., *Computational Aspects of Social Networks*, Springer Optimization and Its
Applications 234, https://doi.org/10.1007/978-3-032-14833-9_9

In the literature, there are different definitions about the adaptive influence maximization problem. In this book, we would like to use the following:

Definition 9.1.1 (*Adaptive Influence Maximization*) Consider a social network $G = (V, E)$ with an information diffusion model m and a feedback model D. Given a positive integer k, deploy at most k seeds into G to maximize the influence spread $\sigma_{m,D}(S)$ where S is the set of seed locations.

Adaptive submodularity and adaptive greedy algorithm are important tools in the study of adaptive optimization problems. Let us learn them and then use them to study the adaptive influence maximization.

9.2 Adaptive Submodularity

Consider a social network $G = (V, E)$ with the deterministic information diffusion model and the full feedback model. Note that each node has two states, active and inactive. Initially, all nodes are in inactive state. Assume that when a seed is placed on a node u, node u is activated with probability p_u.

In fact, in the real world, after receiving a free sample of a product, a person may have two possible attitudes, like the product and dislike the product. This means that when a seed is placed at node u, node u may not be activated for sure. We may assume that such a node u becomes active with certain probability p_u. In this case, the selection of next location for placing a new seed will be affected by outcomes of previous seed-placements. Especially, if u is not activated by the first seed located at u, then u may receive the send seed. In general, a node may receive seeds more than once. Therefore, adaptive distribution and nonadaptive distribution make a difference for the seed-placement.

Thus, each algorithm for the seed placement can be represented by a binary decision tree. If the number of seeds is bounded by k, then the binary decision tree has k levels. Each path from the root to a leaf represents a decision process, or an algorithm outcome. Suppose S is the set of nodes appearing in the path. Then this outcome appears with probability

$$p(\phi, S) = \left(\prod_{v \in S \text{ and } \phi(v)=1} p_v \right) \cdot \left(\prod_{v \in S \text{ and } \phi(v)=0} (1 - p_v) \right)$$

where ϕ is a mapping from S to $\{0, 1\}$ and $\phi(v) = 1$ means that v is active and $\phi(v) = 0$ means that v is inactive. Let $S_\phi = \{v \in S \mid \phi(v) = 1\}$. Let $\sigma_m(S_\phi)$ be the influence spread with initial set S_ϕ of active nodes under deterministic information diffusion model m. Note that S is the definition domain of ϕ, i.e., S can be determined by ϕ. Let $\mathcal{A}$ be an algorithm. Without confusion, we may also denote by $\mathcal{A}$ the set of ϕ appearing in a root-leaf path on the decision tree induced by $\mathcal{A}$. We can obtain an adaptive version of influence maximization problem as follows.

$$\max_{\mathcal{A}} \mathbb{E}_{\phi \in \mathcal{A}}[\sigma(S_\phi)] = \max_{\mathcal{A}} \sum_{\phi \in \mathcal{A}} p(\phi, S)\sigma(S_\phi).$$

This is an adaptive stochastic maximization problem, which can be generalized as follows.

Consider a finite set E of items and a finite set O of states. Each mapping $\phi : E \to O$ is called a *realization* which indicates that each item e is in state $\phi(e)$. Let Φ be the set of all realizations. Assume that a probabilistic distribution $\mathbb{P}$ is already given over Φ. For any subset $D \subseteq E$, any mapping $\psi : D \to O$ is called a *partial realization*. Any partial realization can be represented by a subset of $E \times O$, $\{(e, \psi(e)) \mid e \in D\}$. With this explanation, we can understand the meaning of $\psi \subset \psi'$. In this case, ψ is said to be a *subrealization* of ψ, and ψ' is an extension of ψ from $\mathrm{dom}(\psi)$ to $\mathrm{dom}(\psi')$ where $\mathrm{dom}(\psi)$ denotes the domain of ψ. The probabilistic distribution on partial realization can be induced by $\mathbb{P}$ as follows.

$$\mathbb{P}[\psi] = \sum_{\phi \in \Phi \text{ and } \psi \subset \phi} \mathbb{P}[\phi].$$

Let Ψ be the set of all partial realizations. A *policy* π is a mapping from a subsets of Ψ to E satisfying the following property:

- If $\psi \subset \psi' \in \mathrm{dom}(\pi)$, then $\psi \in \mathrm{dom}(\pi)$.

Each adaptive algorithm is associated with a policy π as follows.

- The algorithm is usually initiated with a partial realization $\psi_0 = \emptyset$, i.e., the partial realization with empty definition domain. In each subsequent ith iteration for $i \geq 1$, if there exists a state x such that $\{\psi_{i-1}, (\pi(\psi_{i-1}), x)\} \in \mathrm{dom}(\pi)$, then set $\psi_i = \{\psi_{i-1}, (\pi(\psi_{i-1}), x)\}$ and the algorithm goes to next iteration. Otherwise, the algorithm stops.

Note that $\psi_i(\pi(\psi_{i-1}))$ has $|O|$ possible values. Thus, each algorithm with policy π can be represented by a decision tree T^π in which each internal node has at most $|O|$ children. An example is shown in Fig. 9.1 for $O = \{0, 1\}$ and $\mathrm{dom}(\pi) = \{\psi \mid |\psi| \leq 3 \text{ and } \{(a, 1), (c, 0)\} \not\subseteq \psi\}$.

In the policy tree T^π, every node represents a partial realization or realization $\psi \in \mathrm{dom}(\pi)$ and every $\psi \in \mathrm{dom}(\pi)$ is associated with a node. Consider a path from the root to a node representing ψ. Suppose $(a_1, s_1, a_2, s_2, ..., a_k)$ is the sequence of node and edge labels alternatively appearing in path. Then $\psi = \{(a_1, s_1), (a_2, s_2), ..., (a_{k-1}, s_{k-1})\}$ and $\pi(\psi) = a_k$.

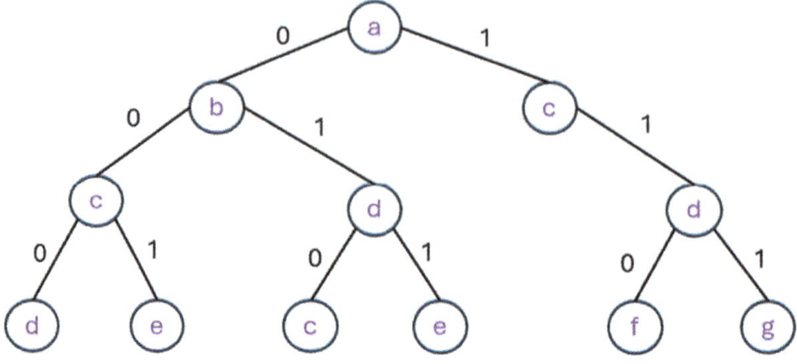

Fig. 9.1 A policy tree T^π has node labels in E and edge labels in O. Each path from the root to a node gives a partial realization. For example, $(a, 0, b, 1, c)$ gives a partial realization $\psi = \{(a, 0), (b, 1)\}$ and $\pi(\psi) = c$

For a policy π and a realization $\phi \in \Phi$, define

$$E(\pi, \phi) = \begin{cases} \mathrm{dom}(\psi) & \psi \text{ is a leaf of } T^\pi \text{ and } \psi \subseteq \phi, \\ \emptyset & \text{otherwise.} \end{cases}$$

To validate the correctness of above definition, let us indicate the following fact.

Lemma 9.2.1 *Let ψ and ψ' be two distinct leaves in T^π. Suppose $\psi \subseteq \phi$ and $\psi' \subseteq \phi'$. Then $\phi \neq \phi'$.*

Proof Let $\zeta \to e$ be the internal node of T^π, at which two paths from the root to ψ and ψ' start to be separated. Then $\psi(e) \neq \psi'(e)$. Hence, $\phi(e) \neq \phi'(e)$. $\qquad\square$

Consider a utility function $f : 2^E \times \Phi \to \mathbb{R}_{\geq 0}$. Define

$$f_{avg}(\pi) = \mathbb{E}_{\phi \in \Phi}[f(E(\pi, \phi), \phi)].$$

We are interested in finding π to maximize $f_{avg}(\pi)$, i.e.,

$$\pi^* = \mathrm{argmax}_\pi \mathbb{E}_{\phi \in \Phi}[f(E(\pi, \phi), \phi)].$$

Usually, this is an intractable problem. However, for f_{evg} with certain properties, some good approximation solutions can be obtained. To describe those properties, let us introduce a notation.

$$\Delta f(e \mid \psi) = \mathbb{E}_\phi[f(\mathrm{dom}(\psi) \cup \{e\}, \phi) - f(\mathrm{dom}(\psi), \phi) \mid \psi \subset \phi \in \Phi].$$

Definition 9.2.2 (*Adaptive Monotone Nondecreasing*) A function $f : 2^E \times \Phi \to \mathbb{R}_{\geq 0}$ is adaptive monotone nondecreasing with respect to distribution $\mathbb{P}[\phi]$ if for any partial realization ψ and $e \in E$, $\Delta f(e \mid \psi) \geq 0$.

Definition 9.2.3 (*Adaptive Submodularity*) A function $f : 2^E \times \Phi \to \mathbb{R}_{\geq 0}$ is adaptive submodular with respect to distribution $\mathbb{P}[\phi]$ if for any two partial realizations ψ and ψ' with $\psi \subset \psi'$, and $e \in E \setminus \mathrm{dom}(\psi')$,

$$\Delta f(e \mid \psi) \geq \Delta f(e \mid \psi').$$

Remark: It is important to note that the adaptive submodularity of a function $f(A, \phi)$ is closely related to the probabilistic distribution of realization *phi*. The same function f can be adaptive submodular in one probabilistic distribution of ϕ and adaptive nonsubmodular in another probabilistic distribution of ϕ. The reader may find such an example in Exercises. In the study of adaptive influence maximization, different information diffusion models and feedback models give different probabilistic distribution of realizations. Therefore, they introduce different outcomes about adaptive submodularity.

We next study some properties of above definitions. Since each algorithm is determined by associated policy π, its associated decision tree can be denoted by T^π. Define $f_{avg}(T^\pi) = f_{avg}(\pi)$. Then we have the following.

Lemma 9.2.4 *A function $f : 2^E \times \Phi \to \mathbb{R}_{\geq 0}$ is adaptive monotone nondecreasing with respect to distribution $\mathbb{P}[\phi]$ if and only if for any two policies π and π' with $T^\pi \subset T^{\pi'}$ (i.e., T^π is a subtree of $T^{\pi'}$),*

$$f_{avg}(T^\pi) \leq f_{avg}(T^{\pi'}).$$

Proof Suppose $T^{\pi'}$ contains exactly one more internal node than T^π. Assume this node in $T^{\pi'}$ has label a and the path from the root to this node induces a partial realization ψ. Then, we have

$$f_{avg}(T^{\pi'}) - f_{avg}(T^\pi) = \mathbb{P}[\psi]\Delta f(a \mid \psi).$$

The sufficiency can be obtained from this equality immediately. For the necessity, we only need to note that when $T^{\pi'}$ contains more than one internal nodes than T^π, those internal nodes can be added to T^π one by one. $\qquad\square$

Definition 9.2.5 (*Concatenation*) Consider two policies π and π'. Their concatenation $\pi @ \pi'$ is defined to be the policy obtained by running π until completion and then running π'. π' runs from a fresh start and ignore the information collected during the period of running π. An example is as shown in Fig. 9.2.

Clearly, T^π is a subtree of $T^{\pi @ \pi'}$. Therefore, we have the following.

Corollary 9.2.6 *If a function $f : 2^E \times \Phi \to \mathbb{R}_{\geq 0}$ is adaptive monotone nondecreasing with respect to distribution $\mathbb{P}[\phi]$, then*

$$f_{avg}(T^\pi) \leq f_{avg}(T^{\pi @ \pi'}).$$

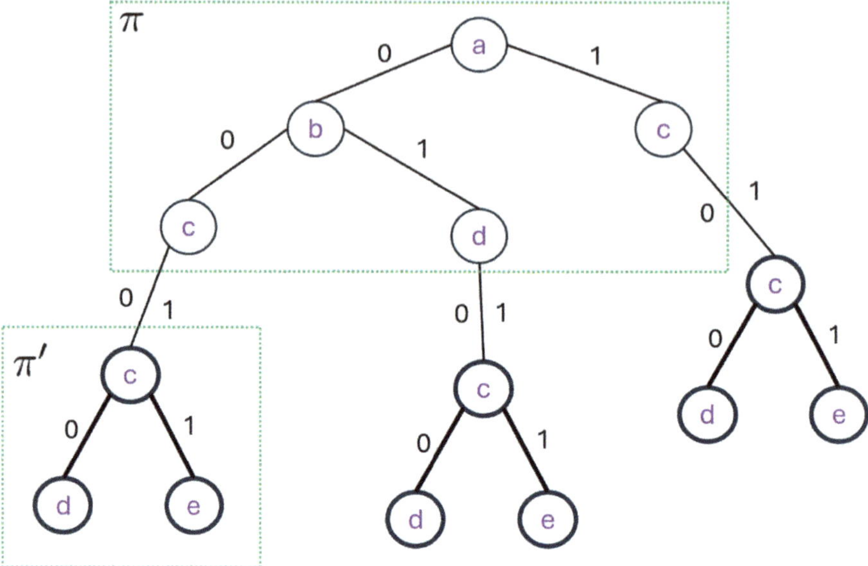

Fig. 9.2 Tree concatenation $T^{\pi@\pi'}$

Definition 9.2.7 (*Collapse*) Collapse is an operation to obtain a tree $T^{\pi'}$ from tree T^{π} as follows: Consider a internal node x in T^{π} and the subtree T_x rooted at x. Let $D(x)$ denote the set of all leaves of T_x. Then replace subtree T_x by the set $D(x)$ (Fig. 9.3). Tree $T^{\pi'}$ represents a random policy π'. Let ψ_x denote the partial realization induced by the path from the root to node x in T^{π}. Then policy π' at partial realization ψ_x will select an element $z \in D(x)$ with probability $\mathbb{P}[\psi_z \mid \psi_x]$.

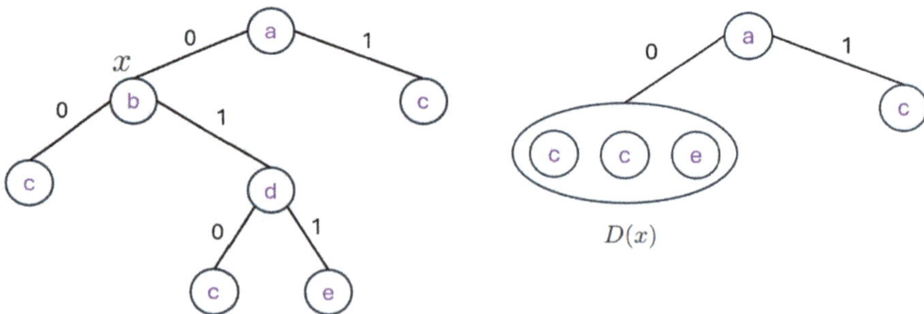

Fig. 9.3 Collapse: a subtree T_x is collapsed to a super node $D(x)$

For $0 \le i < j$, let $T_{[i]}$ denote the tree obtained from T by cutting off the part below level i and $T_{[i] \cup \{j\}}$ the tree obtained from $T_{[i]}$ through collapses of subtrees rooted at level $i + 1$.

Corollary 9.2.8 *If a function $f : 2^E \times \Phi \to \mathbb{R}_{\ge 0}$ is adaptive submodular with respect to distribution $\mathbb{P}[\phi]$, then for every policy π and for all $0 \le i < j$,*

$$f_{avg}(T_{[j]}^{\pi}) - f_{avg}(T_{[j-1]}^{\pi}) \le f_{avg}(T_{[i] \cup \{j\}}^{\pi}) - f_{avg}(T_{[i]}^{\pi}).$$

For a subset I of internal nodes in T^{π}, denote $D(I) = \cup_{x \in I} D(x)$ and denote by $T_{D(I)}^{\pi}$ the tree obtained from T^{π} by deleting nodes in $D(I)$. Then we have the following.

Lemma 9.2.9 *A function $f : 2^E \times \Phi \to \mathbb{R}_{\ge 0}$ is adaptive submodular with respect to distribution $\mathbb{P}[\phi]$ if and only if for every policy π and every policy π' obtained from π through a sequence of collapses,*

$$f_{avg}(T^{\pi}) - f_{avg}(T_{D(I)}^{\pi}) \le f_{avg}(T^{\pi'}) - f_{avg}(T_{D(I)}^{\pi'})$$

where I is a subset of internal nodes of T^{π}.

Proof Assume $T^{\pi'}$ is obtained from T^{π} through a collapse on subtree T_b. Let c be an internal node of T_b such that its all children are leaves. Then we have

$$f_{avg}(T^{\pi}) - f_{avg}(T_{D(c)}^{\pi}) = \mathbb{P}[\psi_c]\Delta_{\psi_c}(c),$$
$$f_{avg}(T^{\pi'}) - f_{avg}(T_{D(c)}^{\pi'}) = \mathbb{P}[\psi_c]\Delta_{\psi_b}(c).$$

The sufficiency of the condition follows immediately from those two equalities. The necessity can be proved by applying these equalities several times. $\qquad\square$

9.3 Adaptive Greedy

Now, we consider an adaptive submodular maximization problem as follows: Suppose that $f : 2^E \times \Phi \to \mathbb{R}_{\ge 0}$ is an adaptive monotone nondecreasing and adaptive submodular function with respect to distribution $\mathbb{P}[\phi]$. The problem is to maximize $f_{avg}(\pi)$ under constraint $|E(\pi, \phi)| \le k$, i.e.,

$$\max f_{avg}(\pi) = \mathbb{E}_{\phi \in \Phi}[f(E(\pi, \phi), \phi)]$$
$$\text{subject to} \quad |E(\pi, \phi)| \le k,$$

where constant k is a positive integer. A greedy algorithm is given in Algorithm 43.

Algorithm 43 Adaptive Greedy Algorithm.

Input: An adaptive monotone nondecreasing and adaptive submodular function $f : 2^E \times \Phi \to \mathbb{R}_{\geq 0}$ with respect to distribution $\mathbb{P}[\phi]$.
Output: A greedy policy π.
1: $\psi_0 \leftarrow \emptyset$.
2: **for** $i = 1$ to k **do**
3:     find e to maximize $\Delta_{\psi_{i-1}} f(e)$;
4:     set $\pi(\psi_{i-1}) \leftarrow e$ and $\psi_i \leftarrow \{\psi_{i-1}, (e, x)\}$ for $x \in O$ with probability $\mathbb{P}[\{\psi_{i-1}, (e, x)\}]$;
5: **end for**
6: **return** π.

Analysis is presented in the following theorem.

Theorem 9.3.1 *Consider an adaptive monotone nondecreasing and submodular function $f : 2^E \times \Phi \to \mathbb{R}_{\geq 0}$ with respect to distribution $\mathbb{P}[\phi]$. Let π be a greedy policy obtained by Algorithm 43. Assume that π^* is the optimal policy. Then*

$$f_{avg}(\pi) \geq (1 - e^{-1}) f_{avg}(\pi^*).$$

Proof Denote $T = T^\pi$ and $T^* = T^{\pi^*}$. Then we have the following.

$$f_{avg}(T^*) \leq f_{avg}(T_{[i]} @ T^*)$$

$$= f_{avg}(T_{[i]}) + \sum_{j=1}^{k} (f_{avg}(T_{[i]} @ T^*_{[j]}) - f_{avg}(T_{[i]} @ T^*_{[j-1]}))$$

$$\leq f_{avg}(T_{[i]}) + \sum_{j=1}^{k} (f_{avg}((T_{[i]} @ T^*)_{[i]\{i+j\}}) - f_{avg}(T_{[i]}))$$

$$\leq f_{avg}(T_{[i]}) + k(f_{avg}(T_{[i+1]}) - f_{avg}(T_{[i]})).$$

Denote $\alpha_i = f_{avg}(T^*) - f_{avg}(T_{[i]})$. Then we have

$$\alpha_i \leq k(\alpha_i - \alpha_{i+1}).$$

Thus,

$$\alpha_{i+1} \leq \alpha_i (1 - \frac{1}{k}) \leq e^{-1/k} \cdot \alpha_i.$$

Therefore,

$$\alpha_k \leq e^{-1} \alpha_0 = e^{-1} \cdot f_{avg}(T^*).$$

That is,

$$f(T) \geq (1 - e^{-1}) f(T^*).$$

$\square$

Now, we return to the adaptive influence maximization problem for social network with deterministic diffusion model and full feedback model. To formulate it as a special case of adaptive submodular maximization. We may consider the element set E consisting of all nodes in the social network and the state set contains two states, active and inactive. For each subset D of elements and a realization ϕ, define

$$f(D, \phi) = \sigma_m(D_\phi)$$

where m is the diffusion model and $D_\phi = \{u \in D \mid u \text{ is active}\}$.

Lemma 9.3.2 *Let $\sigma_m(S)$ denote the influence spread of seed set S under information diffusion model m and nonadaptive feedback. Let $\sigma_{m,full}(S)$ denote the influence spread of seed set S under information diffusion model m and full feedback. Then, for every seed set S, $\sigma_m(S) = \sigma_{m,full}(S)$.*

Proof Argument is similar to the proof of Lemma 2.4.5. $\qquad\qquad\qquad\qquad\square$

Lemma 9.3.3 *If the influence spread σ_m is monotone nondecreasing and submodular, then f defined as above is adaptive monotone nondecreasing and adaptive submodular under the full feedback model.*

Proof For $D \subset D'$, we have $D_\phi \subset D'_\phi$. Since σ_m is monotone nondecreasing, we have $f(D, \phi) \leq f(D', \phi)$. Therefore,

$$\Delta f(u \mid \psi) = \mathbb{E}[f(\text{dom}(\psi) \cup \{u\}, \phi) - f(\text{dom}(\psi), \phi) \mid \psi \subseteq \phi] \geq 0,$$

that is, f is adaptive monotone nondecreasing.

Next, consider two partial realizations ψ and ψ' with $\psi \subset \psi'$. Then, $\text{dom}(\psi) \subset \text{dom}(\psi')$. Since σ_m is submodular, we have

$$f(\text{dom}(\psi) \cup \{u\}, \phi) - f(\text{dom}(\psi), \phi) \geq f(\text{dom}(\psi') \cup \{u\}, \phi) - f(\text{dom}(\psi'), \phi).$$

Therefore,

$$\begin{aligned}
\Delta f(u \mid \psi) &= \mathbb{E}[f(\text{dom}(\psi) \cup \{u\}, \phi) - f(\text{dom}(\psi), \phi) \mid \psi \subseteq \phi] \\
&\geq \mathbb{E}[f(\text{dom}(\psi') \cup \{u\}, \phi) - f(\text{dom}(\psi'), \phi) \mid \psi' \subseteq \phi] \\
&= \Delta f(u \mid \psi').
\end{aligned}$$

Moreover,

$$\mathbb{E}[f(\text{dom}(\psi) \cup \{u\}, \phi) - f(\text{dom}(\psi), \phi) \mid \psi \subseteq \phi \text{ and } \psi' \not\subseteq \phi] \geq 0.$$

Therefore,

$$\Delta_\psi(u) \geq \Delta_{\psi'}(u),$$

that is, f is adaptive submodular. $\qquad\qquad\qquad\qquad\qquad\qquad\qquad\qquad\square$

The following follows immediately from the lemma.

Theorem 9.3.4 *The adaptive influence maximization problem has a greedy approximation with performance ratio $1 - e^{-1}$ under the full feedback model.*

9.4 Golovin-Kraus Conjecture

There is an interesting conjecture about adaptive influence, which induced a sequence of research outcomes. Let us contribute this section to this conjecture. To introduce it, let us consider a (directed) social network (V, E) with IC model and myopic feedback model.

To formulate adaptive influence, let O denote the set of states. Each state is a subset of edges going out from a node, which represents the myopic feedback when a node is activated. In fact, when a node u is activated by a seed, the myopic feedback is that every out-edge from u has its state explored to be active or inactive. Therefore, the set of all active out-edges may represent the myopic feedback while all inactive out-edges can be treated as disappeared.

With this formulation, the influence spread is not adaptive submodular. This important fact is point out by Golovin and Kraus in 2011.

Counterexample 9.4.1 *(Golovin and Kraus [142])* Consider a social network with node set $V = \{u, v, w\}$ and edge set $E = \{(u, v), (v, w)\}$. Moreover, two edges have probabilities $p_{uv} = 1$ and $p_{vw} = 1 - \varepsilon$, respectively. Let $\psi = \{(u, \{(u, v)\})\}$ and $\psi' = \{(u, \{(u, v)\}), (v, \emptyset)\}$. Then, $\psi \subset \psi'$. Moreover,

$$\Delta\sigma(w \mid \psi) = \varepsilon < \Delta\sigma(w \mid \psi') = 1.$$

Therefore, the influence spread σ is not adaptive submodular.

Although the influence spread is not adaptive submodular, Golovin and Kraus [142] still made the following conjecture.

Conjecture 9.4.2 *(Golovin and Kraus [142])* The greedy policy still obtains a constant factor approximation even in the myopic feedback model.

This conjecture was settled by Peng and Chen [327] in 2019, through the study of adaptivity gap defined as follows.

Definition 9.4.3 *(Adaptivity Gap [47, 49])* The adaptivity gap in the IC model is defined as the supremum for the ratio of the influence spreads between the optimal adaptive policy and the optimal nonadaptive policy, over all possible social network G and seed budget k, i.e.,

$$\sup_{G,k} \frac{opt_a(G, k)}{opt_n(G, k)}.$$

They proved the following.

Theorem 9.4.4 (Peng and Chen [327]) *Under the IC model with myopic feedback, the adaptivity gap for the influence maximization problem is at most 4.*

Corollary 9.4.5 (Peng and Chen [327]) *The adaptive greedy produces a $\frac{1}{4}(1 - e^{-1})$-approximate solution for the adaptive influence maximization under the IC model with myopic feedback.*

9.5 Adaptivity via Non-adaptivity

In [327], the following result is also established.

Theorem 9.5.1 *Under the IC model with myopic feedback, the adaptivity gap for the influence maximization problem is at least $e/(e - 1)$.*

This means that for any $\varepsilon > 0$, there exist G and k such that

$$\frac{opt_a(G, k)}{opt_n(G, k)} > e/(e - 1) - \varepsilon.$$

Actually, in the sketch proof of Theorem 9.5.1, some bipartite graphs are constructed which are claimed to meet required property.

However, the truth of Theorem 9.5.1 could be suspected due to the following results in [85].

Theorem 9.5.2 (Du et al. [85]) *Under IC model, the influence spread is independent from feedback model, i.e., its value is independent from the feedback model, only dependent to seed locations.*

Proof Let us first consider a simple case that the social network is with deterministic model, i.e., for every edge (u, v), $p_{uv} = 1$. In this case, at the end of information diffusion process a node v becomes active if and only if there exists a path from a seed u to node v. Therefore, the set of active nodes at the end of information diffusion process depends on only seed locations, i.e., it does not matter whether seed allocation is nonadaptive or adaptive, or what feedback model is.

Next, consider the general case of a social network with IC model. An edge (u, v) is said to be *alive* if v is activated by v through edge (u, v) during information diffusion process. An edge (u, v) is said to be *blocked* if (u, v) is not alive. All alive edges induce a subgraph H of social network G, called a *realization*. Then, H is realized with probability

$$\Pr[H] = \left(\prod_{(u,v):\text{alive}} p_{uv} \right) \left(\prod_{(u,v):\text{blocked}} (1 - p_{uv}) \right).$$

Let $\mathcal{G}$ be the set of all realizations. Let $\sigma_{IC}(S)$ denote the influence spread with respect to seed set S (together with their locations). Let $\sigma^H(S)$ denote the influence spread for subgraph H with deterministic model with respect to seed set S. Then, we have

$$\sigma_{IC}(S) = \sum_{H \in \mathcal{G}} \Pr[H] \sigma^H(S). \tag{9.1}$$

(See in the proof of Lemma 2.1.6.) Therefore, the influence spread for the social network with IC model depends on only seed locations, i.e., it does not matter whether seed allocation is nonadaptive or adaptive. $\qquad\square$

The following follows immediately from above theorem.

Corollary 9.5.3 *Under IC model and myopic feedback,*

$$\frac{opt_a(G, k)}{opt_n(G, k)} = 1$$

for any G and k.

By Theorem 2.4.8, Theorem 9.5.2 can be extended to General Triggering Model.

Theorem 9.5.4 (Du et al. [85]) *Under Gtrig model, the influence spread is independent from feedback model, i.e., its value is independent from the feedback model, only dependent to seed locations.*

By a different argument given in the proof of Lemma 2.4.5, Theorem 9.5.2 can be further extended to general threshold model. In fact, Lemma 2.4.5 (Piecemeal Growth) indicates that whether seeds are distributed at the beginning (i.e., in nonadaptive manner) or one by one (i.e., in adaptive manner), the outcome of influence is the same under the general threshold information diffusion model. Furthermore, this extension can reach general cascade model.

Theorem 9.5.5 *Under any general threshold model or general cascade model, the influence spread is independent from feedback model, i.e., its value is independent from the feedback model, only dependent to seed locations.*

How do we understand results in this section? Recall the study of adaptive influence maximization during last more than ten years: During this period, other than those claimed in the proof of the lower bound for adaptivity gap (the detail is not given in [327]), we never find a social network with IC model and a seed set so that the influence spread gives different values if seeds in the set are selected nonadaptively and adaptively. This fact has implicitly told us the existence of those results in this section, i.e., the adaptivity gap for the influence maximization problem is equal to one.

9.6 Dynamic Independent Cascade

The dynamic independent cascade (DIC) model is specially for adaptive influence, introduced in 2017 [402]. It is based on IC model with the following differences.

- When a seed is allocated at a node x, x is activated with probability p_x. This means that with probability $1 - p_x$, x keeps inactive, so that x may receive a seed, again and actually, possibly more than once until no seed is available.
- Each edge (x, y) is associated possibly more than one probabilities, say $p_{xy}^{(1)}, ..., p_{xy}^{(h)}$ for $h \geq 1$. When x has a chance to influence y through edge (x, y), the success probability selects one of those probabilities evenly at random.

Clearly, above differences make that under DIC model, nonadaptive seed distribution is meaningless. However, it indicates in [85] that an auxiliary graph G' can be constructed such that adaptive influence on graph G under DIC model and cardinality constraint is equivalent to the auxiliary graph G' under triggering model and a knapsack constraint. The construction of G' is as follows.

- For each node $v \in V$, add $k + 1$ new nodes $v_0, v_1, ..., v_k$ and $k + 1$ new edges $(v_0, v), (v_1, v_0), ..., (v_k, v_0)$ (Fig. 9.4). Set $p_{v_0 v} = 1$ and define triggering set at v_0 by

$$
T_{v_0} = \begin{cases}
\{v_1\} & \text{with probability } p_v \\
\{v_2\} & \text{with probability } (1 - p_v)p_v \\
... & \\
\{v_k\} & \text{with probability } (1 - p_v)^{k-1} p_v \\
\emptyset & \text{with probability } (1 - p_v)^k.
\end{cases}
$$

Nodes $v_1, ..., v_k$ represent k possible cases for activating v by seeds. If v is activated by the first seed allocating at v, then v_1 activates v through v_0 and this occurs with probability p_v. If v is not activated by the first seed lying on it and is activated by the second seed, then v_2 actives v through v_0 and this occurs with probability $(1 - p_v)p_v$, and so on. Based on those cases, T_v takes corresponding sets with corresponding probabilities.

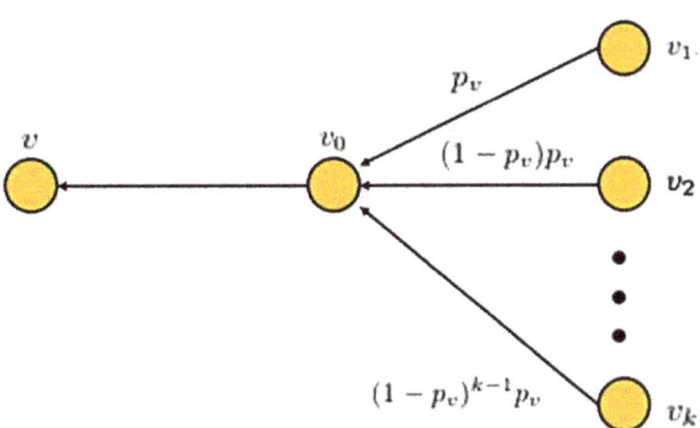

Fig. 9.4 Construction for each node v

- For each edge $(u, v) \in E$ with influence probabilities $p_{uv}^{(1)}, \ldots, p_{uv}^{(h)}$, define

$$p'_{uv} = \frac{1}{h}(p_{uv}^{(1)} + \cdots + p_{uv}^{(h)}).$$

Actually, edge (u, v) becomes alive has only one chance when u is freshly-active. In this case, the success probability for (u, v) selects a value from $\{p_{uv}^{(1)}, \ldots, p_{uv}^{(h)}\}$ evenly at random. Hence, the resulting success probability is p'_{uv}.
- Let $N^-(v)$ denote the set of incoming neighbors of v in G. For $A \subseteq N^-(v)$, define triggering set at v by

$$T_v = \{v_0\} \cup A \text{ with probability } \left(\prod_{u \in A} p'_{uv}\right)\left(\prod_{u \in N^-(v) \setminus A} (1 - p'_{uv})\right).$$

Since $p_{v_0 v} = 1$, T_v must contain v_0. The probability is calculated based on the fact that every node in $N^-(v)$ influences v independently.

Let $STri$ denote the special triggering model described as above. Then, the adaptive influence maximization problem under DIC model can be transferred into the following:

Problem 9.6.1 Let $V^{new} = \{v_1, \ldots, v_k \mid v \in V\}$. Define weight $w(v_i) = i$ for $v_i \in V^{new}$.

$$\begin{aligned} \max \quad & \sigma_{STri}(S) \\ \text{subject to} \quad & S \subseteq V^{new} \\ & \sum_{s \in S} w(s) \leq k \end{aligned}$$

where $\sigma_{STri}(S)$ is the expected number of active nodes in V at the end of information diffusion process with seed set S.

In this section, a new approach is proposed in the study of adaptive influence maximization, by reducing an adaptive optimization problem into a nonadaptive optimization problem. This approach could be extended to other optimization problem on social influence. Here, let us mention an example, the adaptive coupon allocation problem proposed by Yuan and Tang [476]. This problem has a similar taste as the variant DIC model: The probability for a customer to accept a commodity will be increased if a higher discount is offered. Therefore, we may construct a similar auxiliary graph to solve this problem. Actually, approach introduced in this section could open a new direction in the study of influence-based adaptive optimization problems.

9.7 Adaptive Discount Allocation

In viral marketing, the seed represents a free sample or a discount coupon. In the study of nonadaptive influence, when a seed is allocated at a node, the node will be activated definitely. This means that when a free sample is given a person, the person must use it, or when a discount coupon is given a person, the person will make a purchase with the coupon definitely. Of course, this may not happen in the real world. In the study of adaptive influence, the seed is assumed not so powerful. It activates a node with certainly probability. This is more close to the case in the real world.

There is an interesting difference between the free sample and the discount coupon when consider advertise budget in the study of adaptive influence. For free sample, once it is given out, a cost is generated no matter whether the sample will be used or not. However, for discount coupon, a cost is generated only if the coupon is used. In other words, only those used coupons can be included in calculation of budget. This difference introduces a matroid constraint into submodular maximization when transfer adaptive discount allocation to a nonadaptive optimization problem.

Specifically, suppose that there are k coupon values, $c_1, ..., c_k$ and with coupon c_i, each node would make a purchase with probability p_i. In this case, to make an expected transformation, we may add a local structure to every node v as shown in Fig. 9.5.

Let $G' = (V', E')$ be the obtained network. Suppose the viral marketing is studued in social network G with dynamic IC model. Then, this study can be transformed to G' with IC model and the following changes.

- For each edge $(u, v) \in E$ with influence probabilities $p_{uv}^{(1)}, ..., p_{uv}^{(h)}$, edge (u, v) in E' has influence probability

$$p'_{uv} = \frac{1}{h} \cdot (p_{uv}^{(1)} + \cdots + p_{uv}^{(h)}).$$

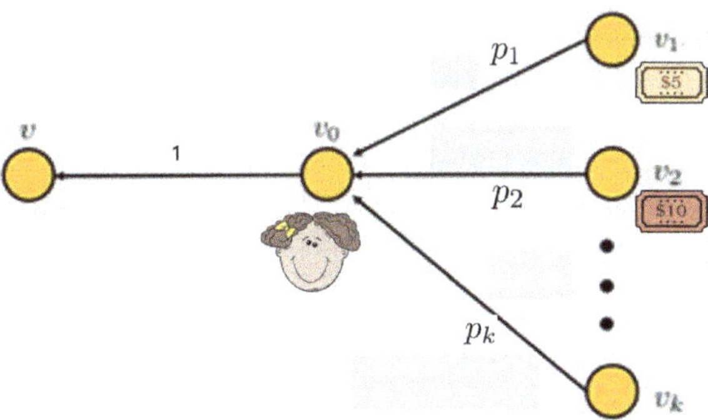

Fig. 9.5 Construction for each node v

- All seeds are allocated on nodes in

$$U = \{v_1, ..., v_k \mid v \in V\}.$$

Let S be the set of seeds. We may consider the following constraints for S.

- **Knapsack Constraint (Budget Bounded)**.

$$\sum_{s \in S} w(s) \le B$$

  where B is the budget upper bound and

$$w(s) = c_i \text{ if } s = v_i \text{ for some } v \in V.$$

- **Matroid Constraint**. For every $v \in V$,

$$|S \cap \{v_1, ..., v_k\}| \le 1,$$

  that is, every person can use at most one coupon. These inequalities give a partition matroid.

There are two possible objective functions for optimization in viral marketing.

- The number of persons who purchased the product. This is a monotone nondecreasing and submodular function with respect to the seed set.
- The revenue. This is a nonmonotone submodular function with respect to the seed set.

Therefore, the adaptive discount allocation can be reduced to a monotone submodular maximization with a knapsack constraint and a matroid constraint or a nonmonotone submodular maximization with a knapsack constraint and a matroid constraint. For these two problems, their approximation solutions are still in research progress. In the remainder of this section, an algorithm is presented to record such a research outcome. This algorithm is to give a $(1 - e^{-2})/2$-approximation with running time $O(n^6)$, for monotone submodular maximization with a knapsack constraint and a matroid constraint.

 This is a local search algorithm with three operation, adding, deleting, and swap. To unify the description about these three operations, a special element ϕ is introduced to satisfying the following properties.

- Cost is zero, i.e., $w(\phi) = c_\phi = 0$.
- For any set A, $A \cup \{\phi\} = A$.

For any set A, let $L(A)$ be the set of swaps involving elements from A. The marginal density profit of a swap (x, y) is defined by

$$\rho_{(x,y)} = \frac{f((A \setminus \{y\}) \cup \{x\}) - f(A)}{c_x}.$$

Consider Algorithm 44. It gives an approximation solution for maximizing a nonnegative, monotone nondecreasing, submodular function $f : 2^U \to \mathbb{R}_{\geq 0}$ subject to the following constraints on set A:

- (Knapsack Constraint) $\sum_{u \in A} c_u \leq B$, where all c_u are positive integers, and
- (Matroid Constraint) $A \in \mathcal{F}$ where $\mathcal{M} = (U, \mathcal{F})$ is a matroid.

This algorithm consists of two parts.

- Let $S^* = \{u_1, u_2, ..., u_p\}$ be an optimal solution of given instance, in the ordering that

$$u_i = \mathrm{argmax}_{u \in S^* \setminus \{u_1, ..., u_{i-1}\}} (f(\{u, u_1, ..., u_{i-1}\}) - f(\{u_1, ..., u_{i-1}\})).$$

Let $Y = \{u_1, u_2\}$. Since S^* is unknown, in the first part, we have to guess u_1 and u_2. In implementation, "guess" will be replaced by "for any two elements".
- In the second part, algorithm starts with $S = Y$ and expand S by applying iterations of greedy swaps while keep knapsack and matroid constraints.

Algorithm 44 Adaptive Greedy Algorithm.

Input: A nonnegative, monotone nondecreasing submodular function $f : 2^U \to \mathbb{R}_{\geq 0}$, a knapsack constraint $\sum_{u \in A} c_u \leq B$ and a matroid $\mathcal{M}$.
Output: An approximation solution A.

1: Let $S^* = \{u_1, u_2, ..., u_p\}$ be an optimal solution.
2: Let $u_1 = \mathrm{argmax}_{u \in S^*}(f(\{u\}) - f(\emptyset))$
3: and $u_2 = \mathrm{argmax}_{u \in S^* \setminus \{u_1\}}(f(\{u, u_1\}) - f(\{u_1\}))$.
4: Guess $Y = \{u_1, u_2\}$.
5:
6: Initially, $S \leftarrow Y$ and $\alpha \leftarrow 1$.
7: **while** $\alpha = 1$ **do**
8:   $\alpha \leftarrow 0$.
9:   $L \leftarrow L(S)$.
10:   **while** $\alpha = 0$ and $L \neq \emptyset$ **do**
11:     Pick and remove from L a swap (x, y) with maximum $\rho_{(x,y)}$.
12:     **if** $(y \notin Y$ and $\rho_{(x,y)} > 0$ and $C(S) - c_y + c_x \leq B)$ **then**
13:       $S \leftarrow (S \setminus \{y\}) \cup \{x\}$
14:       $\alpha \leftarrow 1$
15:     **end if**
16:   **end while**
17: **end while**
18: **return** S.

Next, an analysis on performance ratio is given.

Theorem 9.7.1 (Sarpatwar et al. [352]) *Algorithm 44 produces an approximation solution with performance ratio* $(1 - e^{-2})/2$.

Before giving the proof of Theorem 9.7.1, let us present some lemmas.

Lemma 9.7.2 *Let* B_1 *and* B_2 *be two bases of a matroid* $\mathcal{M}$. *Then, for any* $x \in B_2 \setminus B_1$, *there exists* $y \in B_1 \setminus B_2$ *such that* $(B_1 \setminus \{y\}) \cup \{x\}$ *is still a basis of* $\mathcal{M}$.

Proof Consider a maximal independent set F in $B_1 \cup \{x\}$ such that $x \in F$. By the property of matroid, $|F| = |B_1|$. Let $y \in (B_1 \cup \{x\}) \setminus F$. Then, $(B_1 \setminus \{y\}) \cup \{x\} = F$, which is a basis. $\qquad\square$

Lemma 9.7.3 *Let* S *and* T *be two independent sets of a matroid* $\mathcal{M} = (U, \mathcal{F})$. *Then, there exists a mapping* $b : T \setminus S \to (S \setminus T) \cup \{\phi\}$ *such that for all* $u \in T \setminus S$, $(S \setminus \{b(u)\}) \cup \{u\} \in \mathcal{F}$, *and for all* $y \in S \setminus T$, $|b^{-1}(y)| \leq 1$.

Proof Let B be a basis containing T. By argumentation property, B has $|B| - |S|$ elements which are added to S to form another basis D.

For each element $x \in T \cap (D \setminus S)$, define $b(x) = \phi$.

For each $x \in T \setminus (D \setminus S)$, by Lemma 9.7.2, there exists $y \in D \setminus B$ such that $(D \setminus \{y\}) \cup \{x\}$ is still a basis. Since $D \setminus S \subseteq B$, we must have $y \in S$, which implies $y \in S \setminus T$. Define $b(x) = y$. Then, $(S \setminus \{b(x)\}) \cup \{x\} \in \mathcal{F}$. Replace S by $(S \setminus \{b(x)\}) \cup \{x\}$. Inductively, we can define $b(x)$ for all $x \in T \setminus (D \setminus S)$. $\qquad\square$

Lemma 9.7.4 (Wolsey [436]) *Let* P *and* D *be two positive integers and* $\{\gamma_i \mid i = 1, ..., P\}$ *a set of real-valued numbers. Then,*

$$\frac{\sum_{i=1}^{P} \gamma_i}{\min_{i \in [1..P]}(\sum_{i=1}^{t-1} \gamma_i + D\gamma_t)} \geq 1 - (1 - \frac{1}{D})^P \geq 1 - e^{-P/D}.$$

Proof Denote

$$Z_1 = \sum_{i=1}^{P} \gamma_i$$

$$Z_2 = \min_{i \in [1..P]} (\sum_{i=1}^{t-1} \gamma_i + D\gamma_t).$$

Let $\rho_i = \gamma_i / Z_2$. Then, it suffices to show that the following LP has optimal objective function value is at least $1 - (1 - \frac{1}{D})^P$.

$$\min \quad \sum_{i=1}^{P} \rho_i$$

$$\text{subject to} \quad D \cdot \rho_1 \geq 1$$
$$\rho_1 + D \cdot \rho_2 \geq 1$$
$$\cdots$$
$$\rho_1 + \cdots + \rho_{P-1} + D \cdot \rho_P \geq 1.$$

Its dual LP is as follows.

$$\max \quad \sum_{i=1}^{P} v_i$$

$$\text{subject to} \quad D \cdot v_1 + v_2 + \cdots + v_P \leq 1$$
$$D \cdot v_2 + \cdots + v_P \leq 1$$
$$\cdots$$
$$D \cdot v_P \leq 1$$

Set

$$\rho_i = v_{P-i} = \frac{1}{D}(1 - \frac{1}{D})^{i-1}.$$

Then, all constraints in above LP and dual LP are satisfied with equality sign. Meanwhile

$$\sum_{i=1}^{P} \rho_i = \sum_{i=1}^{P} v_i = 1 - (1 - \frac{1}{D})^P.$$

Therefore, $1 - (1 - \frac{1}{D})^P$ is the optimal objective function value. $\qquad\square$

Lemma 9.7.5 *Consider an element u_ℓ with $\ell \geq 3$ and a set $W \subseteq U \setminus \{u_1, u_2, u_\ell\}$. Then, for any swap (u_ℓ, w) such that $w \in W \cup \{\phi\}$,*

$$f((Y \cup W \setminus \{w\}) \cup \{u_\ell\}) - f(Y \cup W) \leq f(Y)/2/.$$

Proof Note that

$$f((Y \cup W \setminus \{w\}) \cup \{u_\ell\}) - f(Y \cup W)$$
$$\leq f((Y \cup W \setminus \{w\}) \cup \{u_\ell\}) - f(Y \cup W \setminus \{w\})$$
$$\leq f(\{u_\ell\}) - f(\emptyset)$$
$$\leq f(\{u_1\}) - f(\emptyset)$$
$$\leq f(\{u_1\})$$

and

$$f((Y \cup W \setminus \{w\}) \cup \{u_\ell\}) - f(Y \cup W)$$
$$\leq f((Y \cup W \setminus \{w\}) \cup \{u_\ell\}) - f(Y \cup W \setminus \{w\})$$
$$\leq f(\{u_1, u_\ell\}) - f(\{u_1\})$$
$$\leq f(\{u_1, u_2\}) - f(\{u_1\}).$$

Adding two inequalities together, we obtain the inequality in the lemma. $\square$

Now, it is ready to prove Theorem 9.7.1.

Proof (Theorem 9.7.1) Let S be the solution obtained by Algorithm 44. Recall that S^* be the optimal solution. Suppose the guess of u_1 and u_2 are correct. By Lemma 9.7.3, there exists a mapping b from $S^* \setminus S$ to $(S \setminus S^*) \cup \{\phi\}$ such that for any $x \in S^* \setminus S$, $(Ssetminus\{b(x)\}) \cup x$ is independent and for any $y \in S \setminus S^*$, $|b^{-1}(y)| \leq 1$. Next, the proof is divided into two cases.

Case 1. For every $x \in S^* \setminus S$, swap $(x, b(x))$ is rejected at the last iteration duo to $\rho_{(x,b(x))} \leq 0$, not due to $C(S) - c_{b(x)} + c_x > B$.

In this case, we have

$$f(S^*) - f(S) \leq \sum_{x \in S^* \setminus S} (f(S \cup \{x\}) - f(S))$$
$$\leq \sum_{x \in S^* \setminus S} (f((S \setminus \{b(x)\}) \cup \{x\}) - f(S \setminus \{b(x)\}))$$
$$\leq \sum_{x \in S^* \setminus S} (f(S) - f(S \setminus \{b(x)\}))$$
$$\leq f(S).$$

The first two inequalities are duo to monotone submodular property. The third inequality is because of $\rho_{(x,b(x))} \leq 0$. The fourth inequality is obtained by the following argument: Since for any $y \in S \setminus S^*$, $|b^{-1}(y)| \leq 1$, we have that for $x, x' \in S^* \setminus S$ with $x \neq x'$, if $b(x) \neq \phi$ and $b(x') \neq \phi$, then $b(x) \neq b(x')$. Let $\{b(x) \in S \setminus S^* \mid x \in S^* \setminus S\} = \{u_1, u_2, ..., u_K\}$ where $K \leq |S^* \setminus S|$. Then,

$$\sum_{x \in S^* \setminus S} (f(S) - f(S \setminus \{b(x)\})) \leq \sum_{i=1}^{K} (f(\{u_1, ..., u_i\}) - f(\{u_1, ..., u_{i-1}\}))$$
$$\leq f(\{u_1, ..., u_K\}) - f(\emptyset)$$
$$\leq f(\{u_1, ..., u_K\})$$
$$\leq f(S).$$

Case 2. There exists a swap $(x, b(x))$, for $x \in S^* \setminus S$, which is rejected in the last iteration, due to violation of the knapsack constraint.

Let S^ℓ, for $\ell \geq 0$, denote the subset of elements in S right after the ℓ'th iteration of the algorithm. Clearly, $S^0 = Y$. Let b_ℓ be the mapping from $S^* \setminus S^\ell$ to $S^\ell \setminus S^*$, satisfying conditions claimed in Lemma 9.7.3. Note that for each iteration ℓ except the last one, a swap (x_ℓ, y_ℓ) is carried out with $\rho_\ell = \rho_{(x_\ell, y_\ell)} > 0$.

Let $t * +1$ be the first iteration in which a swap $(u^*, b_{t^*}(u^*))$, for $u^* \in S^* \setminus S^{t^*+1}$ is considered and is rejected due to violation of knapsack constraint. Such iteration exists because it can be deduced from the assumption of Case 2. By choice of t^*, for $t = 0, 1, ..., t^* - 1$ and for any $x \in S^* \setminus S^t$, swap $(x, b_{t+1}(x))$ satisfies the knapsack constraint and hence, we must have $\rho_{(x, b_{t+1}(x))} \leq \rho_{t+1}$.

Define $g(A) = f(A) - f(Y)$. Then $g(A)$ is monotone nondecreasing and submodular. Therefore,

$$
\begin{aligned}
g(S^*) - g(S^t) &\leq \sum_{x \in S^* \setminus S^t} \left(g(S^t \cup \{x\}) - g(S^t) \right) \\
&\leq \sum_{x \in S^* \setminus S^t} \left(g((S^t \{b_t(x)\}) \cup \{x\}) - g(S^t \setminus \{b_t(x)\}) \right) \\
&= \sum_{x \in S^*} \left(g(S^t) - g(S^t \setminus \{b_t(x)\}) \right) \\
&\quad + \sum_{x \in S^* \setminus S^t} \left(g((S^t \setminus \{b_t(x)\}) \cup \{x\}) - g(S^t) \right) \\
&\leq g(S^t) + \sum_{x \in S^* \setminus S^t} \left(g((S^t \setminus \{b_t(x)\}) \cup \{x\}) - g(S^t) \right).
\end{aligned}
$$

The first two inequalities follow from the monotone nondecreasing and submodular properties of g. The third inequality is obtained from the same argument as that for the fourth inequality in Case 1. Note that

$$
\begin{aligned}
g((S^t \setminus \{b_t(x)\}) \cup \{x\}) - g(S^t) &= f((S^t \setminus \{b_t(x)\}) \cup \{x\}) - f(S^t) \\
&\leq c_x \cdot \rho_{(x, b_t(x))} \\
&\leq c_x \cdot \rho_{t+1},
\end{aligned}
$$

and

$$
\sum_{x \in S^* \setminus S^t} c_x B - c(Y).
$$

Therefore,

$$
\sum_{x \in S^* \setminus S^t} \left(g((S^t \setminus \{b_t(x)\}) \cup \{x\}) - g(S^t) \right) \leq (B - c(Y))\rho_{t+1}.
$$

It follows that

$$g(S^*) \le 2(g(S^t) + \frac{B''}{2}\rho_{t+1}) \tag{9.2}$$

where $B'' = B - c(Y)$.

Note that in each iteration t, for $t = 1, 2, ..., t^*$, the algorithm performs a swap (x_t, y_t). Thus,

$$(f(S^t) - f(S^{t-1}))/c_{x_t} = \rho_t.$$

Define $B_0 = 0$, $B_t = \sum_{h=1}^t c_{x_h}$ for $t = 1, ..., t^*$, and $B_{t^*+1} = B_{t^*} + c_{u^*}$. Then, the following facts can be obtained.

- By assumption on $t^* + 1$, $B_{t^*+!} > B - c(Y) = B''$.
- Note that

$$g(S^{t^*}) = \sum_{t=1}^{t^*}(g(S^t) - g(S^{t-1}))$$

$$= \sum_{t=1}^{t^*} \rho_t c(x_t)$$

$$= \sum_{t=1}^{t^*} \rho_t (B_t - B_{t-1}).$$

Denote $\rho_{t^*+1} = \rho_{(u^*, b_{t^*}(u^*))}$. Then $g((S^{t^*} \setminus \{b_{t^*}(u^*)\}) \cup \{u^*\}) - g(S^{t^*}) = \rho_{t^*+1}c_{u^*} = \rho_{t^*+1}(B_{t^*+1} - B_{t^*})$. Therefore,

$$g((S^{t^*} \setminus \{b_{t^*}(u^*)\}) \cup \{u^*\}) = \sum_{t=1}^{t^*+1} \rho_t (B_t - B_{t-1}).$$

Define $\gamma_j = \rho_t$ for $j = B_{t-1} + 1, ..., B_t$. Denote $B' = B_{t^*+1}$. Then,

$$g((S^{t^*} \setminus \{b_{t^*}(u^*)\}) \cup \{u^*\}) = \sum_{j=1}^{B'} \gamma_j.$$

We also have

$$g(S^t) = \sum_{j=1}^{B_t} \gamma_j.$$

Therefore,

$$\min_{s\in[1..B']}\left\{\sum_{j=1}^{s}\gamma_j + \frac{B''}{2}\gamma_s\right\} = \min_{t\in[1..t^*]}\left\{\sum_{j=1}^{B_t}\gamma_j + \frac{B''}{2}\gamma_{B_t+1}\right\}$$

$$= \min_{t\in[1..t^*]}\left\{g(S^t) + \frac{B''}{2}\gamma_{B_t+1}\right\}$$

$$\geq g(S^*)/2.$$

The inequality is due to (9.2).

By Lemma 9.7.4,

$$\frac{g((S^{t^*}\setminus\{b_{t^*}(u^*)\})\cup\{u^*\})}{g(S^*)} \geq \frac{\sum_{j=1}^{B'}\gamma_j}{2\cdot\min_{t\in[1..t^*]}\left\{g(S^t) + \frac{B''}{2}\gamma_{B_t+1}\right\}}$$

$$\geq \frac{1}{2}\cdot(1 - (1 - \frac{2}{B''})^{B'})$$

$$\geq \frac{1}{2}(1 - e^{-2\cdot\frac{B'}{B''}})$$

$$\geq \frac{1}{2}(1 - e^{-2}).$$

By Lemma 9.7.5,

$$f(S^{t^*}) = f(Y) + g(S^{t^*})$$

$$= f(Y) + g((S^{t^*}\setminus\{b_{t^*}(u^*)\})\cup\{u^*\})$$

$$- (g((S^{t^*}\setminus\{b_{t^*}(u^*)\})\cup\{u^*\}) - g(S^{t^*}))$$

$$\geq f(Y) + \frac{1 - e^{-2}}{2}g(S^*) - \frac{f(Y)}{2}$$

$$\geq \frac{1 - e^{-2}}{2}f(S^*).$$

$\square$

For running time, there is an improvement as follows.

Theorem 9.7.6 (Sarpatwar et al. [352]) *Keeping the performance ratio unchanged, Algorithm 44 can be modified to run in $\tilde{O}(n^6)$ time where $\tilde{O}$ means that a poly-logarithmic factor is hidden.*

Proof Let $S^* = \{u_1, u_2, ..., u_p\}$. By Lemma 9.7.5,

$$f(S^*) = f(Y) + \sum_{i=3}^{p}(f(\{u_1, u_2, ..., u_i\}) - f(\{u_1, u_2, ..., u_{i-1}\}))$$

$$\leq f(Y) + (p-2) \cdot \frac{f(Y)}{2}$$

$$\leq \frac{n}{2} \cdot f(Y).$$

To improve the running time, we modify Algorithm 44 by inserting a condition for performing a swap. Let ε be a positive number which will be determined later. The condition is that a swap (not with ϕ) is allowed to perform only if it can increase the value of $f(\cdot)$ by a factor of at least $1 + \frac{\varepsilon}{n^2}$. This condition would limit the number of swaps, which is upper-bounded by

$$\frac{\ln(n/2)}{\ln(1 + \frac{\varepsilon}{n^2})} \leq (\ln n) \cdot \frac{n^2}{\varepsilon}.$$

For analysis of Case 1, note that for any $x \in S^* \setminus S$, swap $(x, b(x))$ is rejected due to $\rho_{(x,b(x))} < 0$ or $f((S \setminus \{b(x)\}) \cup \{x\}) < (1 + \frac{\varepsilon}{n^2})f(S)$. Therefore, the modification is as follows.

$$f(S^*) - f(S) \leq \sum_{x \in S^* \setminus S} (f(S \cup \{x\}) - f(S))$$

$$\leq \sum_{x \in S^* \setminus S} (f((S \setminus \{b(x)\}) \cup \{x\}) - f(S \setminus \{b(x)\}))$$

$$\leq \sum_{x \in S^* \setminus S} ((1 + \frac{\varepsilon}{n^2})f(S) - f(S \setminus \{b(x)\}))$$

$$\leq \frac{\varepsilon}{n} + \sum_{x \in S^* \setminus S} (f(S) - f(S \setminus \{b(x)\}))$$

$$\leq (1 + \frac{\varepsilon}{n})f(S).$$

Thus, $f(S^*) \leq (2 + \frac{\varepsilon}{n})f(S)$. To keep the performance ratio $1 - e^{-2}$, it is sufficient to choose ε such that

$$\frac{1}{2 + \frac{\varepsilon}{n}} \geq 1 - e^{-2}$$

that is,

$$\varepsilon \leq \frac{2n}{e^2 - 1}.$$

For analysis in Case 2, a similar modification can be made. Let C be the subset of $x \in S^* \setminus S$ such that

$$g((S^t \setminus \{b_t(x)\}) \cup \{x\}) < (1 + \frac{\varepsilon}{n^2})g(S^t).$$

Then,

$$
\begin{aligned}
g(S^*) - g(S^t) &\leq \sum_{x \in S^* \setminus S^t} (g(S^t \cup \{x\}) - g(S^t)) \\
&\leq \sum_{x \in S^* \setminus S^t} (g((S^t\{b_t(x)\}) \cup \{x\}) - g(S^t \setminus \{b_t(x)\})) \\
&\leq \sum_{x \in C}(1 + \frac{\varepsilon}{n^2})g(S^t) \\
&\quad + \sum_{x \in S^* \setminus S^t} (g((S^t\{b_t(x)\}) \cup \{x\}) - g(S^t \setminus \{b_t(x)\})) \\
&\leq \cdots \\
&\leq (2 + \frac{\varepsilon}{n})g(S^t) + (B - c(Y))\rho_{t+1}.
\end{aligned}
$$

With this modification, along the proof line in Case 2, we can finally obtain

$$f(S^{t^*}) \geq \frac{f(Y)}{2 + \varepsilon/n)e^2} + \frac{1}{2 + \varepsilon/n} \cdot (1 - e^{-2})f(S^*).$$

It follows that to keep the approximation performance $1 - e^{-2}$, it is sufficient to choose $\varepsilon \leq \frac{4}{e^2 - 1}$.

Now, guessing Y spends $O(n^2)$ time. For each guessed Y, there are $\tilde{O}(n2)$ successful swaps. Each such swap needs $O(n^2)$ time to carry out. Therefore, the total running time is $\tilde{O}(n^6)$. $\qquad\square$

9.8 Adaptive Rumor Blocking

There are three approaches to block the rumor, cutting nodes, cutting edges, and placing protectors. We may also consider to distribute the node-cut, the edge-cut, and the protector in a adaptive manner. How to distribute them to maximize the number of protected nodes, or to minimize the number of rumor-influenced nodes? To study this problem, the following results from techniques that we used in previous sections.

Theorem 9.8.1 *Consider a social network $G = (V, E)$ with IC model. Suppose that G contains a set of nodes spreading a rumor. Then the nonadaptive distribution is the best way for placement of given number of node-cuts, edge-cuts, and protectors among all feedback models.*

Proof Let us first consider a special case of IC model, the deterministic model. Let R be the set of nodes which are able to spread a rumor. Initially, $R \neq \emptyset$. As information diffusion process is going step by step, R is growing. Suppose we place a protector p in G. Then p can protect those nodes each of which is closer to p than to R. Clearly, the number of such nodes is getting smaller if p is placed later because R is growing. Therefore, placing p at the first step is the best in order to maximize the number of protected nodes or to minimize the number of rumor-infected nodes. A similar argument can be applied to the node-cut and the edge-cut. This means that the theorem holds for the deterministic model.

Let R_0 be the initial set of R. Let A be the set of node-cuts, edge-cuts, and protectors. Denote by $\sigma(A, R_0)$ the expected number of protected nodes (or the expected number of rumor-influenced nodes). Then,

$$\sigma(A, R_0) = \sum_{X \subseteq G} \Pr[X]\sigma^X(A, R_0)$$

where X is over all subgraphs of G, $\Pr[X]$ is the probability that X is realized, $\sigma^X(A, R_0)$ denotes the number of protected nodes (or the number of rumor-influenced nodes) in subgraph X under deterministic model. Through this equality, above proved properties for deterministic model can be extended to the IC model. $\qquad\square$

9.9   Adaptive Nonsubmodular

In Chap. 5, we studied several nonsubmodular optimization problems. Each of them has an adaptive variation, i.e., an adaptive nonsubmodular optimization problem. For those problems, the sandwich method is often extended and utilized. Here, let us study an example, a profit maximization in [126]. This work compared the nonadaptive version with adaptive version, which can make us get better understanding their difference and relationship. Let us start from description of its nonadaptive version.

Consider a social network $G = (V, E)$ with the IC model. Let $c : V \times V \to \mathbb{R}_{\geq 0}$ be a profit function such that $c(u, v) = c(v, u)$. For any seed set S, denote by $I(S)$ the set of all active nodes at end of the diffusion process. For any node subset A, let $P(A)$ denote the set of unordered pair of nodes in A. Define

$$f(S) = \mathbb{E}[\sum_{(u,v) \in P(I(S))} c(u, v)].$$

Problem 9.9.1 (*Nonadaptive Profit Maximization*) Given a positive integer k, we are interested in studying

$$\max \quad f(S)$$
$$\text{subject to} \quad |S| \le k.$$

To see adaptive version, consider a social network $G = (V, E)$ with the IC model and the full feedback. Let $I(S, \phi)$ denote the set of nodes activated by seed set S under realization ϕ. Define

$$f(S, \phi) = \sum_{(u,v) \in P(I(S,\phi))} c(u, v).$$

For a policy π, let $N(\pi, \phi)$ denote the seed set selected by policy π under realization ϕ. The expected profit of π is defined by

$$f_{avg}(\pi) = \mathbb{E}[f(N(\pi, \phi), \phi)]$$

where the expectation is taken over the probability distribution $p(\phi)$ of ϕ.

Problem 9.9.2 (*Adaptive Profit Maximization*) Given a positive integer k, we are interested in studying

$$\max \quad f_{avg}(\pi)$$
$$\text{subject to} \quad |N(\pi, \phi)| \le k \text{ for all } \phi.$$

Define
$$U(S, \phi) = \sum_{v \in I(S,\phi)} w(v) \tag{9.3}$$

where
$$w(v) = \frac{1}{2} \sum_{u \in V} c(v, u).$$

Also, define
$$L(S, \phi) = \sum_{\{u,v\} \in \cup_{x \in S} P(I(\{x\},\phi))} c(u, v). \tag{9.4}$$

Theorem 9.9.3 *For any ϕ, $L(S, \phi) \le f(S, \phi) \le L(S, \phi)$.*

Theorem 9.9.4 *$U(S, \phi)$ and $L(S, \phi)$ are adaptive monotone nondecreasing and adaptive submodular under the full feedback model.*

In this research direction, there exist many interesting research issues, such as extension of DS-decomposition and related sandwich theorem.

Exercises

1. Let $U(S, \phi)$ be the function defined by (9.3) in Sect. 9.9. Show that for fixed realization ϕ, $U(S, \phi)$ is monotone nondecreasing and submodular with respect to S.
2. Let $I(S, \phi)$ denote the set of all activated nodes when S is the set of selected seeds under the realization ϕ. Suppose that in the full feedback model, ϕ and ψ are realization and partial realization, respectively such that $\phi \sim \psi$. Show that

$$I(dom(\psi), \phi) = \cup_{x \in dom(\psi)} I(x, \phi).$$

3. Let $\sigma_m(S)$ denote the influence spread of seed set S under information diffusion model m and nonadaptive feedback. Let $\sigma_{m, full}(S)$ denote the influence spread of seed set S under information diffusion model m and full feedback. Show that for every seed set S, $\sigma_m(S) = \sigma_{m, full}(S)$.
4. As shown in Fig. 9.6, a social network consists of three nodes.
   Let ϕ be the realization such that all three nodes are active. Show that under the IC model and seed set $\{2, 3\}$, we have

$$\Pr[\phi] = p_{21} + p_{31} - p_{21} \cdot p_{31}$$

   and under the LT model and seed set $\{2, 3\}$, we have

$$\Pr[\phi] = p_{21} + p_{31}.$$

5. Let $L(S, \phi)$ be the function defined by (9.3) in Sect. 9.9. Show that for fixed realization ϕ, $L(S, \phi)$ is monotone nondecreasing and submodular with respect to S.
6. Consider Counterexample 9.4.1. Compute $\Delta\sigma(w \mid \psi)$ and $\Delta\sigma(w \mid \psi')$ under the full feedback model.
7. Consider a social network with the IC model. Let $\sigma(A, \phi)$ be the number of active nodes in set A under realization ϕ. Show that $\sigma(A, \phi)$ is adaptive monotone nondecreasing and adaptive submodular under full feedback model.

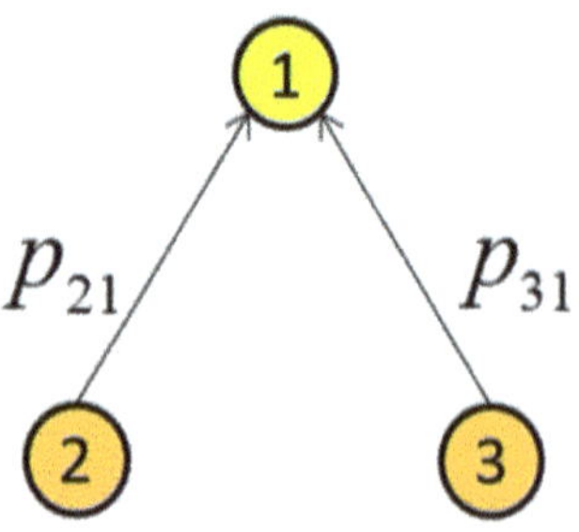

Fig. 9.6 A social network with three nodes

8. Using Counterexample 9.4.1, find, in the proof of Lemma 9.3.2, where the assumption of the full feedback model is used?

9. Consider function $f : 2^E \times \Phi \to \mathbb{R}_{\geq 0}$. Suppose that $f(D, \phi)$ is monotone nondecreasing and submodular with respect to D. Show that f is adaptive monotone nondecreasing and adaptive submodular under any probabilistic distribution of realization ϕ.

10. Show that for any function $f : 2^E \times \Phi \to \mathbb{R}$ with probabilistic distribution $p(\phi)$, there are two adaptive monotone nondecreasing and submodular function $g : 2^E \times \Phi \to \mathbb{R}$ and $h : 2^E \times \Phi \to \mathbb{R}$ under the same distribution $p(\phi)$ such that $f = g - h$.

11. Let $I(S)$ denote the influence spread of seed set S under the IC model and nonadaptive feedback. Let $I_{full}(S)$enote the influence spread of seed set S under the IC model and full feedback. Show that for every seed set S, $I(S) = I_{full}(S)$.

12. (Adaptivity Gap [327]) The *adaptivity gap* in the IC model is defined as follows:

$$\sup_{G,k} \frac{opt_a(G, k)}{opt_n(G, k)}$$

where G and k are over all possible social networks and seed budgets, $opt_a(G, k)$ is the objective function value of optimal solution for adaptive influence maximization on instance (G, k) and $opt_n(G, k)$ is the objective function value of optimal solution for nonadaptive influence maximization on the same instance (G, k). Show that this gap is equal to one.

13. Let S and T be two independent sets of a matroid $\mathcal{M}$ and $|S| = |T|$. Show that there exists a one-to-one onto mapping τ from $S \setminus T$ to $T \setminus S$ such that for any $x \in S \setminus T, (S \setminus \{x\}) \cup \{\tau(x)\}$ is independent.

14. Consider a nonnegative monotone nondecreasing submodular function $f : 2^U \to \mathbb{R}_{\geq 0}$. Let S and T be two subsets. Suppose b is a one-to-one mapping from $S \setminus T$ to $T \setminus S$. Then

$$\sum_{x \in S\setminus T} (f(T) - f(T \setminus \{b(x)\})) \leq f(T).$$

15. (Sarpatwar et al. [352]) Algorithm 44 can be extended to Algorithm 45 for monotone submodular function maximization with a knapsack constraint and k matroid constraints. In this algorithm, a k-swap is to add one element and meanwhile, to remove k elements.

Algorithm 45 Adaptive Greedy Algorithm.

Input: A nonnegative, monotone nondecreasing submodular function $f : 2^U \to \mathbb{R}_{\geq 0}$, a knapsack constraint $\sum_{u \in A} c_u \leq B$ and k matroid $\mathcal{M}_i = (U, \mathcal{F}_i$ for $i = 1, ..., k$.
Output: An approximation solution A.

1: Let $S^* = \{u_1, u_2, ..., u_p\}$ be an optimal solution.
2: Let $u_1 = \text{argmax}_{u \in S^*}(f(\{u\}) - f(\emptyset))$
3: and $u_2 = \text{argmax}_{u \in S^* \setminus \{u_1\}}(f(\{u, u_1\}) - f(\{u_1\}))$.
4: Guess $Y = \{u_1, u_2\}$.
5:
6: Initially, $S \leftarrow Y$ and $\alpha \leftarrow 1$.
7: **while** $\alpha - 1$ **do**
8:    $\alpha \leftarrow 0$.
9:    $L \leftarrow L_k(S)$. % the set of all k-swaps %
10:    **while** $\alpha = 0$ and $L \neq \emptyset$ **do**
11:      Pick and remove from L a k-swap $(x, \bar{y})$ with maximum $\rho_{(x,\bar{y})} = \frac{f((S \setminus \bar{y}) \cup \{x\}) - f(S)}{c_c}$.
12:      **if** $(y \notin Y$ and $\rho_{(x,\bar{y})} > 0$ and $C(S) - c(\bar{y}) + c_x \leq B)$ **then**
13:        $S \leftarrow (S \setminus \bar{y}) \cup \{x\}$
14:        $\alpha \leftarrow 1$
15:      **end if**
16:    **end while**
17: **end while**
18: **return** S.

Show the following for analysis of Algorithm 45:

(a) Let S and T be two independent sets in the intersection of k matroids. Let $[S \setminus T]^{\leq k}$ denote the collection of subsets of $S \setminus T$ such that each subset has size at mot k. Then, there exists a mapping from $T \setminus S$ to $[S \setminus T]^{\leq k}$ such that each element of $S \setminus T$ appears in at most k mapped subsets.

(b) Consider an element $u_\ell \in S^*$, for $\ell \geq 3$, and a subset $W \subseteq U \setminus \{u_1, u_2, u_\ell\}$. Then, for any k-swap $(u_\ell, \bar{w})$ with $\bar{w} \in [W]^{\leq k}$,

$$f(Y \cup W \setminus \bar{w}) \cup \{u_\ell\}) - f(Y \cup W) \leq f(Y)/2.$$

(c) Let $b : S^* \setminus S \to [S \setminus S^*]^{\leq k}$ be the mapping mentioned in (a). Then,

$$\sum_{x \in S^* \setminus S} (f(S) - f(S \setminus b(x))) \leq k \cdot f(S).$$

(d) Show that for any fixed $k \geq 1$, Algorithm 45 produces an approximation solution with performance ratio $\frac{1 - e^{-(k+1)}}{k+1}$.

(e) Keeping the performance ratio unchanged, Algorithm 45 can be modified to run in $\tilde{O}(n^{k+5})$ time.

Historical Notes

A lot of efforts have been made in the recent study of adaptive construction of objects and adaptive process of information [126, 141, 166, 167, 222, 244, 382, 383]. Adaptive influence maximization is an important example [48, 49, 170, 176, 353, 358, 370, 373, 380, 395, 397, 402, 477]

Golovin and Krause [141] introduced the concept of adaptive monotone nondecreasing and the concept of adaptive submodular. Using those concepts, they established the performance $1 - e^{-1}$ of adaptive greedy algorithm for adaptive submodular maximization. As an example, they showed the adaptive monotone nondecreasing and adaptive submodularity for adaptive influence maximization under the IC model and the full feedback model. Moreover, in [142], they indicated, by presenting a counterexample, that for adaptive influence maximization under the IC model and the myopic feedback, the adaptive submodularity does not hold. Furthermore, they conjectured that in this case, the adaptive greedy algorithm is still a constant-approximation for adaptive influence maximization.

Salha et al. [353] modify IC model such that edges can be reactivated at each time step. With this modified IC model, they show that myopic feedback is adaptive submodular and hence, the adaptive greedy algorithm gives a constant-approximation. To attack the conjecture, Tong and Wang [395, 396] introduce a parameter, the regret ratio, and with this parameter, established the approximation performance of adaptive greedy algorithm. Yuan and Tang [477] introduced a parameter into adaptive greedy algorithm and established the performance ratio with this parameter for partial feedback, which includes myopic feedback as a special case. The conjecture of Golovin and Krause was proved by Peng and Chen [327]. They introduced the concept of adaptivity gap and showed an upper bound 4 and a lower bound $e/(e-1)$ for the adaptivity gap. The conjecture can be proved by using the upper bound.

However, the lower bound contradicts a recent result presented by Du et al. [85]. Actually, the lower bound $e/(e-1) > 1$ means that there exist a social network G and a budget k such that $opt_a(G, k) > opt_n(G, k)$ where $opt_a(G, k)$ $(opt_n(G, k))$ is the influence spread of optimal adaptive (nonadaptive) selection of seed set. But, in [85], it is showed that under IC model, the influence spread of a seed set is independent from the selection manner, i.e., it does not matter whether the seed set is selected adaptively or nonadaptively. This implies that $opt_a(G, k) = opt_n(G, k)$ for any G and k. This seems surprising. Solving this contradiction may give us a deeper understanding the social influence.

Tong et al. [402] proposed the dynamic IC (DIC) model. Under DIC model, seeds have to be allocated adaptively since a location can receive seeds more than once. Du et al. [85] showed that adaptive influence maximization under DIC model can be transformed into a nonadaptive influence maximization in a auxiliary graph under triggering model. Such a method has potential applications in other optimization problems, such as adaptive discount allocation [476] and adaptive revenue (profit) maximization [126, 166]. Adaptive revenue maximization is an example of non-monotone adaptive submodular maximization [145].

In the proof of solution for Golovin-Krause conjecture, Peng and Chen [327] claimed that under myopic feedback, adaptive greedy algorithm produces a solution which is $(1 - e^{-1})$-approximation for optimal nonadaptive influence maximization. By result of Du et al. [85], it is also a $(1 - e^{-1})$-approximation for adaptive influence maximization under myopic feedback model; this is a adaptive nonsubmodular case. Does this mean that the current definition of adaptive submodularity is improper? Therefore, the study on adaptive influence maximization has a broader impact on the study of adaptive submodular maximization [119, 121, 171, 368].

It is interesting to note that adaptive discount allocation can be reduced to a monotone submodular function maximization with a knapsack constraint and a matroid constraint. Such a maximization problem has been studied in [10, 38, 39, 352]. Approximation performance ratio can be close to $1 - e^{-1}$ as computational time goes up rapidly. The approximation algorithm and its analysis in Sect. 9.7 come from [352]. It is still worth making efforts on practical approximation solution, faster running time with reasonable performance.

Chapter 10
Robust and Machine Learning

Computers are able to see, hear and learn. Welcome to the future.
—Dave Waters

In previous chapters, many uncertainties were ignored in the study of social networks. In this chapter, they will be considered. Employed methodologies are robust optimization and machine learning.

10.1   Robust Influence Maximization

In the study of social networks, usually a social network is given together with an information diffusion model and all parameters in the model are also given. However, in the real world, those parameters may not to be known exactly, i.e., there exist uncertainties about parameters.

Chen et al. [45] study the influence maximization in social networks $G = (V, E)$ with IC model. However, influence probability at each edge e can be varied within an interval $[l_e, r_e]$. Let $\Theta = \prod_{e \in E}[l_e, r_e]$ be the parameter space. Denote $\theta^- = (l_e)_{e \in E}$ and $\theta^+ = (r_e)_{e \in E}$. Since true probability $\theta \in \Theta$ is unknown, they employ the sandwich method (Algorithm 46) to find solution, where $\sigma_\theta(S)$ denotes the expected number of influenced nodes when S is the set of seeds.

© The Author(s), under exclusive license to Springer Nature Switzerland AG 2026
W. Wu et al., *Computational Aspects of Social Networks*, Springer Optimization and Its
Applications 234, https://doi.org/10.1007/978-3-032-14833-9_10

Algorithm 46 Greedy Algorithm.

Input: A social network $G = (V, E)$, an integer $k > 0$, and an influence probability space Θ,
Output: Seed location set S.

1: Use greedy algorithm to find a $(1 - 1/e)$-approximation S^- for IC model with parameter θ^-.
2: Use greedy algorithm to find a $(1 - 1/e)$-approximation S^+ for IC model with parameter θ^+.
3: $S = \text{argmax}(\sigma_{\theta-}(S^-), \sigma_{\theta-}(S^+))$.
4: **return** S.

For the performance of this algorithm, we have the following.

Theorem 10.1.1 *Consider a social network $G = (V, E)$ with IC model and parameter space Θ. Given a budget limit $k > 0$, Algorithm 46 produces a solution (seed set of size k) S satisfying*

$$g(\Theta, S) = \min_{\theta \in \Theta} \frac{\sigma_\theta(S)}{\sigma_\theta(S_\theta^*)} \geq \alpha(\Theta) \cdot \left(1 - \frac{1}{e}\right),$$

where

$$S_\theta^* = argmax_{S \subseteq V, |S| = k} \sigma_\theta(S)$$

and

$$\alpha(\Theta) = \frac{\sigma_{\theta-}(S)}{\sigma_{\theta+}(S^+)}.$$

Proof Clearly, $\sigma_\theta(S)$ is monotone nondecreasing with respect to θ for fixed S. Therefore,

$$\sigma_\theta(S_\theta^*) \leq \sigma_{\theta+}(S_\theta^*) \leq \sigma_{\theta+}(S_{\theta+}^*) \leq \frac{\sigma_{\theta+}(S^+)}{1 - 1/e},$$

where S_θ^* is an optimal solution for parameter θ. It implies that

$$g(\Theta, S) \geq \min_{\theta \in \Theta} \frac{\sigma_\theta(S)}{\sigma_{\theta+}(S^+)}\left(1 - \frac{1}{e}\right) = \frac{\sigma_{\theta-}(S)}{\sigma_{\theta+}(S^+)}\left(1 - \frac{1}{e}\right) = \alpha(\Theta)\left(1 - \frac{1}{e}\right).$$

$\square$

10.2　Uncertainty of Rumor Source

In the study of rumor blocking, the location of rumor source is often supposed to be already known. However, the rumor source detection is a quite hard research problem. Hence, in the real world, one may know only partial information about rumor source. Therefore, robust rumor blocking is a pretty practical research problem. For partial information, usually a set of suspicious nodes is known. About suspicious nodes, assumption has the following possibilities.

- The probability of each suspicious node is known exactly [369].
- The probability of each suspicious node is known to lie in an interval [369].
- It is known that there are at most r rumor sources.

In the literature, there are several different objective functions in optimization problems about robust rumor blocking. Different objective functions give problems with different computational complexities. In the following, let us mention a few examples.

Shi et al. [369] formulated three optimization problems. To describe them. Let V_s be the set of suspicious nodes. Each suspicious node may be turned into a rumor source node in a step before step $T + 1$. Let S_i denote the set of suspicious nodes which are turned into rumor sources at step i for $1 \le i \le T$. Denote $M = \{S_1, S_2, ..., S_T\}$. For any protector set P, define

$$R_g(P, M) = \{v \mid \exists u \in P, \forall i \in [T], \forall u' \in S_i, d_g(u, v) < d_g(u', v)\},$$

where g is a subgraph consisting of alive edges and $d_g(u, v)$ is the distance between u and v in g, Furthermore, define

$$f(P) = \mathbb{E}[R_g(P, M)],$$

where the expectation is taken over all possible alive subgraphs and possible M. The first problem formulated in [369] is the following.

Problem 10.2.1 Given a suspicious node set V_s with probability p_v for each $v \in V_s$, and an integer $k > 0$, find a set $P \subseteq V \setminus V_s$ with $|P| \le k$ to maximize $f(P)$.

Note that $f(P)$ is monotone nondecreasing and submodular. Therefore, there exists a reverse sampling algorithm to produce a $(1 - 1/e - \varepsilon)$-approximation solution. Shi et al. [369] modified an algorithm [392] to design such an algorithm.

The second problem and the third problem in [369] can be described as follows.

Problem 10.2.2 A T-prob setting is a set of probabilities for nodes in V_s. Given a set $\mathcal{T}$ of T-settings, find a seed set S such that the weighted summation of the preventing function over all T-settings can be maximized, i.e.,

$$\max H(S) = \sum_{t \in \mathcal{T}} w[t] \frac{f_t(S)}{f_t(S_t^g)},$$

where $f_t(\cdot)$ is the expected misinformation reduction under T-prob setting t, $w[t]$ its weight, and S_t^g is the solution returned by a greedy algorithm for Problem 10.2.1.

Problem 10.2.3 Let $\mathcal{T} = \{[l_v, r_v] \mid v \in V_s\}$ where each $[l_v, r_v]$ gives $v \in V_s$ a T-prob range. This problem aims to find a set $S \subseteq V \setminus V_s$ of k nodes that maximizes the robust ratio:

$$S = \operatorname{argmax}_{S \subseteq V \setminus V_s, |S| = k} g(\mathcal{T}, S),$$

where

$$g(\mathcal{T}, S) = \min_{t \in \mathcal{T}} \frac{f_t(S)}{f_t(S_t^*)}.$$

Approximation solutions for these three problems can be found in exercise of this chapter.

In [495], Zhu et al. assume that locations of rumor source are unknown, but it is known that the number of rumor sources is at most ℓ. Consider a social network $G = (V, E)$ with the IC model, Let R be a set of rumor sources and S a set of locations for protectors. Let $\sigma_r(S, R)$ denote the expected number of nodes that are activated by rumor source set R when S is selected to be allocated by protectors. Since R is not exactly known before S is selected, the worst case has to be considered. Therefore, minimization is made on $f(S) = \max_R \sigma_r(R, S)$.

Theorem 10.2.4 (Zhu et al. [495]) $f(S)$ *is neither submodular nor supermodular.*

Proof Consider Fig. 10.1 and $\ell = 1$. Suppose that the positive cascade has the higher priority.

Let $A = \emptyset$ and $B = \{v_9\}$. Then,

$$f(A) = \max_{|R| \leq 1} \sigma_r(R, \emptyset) = \sigma_r(\{v_1\}, \emptyset) = 8$$

and

$$f(B) = \max_{|R| \leq 1} \sigma_r(R, \{v_9\}) = \sigma_r(\{v_1\}, \{v_9\}) = 5.$$

Moreover,

$$f(A \cup \{v_1\}) = \sigma_r(\{v_9\}, \{v_1\}) = 4$$

and

$$f(B \cup \{v_1\}) = \sigma_r(\{v_2\}, \{v_9, v_1\}) = 2.$$

Fig. 10.1 Counterexample

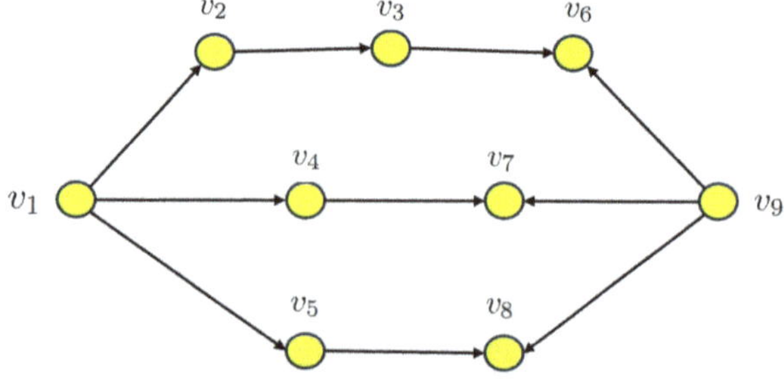

Thus,

$$f(A \cup \{v_1\}) - f(A) = -4 < -3 = f(B \cup \{v_1\}) - f(B).$$

Therefore, $f(S)$ is not submodular.

To see that $f(S)$ is not supermodular, let $A = \{v_1\}$ and $B = \{v_1, v_9\}$. Then $f(A) = 4$ and $f(B) = 2$. Moreover,

$$f(A \cup \{v_2\}) = \max_{|R| \leq 1} \sigma_r(R, \{v_1, v_2\}) = \sigma_r(\{v_9\}, \{v_1, v_2\}) = 4$$

and

$$f(B \cup \{v_2\}) = \max_{|R| \leq 1} \sigma_r(R, \{v_1, v_2, v_9\}) = \sigma_r(\{v_3\}, \{v_1, v_2, v_9\}) = 1.$$

Therefore,

$$f(A \cup \{v_2\}) - f(A) = 0 > -1 = f(B \cup \{v_2\}) - f(B).$$

Hence, $f(S)$ is not supermodular. $\qquad\square$

Theorem 10.2.5 *Computing $\sigma_r(R, S)$ is #P-hard.*

Proof When $S = \emptyset$, $\sigma_r(R, \emptyset)$ is exactly the expected number of nodes influenced by R. Hence, computing it in the IC model is #P-hard. $\qquad\square$

Zhu et al. [495] employed a parameterized method to give a solution. In [498], Zhu et al. discovered some new properties on $\sigma_r(R, S)$.

Theorem 10.2.6 *Given R, denote $h(S) = \sigma_r(R, S)$. Then the following holds.*

(a) $h(S) = 0$ for $|R| \leq |S|$.
(b) $h(S)$ is monotone nonincreasing.
(c) $h(S)$ is supermodular.

Proof (a) Since R is given and $|R| \leq |S|$, protectors can be placed to cover all rumor locations.

For (b) and (c), note that

$$h(S) = \sum_{g \in \mathcal{G}} \Pr[g] \cdot h^g(S),$$

where $\mathcal{G}$ is the set of realizations, $\Pr[g]$ is the probability that g is realized, and $h^g(S)$ is the function $h(S)$ for given graph g with deterministic model. It is sufficient to prove (b) and (c) in deterministic model.

In deterministic model, a node v is influenced by the rumor if and only if

$$dist(v, R) < dist(v, S)$$

since positive cascade has higher priority.

(b) Clearly, if $S \subset S'$, then

$$dist(v, R) < dist(v, S') \leq dist(v, S).$$

Hence,

$$h(S) \geq h(S').$$

(c) Consider $u \in V \setminus S'$. For protector set $S' \cup \{u\}$, v is influence by the rumor if and only if

$$dist(v, R) < dist(v, S' \cup \{u\}).$$

Since $dist(v, S') \geq dist(v, S' \cup \{u\})$, we have

$$h(S') \geq h(S' \cup \{u\}).$$

Thus

$$\Delta_u h(S') \leq 0.$$

Note that $-\Delta_u(S')$ is the number of nodes v such that

$$dist(v, R) < dist(v, S') \text{ and } dist(v, R) \geq dist(v, S' \cup \{u\}).$$

Similarly, $-\Delta_u h(S)$ is the number of nodes v such that

$$dist(v, R) < dist(v, S) \text{ and } dist(v, R) \geq dist(v, S \cup \{u\}).$$

Since $S \subset S'$, we have

$$dist(v, R) < dist(v, S') \Rightarrow dist(v, R) < dist(v, S')$$

and

$$dist(v, R) \geq dist(v, S \cup \{u\}) \Rightarrow dist(v, R) \geq dist(v, S' \cup \{u\}).$$

Therefore,

$$\Delta_u h(S) \leq \Delta_u h(S').$$

$$\square$$

By this theorem, $h(\emptyset) - h(S)$ is monotone nondecreasing and submodular. Zhu et al. [498] The problem of maximizing $h(\emptyset) - h(S)$ and design a randomized algorithm with reverse sampling.

10.3 Uncertainty of Price

Price is an important issue in the study of marketing. There are two important research directions on product price in the literature.

- Predict customer's acceptable price with historical data.
- Determine marketing price to maximize the revenue with various information on marketing environment and customers' behavior.

In general, this research subject belongs management science and many researchers and publications in the area of general marketing as a part of management science. Only a few research papers appear in the study of computational social networks. To see a brief scene, let us mention three research works as follows.

Zhu et al. [504] consider two stage process for product adoption. In the first stage, customers explore the product and may be influenced by the features of the product. In the second stage, customers decide to buy the product if price is acceptable. Each customer has his/her own acceptable price based on his/her evaluation on the product and his/her economic situation. An acceptable price is not easy to know exactly in marketing. However, from historical data, a distribution can be learned, so it can be an important fact for the seller to determine the marketing price based on the prediction of the acceptable price.

Determination of marketing price is an interesting and very complicated problem. It is usually based on the study of customers' behavior. There exist a lot of publications on this topic [259, 475]. To capture the click behavior of a customer, Tang and Yuan [381] proposed "consider-then-choose" model consisting of two phases.

- Phase 1: The customer browses a sequence of products to form a consideration set. In order to capture customer's browsing behavior, Tang and Yuan [381] proposed a cascade browse model for this phase.
- Phase 2: The customer makes a purchase of product selected from consideration set. To capture the customer's purchase behavior, they employed the MNL model, which is related to standard assortment pricing [6, 189, 299], but with a few differences, e.g., they use a combination of cascade browse model and MNL model.

In each phase, Tang and Yuan formulate an optimization problem and provided an efficient approximation (See them in exercises in this chapter.).

The acceptable price is determined by the buyer and the marketing price is determined by the seller. Feldman et al. [106] consider them as two sides of a two-stage game. The situation is that a seller faces the pricing problem for a public product. There are several potential buyers on a relationship network. By public, it means that when a buyer purchases the product, not only he gets benefit but also his neighbors in a network get benefit. In the first stage, the seller announces a price for the product. In the second stage, all potential buyers (each having a private value for the product) make decision simultaneously whether they will purchase the product or not.

10.4 Hidden Links

In the real world, although many connections between nodes in a social network have been identified with certain technologies, there still exist a large number of hidden connections due to individual privacy. This makes hidden link prediction an important research issue in the study of computational social networks [61, 158, 180, 260, 388, 494].

Ma et al. [279, 280] used a game model. This is a quite interesting approach, which ingeniously uses the property of Nash equilibrium. Consider a social network $G = (V, E)$ and a community partition $(C_1, C_2, ..., C_k)$. Let us trace back to the community formation process, which may help us explore the hidden links.

Initially, each individual represents a community. Later, they will get together to form larger communities. The decision for each individual to join a community is made by himself/herself according to certain judgment (k-hop community loyalty), that is, an individual will join a community which has the highest k-hop community loyalty. This process is terminated according to the Nash equilibrium condition. This condition indicates that when the community structure is established, each individual has more neighbors in the community than the neighbors in other communities. If this condition does not hold, then there are hidden links within the community.

Let us give formal definition of k-hop community loyalty.

Definition 10.4.1 (*k-hop community loyalty*) Let k be a positive integer. Given a community C_j and an individual $i \in C_j$, the k-hop community loyalty of i to C_j is defined to be

$$\mathrm{CL}_k(i, C_j) = \alpha_1 w_1 + \cdots + \alpha_k w_k,$$

where

$$\alpha_p = \frac{|N_{i,j,p}^{in}|}{|N_{i,j,p}^{in}| + |N_{i,j,p}^{out}|},$$

where $N_{i,j,p}^{in}$ and $N_{i,j,p}^{out}$ are sets of p-hop neighbors of i, inside community C_j and outside community C_j, respectively; w_p is a coefficient which indicates the importance of k-hop neighbors of u_i. w_p decreases exponentially as p increases [228]. For example, as indicated in [211], $w_p = (1\%)^p$.

For $t \neq j$, $\mathrm{CL}_k(i, C_t)$ is defined similarly.

With above definition of k-hop community loyalty, Nash equilibrium condition can be described as follows.

Definition 10.4.2 (*Nash Equilibrium between Communities*) Given a social network $G = (V, E)$, a community partition $(C_1, ..., C_m)$ is said to be in Nash equilibrium if for any community C_t and any $i \in C_j$,

$$\forall 1 \leq j \neq t \leq m, \mathrm{CL}_k(i, C_j) \leq \mathrm{CL}_k(i, C_t).$$

A formal definition of hidden links and its corresponding mining problem are given in the following.

Definition 10.4.3 (*Hidden Links*) In the real world, some relations (links) between individuals may be lost when information is gathered and recorded. Those relations are called hidden links.

Definition 10.4.4 (*Active Individual*) Given a social network $G = (V, E)$ with a community partition $(C_1, ..., C_m)$, an individual $i \in C_j$ is said to be active if there exists t such that $\mathrm{CL}_k(i, C_t) > \mathrm{CL}_k(i, C_j)$.

Problem 10.4.5 (*Hidden Links Mining*) Given a social network $G = (V, E)$ with a community partition $(C_1, ..., C_m)$, find the active individuals that destroy Nash equilibrium of the community structure, and mine possible hidden links. (Those links connect them to other individuals in the same community.)

Ma et al. [279, 280] proposed Algorithms 47 and 48 to find active individuals and to mine hidden links, respectively.

Algorithm 47 Locate Active Individuals.

Input: A social network $G = (V, E)$ with a community partition $(C_1, ..., C_m)$, coefficients $w_1, ..., w_k$ where k is predetermined.
Output: Active individual set $S \subseteq V$.
1: $S \leftarrow \emptyset$.
2: **for** $j = 1$ to m **do**
3:    **for** $i \in C_j$ **do**
4:       $\mathrm{Act}(i) = \max_{t \neq j}(\mathrm{CL}_k(i, C_t) - \mathrm{CL}_k(i, C_j))$;
5:       **if** $\mathrm{Act}(i) > 0$ **then**
6:          $S \leftarrow S \cup \{i\}$
7:       **end if**
8:    **end for**
9: **end for**
10: **return** S.

Algorithm 48 Search Hidden Links.

Input: The active individual set S, the total number of added links AL.
Output: Hidden link set HL.

1: $L \leftarrow 0, HL \leftarrow \emptyset$.
2: **while** $L < AL$ **do**
3:    **if** $S \neq \emptyset$ **then**
4:      $(p, q) = \text{argmax}_{i,i'}(\text{Act}(i) + \text{Act}(i'))$ where i and i' belong to the same community and at least one in S;
5:    **else**
6:      $(p, q) = \text{argmax}_{i,i'}|N(i) \cap N(i')|$ where $i, i' \in V$ and $i \neq i'$;
7:    **end if**
8:    $HL \leftarrow HL \cup \{(p, q)\}$;
9:    $L \leftarrow L + 1$;
10:    Use Algorithm 47 to update S;
11: **end while**
12: **return** HL.

10.5   Machine Learning for Hidden Links

There are also several methods in traditional machine learning for link prediction. Before introduce them, let us give a general discussion on robust optimization and machine learning.

In robust optimization, some input information is assumed not to be unknown exactly, however, they are known to vary in certain domains. However, in machine learning, the input information may be completely unknown, instead, what is known is a sequence of historical data and the input information is hidden in the historical data. For example, in the robust influence maximization, the influence probability on each link is assumed to vary in an interval. However, for machine learning in the influence maximization, the influence probability may be completely unknown, even not know the information diffusion model.

How to hide the information in historical data? If the historical data is only about where seeds are located, then no information on diffusion can be mined out from the data. Therefore, in machine learning method, a system of parameters is usually designed. The historical data contains values of those parameters. Using the data, through training, the relationship between seed locations and those parameters would be released, so that for currently input social network, the solution of influence maximization can be obtained. It is not a easy job to find a good system of parameters for the influence maximization. We will come back to this problem in a later section.

Now, we consider machine learning approach for the hidden links. The system is much simpler. Each methods use a neighborhood overlap measure and statistics. Assume that the likelihood of an edge (u, v) is simply proportional to the measure. The machine learning approach is built based on this assumption. A list of overlap measures is summarized in [173]. They are divided into two groups, the local overlap measure and the global overlap measure.

The Number of Shared Neighbors: Let $N(u)$ denote the set of neighbors of node u. Then this overlap measure is defined by

$$S_{overlap}[u, v] = |N(u \cap N(v)|.$$

The following are two variations.

Resource Allocation (RA) Index: Let d_u denote the degree of node u. This index is defined by

$$S_{RA}[u, v] = \sum_{w \in N(u) \cap N(v)} \frac{1}{d_w}.$$

Adamic-Adar (AA) Index: This index is defined by

$$S_{AA}[u, v] = \sum_{w \in N(u) \cap N(v)} \frac{1}{\log d_w}.$$

Next, we turn to another type of indexes.

Sorensen Index: The Sorensen index is defined by

$$S_{Sorensen}[u, v] = \frac{2\,N(u) \cap N(v)|}{d_u + d_v}.$$

This index normalizes the count of common neighbors by the sum of the node degrees. This normalization prevents highly biased towards predicting edges for nodes with large degrees. Salton index and Jaccard index are proposed for the same purpose,

Salton Index: It is defined by

$$S_{Saltpn}[u, v] = \frac{2\,N(u) \cap N(v)|}{\sqrt{d_u d_v}}.$$

Jaccard Index: This index is defined by

$$S_{Jaccard}[u, v] = \frac{|N(u) \cap N(v)|}{|N(u) \cup N(v)|}.$$

More variations of the local overlap measure can be found in [269]. For the global overlap measure, we may consider Katz index.

Katz Index: It is defined by

$$S_{katz}[u, v] = \sum_{i=1}^{\infty} \beta^i A^i[u, v],$$

where β is a parameter with $0 < \beta < 1$, A^i is the power i of the adjacency matrix of the graph, and $A^i[u, v]$ is the entry at the row and the column labeled with u and v, respectively, which is the number of paths of length at most i from u to v. Clearly, β can be used to control how much weight is given to short versus long paths. If βA has the largest eigenvalue less than one and $(I - \beta A)$ is non-singular. The Katz indetex can be derived as

$$S_{Kats} = \sum_{i=1}^{\infty} \beta^i A^i$$

$$= (I - \beta A)^i - I.$$

The Katz index is highly biased towards predicting edges for nodes with large degrees since high-degree nodes will generally be involved in more paths. Therefore, Leicht et al. [232] propose an improvement by using

$$\frac{A^i}{\mathbb{E}[A^i]}$$

to replace A^i, where the expectation is taken for a random graph to be drawn from graphs with the same set of degrees as given graph. By analytical computation, we can obtain

$$\mathbb{E}[A[u, v]] = \frac{d_u d_v}{2|E|}.$$

In fact, there are d_u edges leaving u and each of these edges has a $\frac{d_v}{2|E|}$ chance of ending at v. Similarly, by analytical computation, we obtain

$$\mathbb{E}[A^2[u, v]] = \frac{d_u d_v}{(2|E|)^2} \sum_{w \in V} (d_w - 1)d_w.$$

Unfortunately, it is not easy to obtain explicit expression by analytical computation for $\mathbb{E}[A^i]$ with $i \geq 3$. But, in general, we may use an approximation value

$$\mathbb{E}[A^i[u, v]] = \frac{d_u d_v \lambda^{i-1}}{2|E|},$$

where λ is the largest eigenvalue of A. Using this expression, the LNH index is expressed as

$$S_{LNH}[u, v] = I[uv] + \frac{2|E|}{d_u, d_v} \sum_{i=0}^{\infty} \beta^i \lambda^{1-i} A^i[u, v],$$

that is, S_{LNH} can be expressed in the following form:

$$S_{LNH} = 2\alpha \lambda |E| D^{-1} (I - \frac{\beta}{\lambda} A)^{-1} D^{-1},$$

where D is a matrix with node degree on the diagonal and 0 at other entries, called the degree matrix of A.

Random Walk Method: Another method, in the class of global overlap measurements, is to consider random walks, instead of exact counts of paths over the graph. For example, define an index by the following:

$$S_{RW}[u, v] = q_u[v] + q_v[u],$$

where $q_u[v]$ is the stationary probability that random walk, starting at node u, visits node v.

10.6 Clustering and Community Detection

Clustering is a typical method for unsupervised learning. Community detection is its extension for dealing with graph data. In Chap. 1, we already see several approaches for community detection. In this section, we touch a new one, spectral method. To do so, we first introduce the fundamental knowledge on graph Laplacians.

Definition 10.6.1 (*Unnormalized Laplacian*) The unnormalized Laplacian of graph $G = (V, E)$ is defined by
$$L = D - A,$$

where A is the adjacency matrix of G and D is the degree matrix.

If G is a simple graph, then its unnormalized Laplacian has the following properties:

- $L = L^T$.
- For any $x \in \mathbb{R}^{|V|}$, $x^T L x \geq 0$.
- L has n ($n = |V|$) nonnegative eigenvalues, $0 \leq \lambda_n \leq \lambda_{n-1} \leq \cdots \leq \lambda_1$.
- The geometric multiplicity of the 0-eigenvalue of L corresponds to the number of connected components of the graph.

Definition 10.6.2 (*Normalized Laplacian*) The unnormalized Laplacian of graph $G = (V, E)$ is defined by
$$L_{sym} = D^{-\frac{1}{2}} A D^{\frac{1}{2}},$$

where A is the adjacency matrix of G and D is the degree matrix.

Definition 10.6.3 (*Random Walk Laplacian*) The random walk Laplacian of graph $G = (V, E)$ is defined by
$$L_{RW} = D^{-1} L,$$

where L is the unnormalized Laplacian of the graph.

L_{sym} and L_{RW} have properties similar to L. Motivated from the fourth property, a clustering method is proposed to be based on the minimization of the following function:

$$\text{cut}(A_1, ..., A_k) = \frac{1}{2} \sum_{k=1}^{K} |\{(u, v) \in E \mid u \in A_k, v \in \bar{A}_k\}|.$$

However, a known problem with this approach is that it tends to simply make clusters that consist of a single node [372]. To make an improvement on this point, the following two functions are proposed.

$$\text{RationCut}(A_1, ..., A_k) = \frac{1}{2} \sum_{k=1}^{K} \frac{|\{(u, v) \in E \mid u \in A_k, v \in \bar{A}_k\}|}{|A_k|}$$

and

$$\text{NCut}(A_1, ..., A_k) \frac{1}{2} \sum_{k=1}^{K} \frac{|\{(u, v) \in E \mid u \in A_k, v \in \bar{A}_k\}|}{\text{vol}(|A_k)},$$

where $\text{vol}(A_k) = \sum_{u \in A_k} d_u$.

10.7   Graph Features

Node-clarification is an important task for learning from graph data. An typical example is the rumor source detection or the effector detection. Given the set of information-effected nodes, find one or more effectors. This task may be done by graph features and statistics. Consider a graph $G = (V, E)$ with adjacency matrix A. The following are examples of node-level features, listed in [173].

- **Node Degree**: For any node $u \in V$, the degree of u is defined by

$$d_u = \sum_{v \in V} A[u, v].$$

- **Eigenvector Centrality**: For any node u, the eigenvector centrality of u is defined by

$$e_u = \frac{1}{\lambda} \sum_{v \in V} A[u, v] e_v.$$

This equation can be rewritten as

$$\lambda e = Ae.$$

It is easy to see from this equation that λ is an eigenvalue and e is an eigenvector.

- **Clustering Coefficient**: For any node u, its local variant of the clustering coefficient is defined by

$$c_u = \frac{|\{(v, w) \in E \mid v, w \in N(u)\}|}{\binom{d_u}{2}}.$$

  Actually, c_u is the ratio between the actual number and the total possible number of triangles with the ego graph of u.
- In above definition of clustering coefficient, instead of triangle, using other motifs or graphlets within a node's ego graph, we would obtain other node-level features.

There are also a few graph-level features mentioned in [173]. They can be used for graph classification.

- **Bag of Nodes**: Each feature of this type is defined by aggregating node-level feature statics. The degree sequence is an example.
- **The Weisfeiler-Lehman Kernel**: This feature is obtained by applying Weisfeiler-Lehman algorithm [434] and kernel [366] to a feature of bag-of-nodes type.
- **Graphlet-Based Method**: This feature is obtained by simply counting the occurrence of different small subgraph structures. For example, count the occurrence of different subgraphs of three nodes.

All machine learning methods described in this chapter are traditional ones. In next chapter, we will study advanced machine learning methods, especially deep learning.

Exercises

1. Robust optimization is a subfield of optimizations, which deals with uncertainty in the data about objective function and constraints. Under this framework, the parameters in the objective and constraint functions are only assumed to belong to certain sets in function space, called *uncertainty sets*. The aim is to final an optimal solution in the worst case, that is, it is optimal no matter what the objective and constraints turn out to be. For example, a robust linear programming is expressed as follows:

$$\min \quad c^T x$$
$$\text{subject to} \quad a_i^T x \leq b_i, \forall a_i \in U_i, \forall b_i \in U_{b_i}, i = 1, ..., m,$$

where $U_{a_i} \subseteq \mathbb{R}^n$ and $U_{b_i} \in \mathbb{R}$ are given uncertainty sets. Note that in this formulation, we assume there is no uncertainty in the objective function. Why, without of generality, we are able to do so?

2. Consider a special case of robust LP described as in Problem 1. Assume that U_{a_i} and U_{b_i} are polyhedra, that is,

$$U_{a_i} = \{a_i \mid D_i i a_i \le d_i$$

and

$$b_i^- \le b_i \le b_i^+.$$

Please show that the uncertainty of b_i can be removed.

3. ([45]) Show that there exist a graph $G = (V, E)$ and a parameter space Θ such that

$$\max_{|S|=k} g(\Theta, S) = O(k/n),$$

where $n = |V|$.

4. ([45]) Show that there exists a graph $G = (V, E)$ and a parameter space $\Theta = \times_{e \in E}[l_e, r_e]$ such that

$$\max_{|S|=k} g(\Theta, S) = O(\log |V|/|V|),$$

where $r_e - l_e \le \delta$ for every $e \in E$ and a constant $\delta = O(1/|V|)$.

5. ([45]) Show that there exists a graph $G = (V, E)$ and a parameter space $\Theta = \times_{e \in E}[le, r_e]$ such that

$$r_e - l_e \le \delta = O(1/\sqrt{n})$$

and

$$\max_{\Omega} \min_{\theta \in \Theta} \mathbb{E}\left[\frac{\sigma_\theta(\tilde{S})}{\sigma_\theta(S_\theta^*)}\right] = O\left(\frac{\log n}{\sqrt{n}}\right),$$

where Ω is any probability distribution over seed sets of size k and $n = |V|$.

6. ([52]) Show that if graph $G = (V, E)$ is with parameter space Θ such that for any $\theta_i, \theta_2 \in \Theta$, $\|\theta_1 - \theta_2\|_\infty \le \delta$, then for any $S \subseteq V$,

$$|\sigma_{theta_1}(S) - \sigma_{theta_2}(S) \le mn\delta,$$

where $n = (V)$ and $m = |E|$.

7. ([45]) Consider a graph $G = (V, E)$ with parameter space $\Theta = \times_{e \in E}[l_e, r_e]$. Show that if there exists $\lambda > 0$ such that $r_e \le (1 + \lambda)l_e$ for every $e \in E$, then for any non-empty set $S \subseteq V$,

$$\frac{\sigma_{\theta^+}(S)}{\sigma_{\theta^-}(S)} \le (1 + \lambda)^n$$

and

$$\max_{|S|=k} g(\Theta, S) \geq (1 + \lambda)^{-n},$$

where $n = (V)$.

8. In the special case of robust LP, described in Problem 2, the robust LP can be transformed into the following form:

$$\min \quad c^T x$$
$$\text{subject to} \quad p_i^T d_i \leq b_i, i = 1, ..., m$$
$$D_i^T p_i = x, i = 1, ..., m$$
$$p_i \geq 0, i = 1, ..., m.$$

9. ([369]) Please design $(1 - 1/e - \varepsilon)$-approximations for Problem 10.2.1, Problem 10.2.2, and Problem 10.2.3, respectively.

10. ([495]) In Sect. 10.2, we consider a directed social network $G = (V, E)$ with the IC model, and a positive integer ℓ. Let R be the set of rumormongers with $|R| \leq \ell$. Let $\sigma_r(R, S)$ be the expected number of nodes that are effected by rumormongers in R, while nodes in set S are located with protectors. Note that locations of nodes in R are unknown when S is selected. We may select S by minimizing

$$f(S) = \max_{R \subseteq V} \sigma_r(R, S).$$

Please prove the following.

(a) min $f(S)$ is NP-hard.
(b) Computing $\sigma_r(R, S)$ is #P-hard.
(c) $\sigma_r(R, S)$ is a polymatroid function with respect to R.
(d) $f(S)$ is monotone nonincreasing.

11. ([279]) Consider a social network $G = (V, E)$ with a community partition $(C_1, ..., C_m)$. Let $S = \{s_1, ..., s_k\}$ be a subset of V. Suppose that the induced graph of S forms a cycle and only odd vertices $s_1, s_3, ..., s_{2i-1}$ are adjacent to $V \setminus S$. Show that S cannot be partitioned into two communities S_1 and S_2 such that for each $u \in S_1$,

$$|N(u) \setminus S_2| > |N(u) \cap S_2|$$

and for each $v \in S_2$,

$$|N(v) \setminus S_1| > |N(v) \cap S_1|,$$

where $N(u)$ denotes the set of neighbors of u.

12. ([279]) Show that for any social network $G = (V, E)$, it is NP-hard to partition G into two communities to reach strong Nash equilibrium with 1-hop community loyalty condition. (The Nash equilibrium is said to be *strong* if the inequality in the definition holds strongly (or strictly), i.e., "$<$" holds instead of "$\leq$.")

13. (Double Sandwich Method [128]) Consider a social network $G = (V, E)$ with the IC model. Each edge e is associated with a probability $p_e \in [l_e, h_e]$. Given a seed set S, $I(S)$ denote the set of active nodes at the end of information process. A profit function is defined by

$$f(S) = \mathbb{E}[\sum_{u,v \in I(S)} c(u, v)],$$

where $c(u, v) > 0$ is the profit contributed by pair $\{u, v\}$. This setting is a combination of problem formulation in Sects. 10.1 and 5.1. Considered problem is to maximize $f(S)$ subject to $|S| = k$ for given $k > 0$. Clearly, this problem is nonsubmodular. Let $\Theta = \prod_{e \in E}[l_e, h_e]$. θ^- (θ^+) denote the choice of $p_e = l_e$ $(p_e = h_e)$ for every $e \in E$. In Sect. 5.1, a sandwich method is already described for the problem when all p_e are already selected. The second sandwich method will follow from the one in Sect. 10.1 based on selection θ^- and θ^+. Please give the detail description of the second sandwich method and analyze it.

14. Let X be a real-valued square matrix. Let λ be the largest eigenvalue of X. Then

$$(I - X)^{-1} = \sum_{i=0}^{\infty} X^i$$

If and only if $\lambda < 1$ and $(I - X)$ is non-singular.

15. Suppose that the expectation is taken for a random graph to be drawn from graphs with the same set of degrees as given graph. By the analytical computation, show that

$$\mathbb{E}[A^2[u, v]] = \frac{d_u d_v}{(2|E|)^2} \sum_{w \in V} (d_w - 1) d_w.$$

16. ([269]) Show that the LNH index S_{LNH} can be expressed in the following form:

$$S_{LNH} = 2\alpha\lambda|E|D^{-1}(I - \frac{\beta}{\lambda}A)^{-1}D^{-1},$$

where D is a matrix with node degree on the diagonal.

17. Let L be the unnormalized Laplacian of graph $G = (V, E)$. Show that for any $x \in \mathbb{R}^{|V|}$, $x^T L x \geq 0$.

18. Suppose that L is the unnormalized Laplacian of graph $G = (V, E)$. Show that the geometric multiplicity of the 0-eigenvalue of L corresponds to the number of connected components of the graph.

19. Consider a graph $G = (V, E)$. Let $B \subset V$. Define that for any $u \in V$,

$$a[u] = \begin{cases} \sqrt{\frac{\bar{B}}{B}} & \text{if } u \in B, \\ -\sqrt{\frac{B}{\bar{B}}} & \text{if } u \in \bar{B}. \end{cases}$$

Show that

$$\mathrm{RatioCut}(B, \bar{B}) = \frac{a^T L a}{|V|},$$

where L is unnormalized Laplacian of graph G.

20. Consider a graph $G = (V, E)$ and its unnormalized Laplacian L. Suppose that the following minimization is over real-valued vectors:

$$\min \quad x^T L x$$
$$\text{subject to} \quad \|x\|^2 = |V|$$
$$\sum_{u \in V} x[u] = 0.$$

Show that the solution of this minimization is given by the eigenvector of the second smallest eigenvalue of the Laplacian.

Historical Notes

Uncertainty exists widely in the real world. However, for simplicity of study, uncertainty is usually reduced at early stage of research. This happens also in the study of computational social networks. When the research gets in certain stage, it involves uncertainty more and more. This chapter selects a few uncertainties and related researches from [41, 132, 144] to introduce those recent efforts.

Robust optimization and machine learning are two popular methods to deal with uncertainties. With robust optimization, partial information on uncertainty is usually assumed to know. For example, when robust influence maximization is studied by Chen et al. [45], the activate probability at each edge in IC model is allowed to vary in an interval. When robust influence maximization is studied by He and Kempe [187], the information diffusion model can be varied among a set of possible models. When robust misinformation prevention is studied by Shi et al. [369], misinformation (rumor) sources are restricted in a set of suspicious nodes.

However, with machine learning method, the information on uncertainty is totally not given. Actually, such information is hidden in historical data. For example, when Tong [390] studies the rumor blocking with machine learning approach, it is assumed that information diffusion model is unknown. Chen et al. [41] divide learning-based influence maximization into two phases.

- Learning Phase: With designed model, train a group of parameters Θ.
- Testing Phase: Given a target social network G and a budget b, use parameters Θ to find solution of influence maximization problem.

The information diffusion model is hidden in learning phase. If training data comes from certain model, then in testing phase, the solution is given to the social network with such a information diffusion model.

In machine learning approach, the important part is the selection of parameters, which regard to machine learning model. For social network problems, it is about graph representation, hence called graph representation learning [41]. More references can be found from survey papers [41, 249] on machine learning approaches in the study of social networks.

Chapter 11
Deep Learning in Social Networks

Interesting in predictive analytics? Then research artificial intelligence, machine learning, and deep learning.

— `SupplyChainToday.com`

11.1 Popularity Prediction

An important class of graph data is the social data, generated from online social networks (OSNs). The social data have been growing rapidly over the internet through various OSNs, including online websites (such as Facebook, LinkedIn, and Research-Gate) and messengers (such as Skype). The widespread use of them leads to an increasing interest in efficiently and correctly discovering important, useful, and implicit information. Efficient techniques for data harnessing about social data are crucial to applications across many domains, including public safety, environment management, election, and viral marketing.

On the other hand, the development of neural networks has blossomed in the last few years, leading to a large number of graph representation learning (GRL) models. Compared with traditional models, GRL methods are often shown to be more effective. In this chapter, we introduce GRL for harnessing social data, involving deep learning and reinforcement learning.

Let us start from a brief survey on GRL methods for popularity prediction problems [41].

Given information I, let P_I^t denote the number of nodes activated by I up to time t. P_I^∞ is simply written as P_I, which is the total number of nodes activated by I until its propagation stops. P_I is referred as the *popularity* of I.

There are two types of popularity predictions. If prediction aims at the exact value of popularity, such as computation of influence spread, then it is called *regression* popularity prediction. Otherwise, the prediction pays attention only on certain information of popularity. For example, given a threshold θ, the prediction is to determine whether the popularity is bigger than θ or not. In such a case, it is called *classification* prediction.

The regression prediction can be classified into three levels, micro-level, meso-level, and macro-level. Furthermore, macro-level and micro-level predictions have two versions, coarse-grained and fine-grained [456].

Note that in the study of computational social networks, we touched many examples of popularity predictions. There are a lot of machine learning methods for popularity predictions, which can be classified to diffusion model-based methods [143, 226, 422], feature-based methods [131, 199, 220, 438, 461, 492], generative process-based methods [233, 234, 290, 362, 490], and GRL methods. In this chapter, we focus only on GRL methods. The main difference between other works and GRL is the way they deal with representing graph structure. In other works, representing graph structure is treated as a pre-processing step, while GRL methods consider it as a machine learning task itself and apply data-driven approaches to learn embeddings that encode information of graph structure [175]. In subsequent sections, we will introduce the framework of GRL and its applications in computational social networks.

11.2   Node Embeddings

The goal of node embeddings is to encode nodes as low-dimensional vectors such that the structure of their local graph neighborhood is summarized. It usually consists of three components:

- **Encoder**: An encoder model maps each node in the graph into a low-dimensional vector, called *embedding*. In other words, the encoder is a function that maps nodes $v \in V$ to vector embeddings $z_v \in \mathbb{R}^d$, that is,

$$\mathrm{ENC}(v) = z_v.$$

In a lot of cases, the encoder function is simply an embedding established based on the node ID, which is called the *shallow embeddings*. It is worth mentioning that in the graph neural network, the encoder may use node features or the local graph structure around each node as input to produce an embedding.

- **Decoder**: A decoder model uses the low-dimensional node embeddings to reconstruct information about each node's neighborhood in the original graph, that is, the decoder is to reconstruct certain structure from the node embeddings generated by the encoder. One often encounters the pairwise decoder, which is defined as follows:

$$\text{DEC}(\text{ENC}(u), \text{ENC}(v)) = \text{DEC}(z_u, z_v) \approx S(u, v),$$

where $S(u, v)$ is the target relationship between u and v.
- **Loss Function**: This function is to measure the discrepancy between decoded value $\text{DEC}(z_u, z_v)$ and the true value $S(u, v)$. For example, in practice, this function might be a mean-square error

$$\ell(\text{DEC}(z_u, z_v), S(u, v)) = \|\text{DEC}(z_u, z_v) - S(u, v)\|^2.$$

The total loss $\mathcal{L}$ is defined to be the collected over a set of training node pairs, $\mathcal{D}$, that is,

$$\mathcal{L} = \sum_{(u,v)\in\mathcal{D}} \ell(\text{DEC}(z_u, z_v), S(u, v)),$$

which can be used for evaluating the quality of selected embedding method.

Next, we move our attention to a specific method, DeepWalk which is a typical one of random walk embeddings. DeepWalk uses an inner-product decoder as follows:

$$\text{DEC}(z_u, z_v) = \frac{e^{z_u^T z_v}}{\sum_{w\in V} e^{z_u^T z_w}} \approx p_{G,T}(v|u),$$

where $p_{G,T}(v|u)$ is the probability of visiting v on a length-T random walk starting from u. The loss function of DeepWalk is defined by

$$\ell(\text{DEC}(z_u, z_v), S(u, v)) = -S(u, v)\log(\text{DEC}(z_u, z_v)).$$

For training, the general strategy is to minimize the following *cross-entropy loss*:

$$\mathcal{L} = \sum_{(u,v)\in\mathcal{D}} -\log(\text{DEC}(z_u, z_v)).$$

The node2vec approach is another method of random walk embeddings. It uses the same decoder as DeepWalk, but a modified loss function in order to improve computation time.

11.3 Graph Neural Networks

The general framework of deep learning (DL) is as follows:

- Data Input: Extract information from inputs.
- Data Representation: Embed this information into some low-dimensional latent space. For example, if the deep learning method is used for learning a function, then it maps a numeric-form graph to low-dimensional embeddings via optimizing over a large number of expressive neural network functions.
- Model Construction: Select a deep learning model.
- Problem Solving: Solve classification or regression prediction problem.

For example, if the deep learning method is used for learning a function, then it maps a numeric-form graph to low-dimensional embeddings via optimizing over a large number of expressive neural network functions.

For model selection, there are many choices. The following are several possible candidates:

- CNN (Convolutional Neural Network) [428]
- GCN (Graph Convolutional Network) [217]
- GAT (Graph Attention Network) [409]
- GNN (Graph Neural Network) [173]
- RNN (Recurrent Neural Network) [97]

While CNN and RNN are well defined for image and text, respectively, GNN is defined for graphs and hence serves very well in the study of computational social networks.

In this section, we introduce GNN. First, let us explain the idea roughly about operation of GNN. GNN is a multi-layer network. When the information passes through each layer, it will be updated. For example, suppose that initially, the information contains neighbors for each node. After passing the first layer, the information is updated to contain the 1-hop local graph structure for every node. After passing the second layer, the information is updated to contain the 2-hop local graph structure for every node, so on so forth, until all layers are passed and enough information is obtained.

The architecture of GNN is a multi-layer perceptron (MLP). The embedding of a graph G is usually generated by flattening the adjacency matrix:

$$z_G = \mathrm{MLP}(A[v_1] \oplus A[v_2] \oplus \cdots \oplus A[v_n]),$$

where $n = |V|$ and A is the adjacency matrix of G. Moreover, operation $\oplus$ is concatenation of row vectors. The embedding of each node contains graph structure, as well as features that one may concern.

Consider a graph $G = (V, E)$ together with a set of node features $X \in \mathbb{R}^{d|V|}$. Let $h_u^{(k)}$ denote the hidden embedding of node u at layer k. During each message-passing iteration, $h_u^{(k)}$ is updated according to information aggregated from u's neighborhood $N(u)$, which can be expressed in the following:

$$h_u^{(k+1)} = \text{UPDATE}(h_u^{(k)}, \text{AGGREGATE}(\{h_v^{(k)} \mid v \in N(u)\}))$$
$$= \text{UPDATE}(h_u^{(k)}, m_{N(u)}^{(k)}),$$

where UPDATE and AGGREGATE are two differentiable functions computed by neural networks. For initial embeddings, usually set $h_u^{(0)} = x_u \forall v \in V$. After the first iteration, every node embedding $h_u^{(1)}$ contains information from its 1-hop neighborhood. After the second iteration, every node embedding $h_u^{(2)}$ contains information from its 2-hop neighborhood. Suppose GNN consists of K layers. After passing the final layer, define where UPDATE and AGGREGATE are two differentiable functions computed by neural networks. For initial embeddings, usually set $h_u^{(0)} = x_u \forall v \in V$. After the first iteration, every node embedding $h_u^{(1)}$ contains information from its 1-hop neighborhood. After the second iteration, every node embedding $h_u^{(2)}$ contains information from its 2-hop neighborhood. In general, after the kth iteration, every node embedding $h_u^{(k)}$ contains information from its k-hop neighborhood. Suppose GNN consists of K layers. After passing the final layer, the node embedding is defined by

$$z_u = h_u^{(K)}.$$

Usually, each node embedding contains two types of information. One is the structure information. For example, $h_u^{(k)}$ may contain information about degrees of all k-hop neighbors. The other is feature-based information.

For functions UPDATE and AGGREGATE, there are many selections with various considerations. For example, set

$$m_{N(u)}^{(k)} = \text{AGGREGARE}(\{h_v^{(k)} \mid v \in N(u)\})$$
$$= \sum_{v \in N(u)} h_v^{(k)}$$
$$\text{UPDATE}(h_u^{(k)}, m_{N(u)}^{(k)}) = \tau(W_{self} h_u^{(k)} + W_{neig} m_{N(u)}^{(k)} + b^{(k)}),$$

where W_{self} and W_{neig} are trainable parameter matrices, $b^{(k)}$ is the bias term, and τ is a nonlinear function. W_{self}, W_{neig}, and $b^{(k)}$ can be shared across all layers and can also be trained separately. For simplicity of expression, we may omit superscripts and bias term, so that above expressions become the following:

$$m_{N(u)} = \sum_{v \in N(u)} h_v$$
$$\text{UPDATE}(h_u, m_{N(u)}) = \tau(W_{self} h_u + W_{neig} m_{N(u)}).$$

GNN with these update formulas is called a basic GNN [292, 354].

For basic GNN, there is another definition in the graph-level form.

$$H^{(k+1)} = \tau(A H^{(k)} W_{self} + H^{(k)}),$$

where A is the adjacency matrix of graph G.

Theoretically, GNN can be derived independently in three ways.

- In theory of graph signal processing, generalize the Euclidean convolutions to the non-Euclidean graph domain [31].
- Based on message passing algorithms for probabilistic inference in graphical models, develop neural message passing approaches to analogy them [72].
- Motivated from connections to the Weisfeiler-Lehman test for graph isomorphism [174].

For above theoretical results, the reader may find a succinct explanation in [173].

11.4 A Fine Design for Influence

Chen et al. [42] give fine-designed solution for influence maximization (IM) in social networks with certain uncertainty, the IM problem is considered as a reinforcement learning (RL) problem for finding the optimal policy of selecting k seeds, which are considered as k action sequence, to maximize the cumulative reward (influence spread). Especially, they define the learning-based influence maximization problem as follows.

Definition 11.4.1 (*Learning-Based IM [42]*) The problem consists of two phases.

- **Learning Phase**: Given a set of training data, a diffusion model ϕ, and the influence spread function $\sigma : S \to \mathbb{R}^+$, construct a parameter function $\hat{Q}(v, S; \theta)$ and train the set of parameters Θ to make parameter function $\hat{Q}(v, S; \Theta)$ approximate $\Delta_v \sigma(S)$ as accurately as possible.
- **Testing Phrase**: Given a target social network G with diffusion model ϕ, the learned parameters Θ and an integer k, solve the influence maximization under budget constraint $|S| \le k$.

The set of training data consists of a sequence of social networks, $\mathcal{G} = \{G_1, G_2, ..., G_t\}$. Each G_i is a smaller one with diffusion model ϕ.

They present a deep learning method for solving the influence maximization. It is a fine design, named ToupleGNN, which has the following framework.

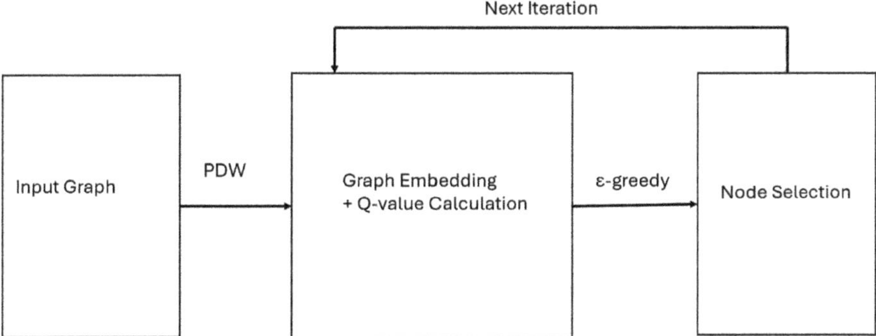

Fig. 11.1   Framework of ToupleGNN

- As shown in Fig. 11.1, given a set of training data they first apply the personalized DeepWalk (PDW) method to get the initial node embedding.
- Then three coupled GNN (ToupleGNN) are designed to complete the following tasks in k iterations, where a coupled GNN is a combination of double deep Q-network (DDQN) [181] and GNN.

  - Utilize ToupleGNN to capture the cascading effect of information diffusion to get node embeddings.
  - Based on node embeddings, Construct the parameterized function $\hat{Q}(v, S; \Theta)$.
  - Use ϵ-greedy to select the next seed set.
  - Use DDQN to learn the parameters.

The reason for using DDQN is that it is not easy to know the exact Q-value in DQN. The DDQN [181] can be used for solving this issue and giving a better performance.

Next, along the framework, we give further explanations.

Node Embedding: Each node embedding contains the information about graph structure and features of influence, for each node u, including the activation state of node u, $X_u \in \mathbb{R}$, the capacity of u to influence other users, $S \in \mathbb{R}^l$, and the tendency of being activated by other users, $T_u \in \mathbb{R}^l$.

Initial Embedding Learning: Initial node embeddings are generated by the personalized DeepWalk (PDW) method as shown in Algorithm 49. It contains two parts: influence context generation (lines 3–9) and parameters learning (lines 10–18), which will be explained below.

Algorithm 49 PDW

Input: A social network $G = (V, E)$ with the IC model.
Output: X_u, S_u, T_u for every node $u \in V$.
1: Initialize X_u, S_u, T_u by Gaussian distribution $N(0, 0.01)$;
2: Initialize $W \leftarrow \emptyset$;
3: **for** each $u \in V$ **do**
4:    $L_u \leftarrow \emptyset, G_u \leftarrow \emptyset, C_u \leftarrow \emptyset$;
5:    $L_u \leftarrow$ Sample αL nodes by RWR starting from u;
6:    $G_u \leftarrow$ Uniformly sample $(1 - \alpha)L$ nodes from $N_{out}^r(u)$;
7:    $C_u \leftarrow L_u \cup G_u$;
8:    Inset (u, C_u) into W;
9: **end for**
10: **for** each $(u, C_u) \in W$ **do**
11:    **for** each $v \in C_u$ **do**
12:      Update X_u, S_u, X_v, T_v;
13:      Sample a set of negative sample N;
14:      **for** each $w \in N$ **do**
15:        Update X_u, S_u, X_w, T_w;
16:      **end for**
17:    **end for**
18: **end for**
19: **return** X_u, S_u, T_u for every $u \in V$.

The first part (influence context generation) is for the computation of three parameters L_u, G_u, C_u, where C_u is the influence context of node u, L_u is the local part of C_u, and G_u is the global part of C_u. Local context L_u is a sampled set of nodes activated by u. To limit the size of L_u, set a threshold αL where L is the length threshold of the node context and $\alpha \in [0, 1]$. L_u is computed by random walk with restart (RWR) strategy. G_u is a sampled set of nodes from the r-hop neighbors of u. After generating L_u, G_u is obtained by randomly sampling $(1 - \alpha)$ nodes from $N_{out}^r(u)$, the set of r-hop out-neighbors of u. At end of the first part, a set of observing contexts, W is obtained.

In the second part (parameters learning), parameters X_u, S_u, X_v, T_v are updated by using W through maximizing the following function:

$$\sum_{(u, G_u) \in W} \sum_{v \in C_u} \log \Pr[v \mid u],$$

where $\Pr[v \mid u]$ is the probability of user v being influenced by user u, which is formulated by

$$\Pr[v \mid u] = \exp(X_u S_u T_v + X_v) / Z(u),$$

where

$$Z(u) = \sum_{w \in V} \exp(X_u S_u T_w + X_w).$$

Fig. 11.2 State GNN

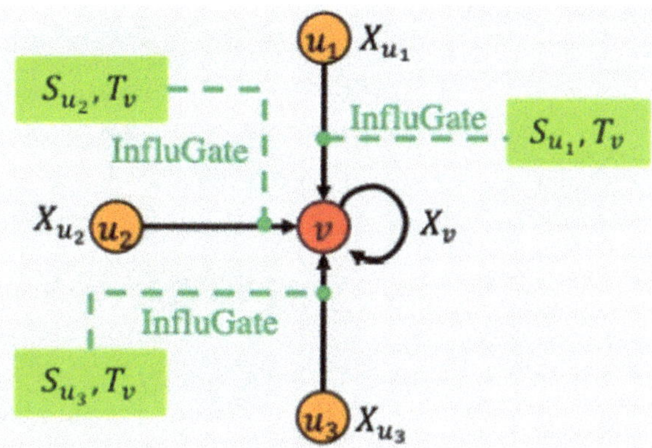

Fig. 11.3 Source GNN

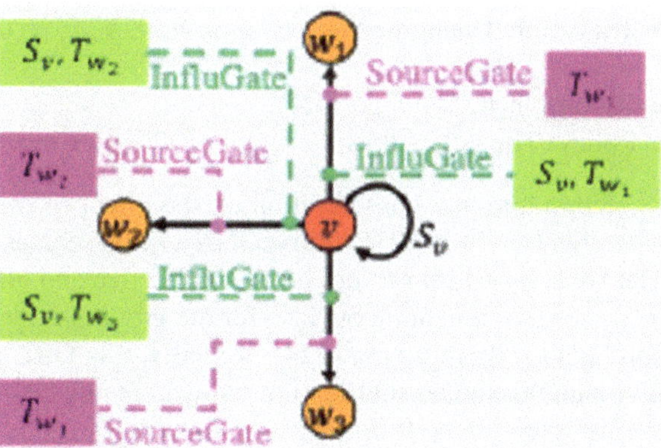

ToupleGNN: Three coupled GNN form the main body of ToupleGNN. They capture the iterative interplay between node state, nodes' influence capacity, and nodes' tendency. They are called state GNN, source GNN, target GNN as show in Figs. 11.2, 11.3, and 11.4, respectively. The reader may find the detail explanation for them in [42].

Reinforcement Learning: What is reinforcement learning (RL)? It is a type of machine learning where intelligent agent can take actions based on the current state and the interaction with environment to maximize the total reward received. The influence maximization can be naturally formulated as an RL problem as follows:

- Action: Let S_t denote the current seed set. The action is to select a node $u \in \bar{S}_t$ as next seed, represented by the embedding of u.
- State: The current state is represented by a sequence of actions of selecting seeds in S_t.

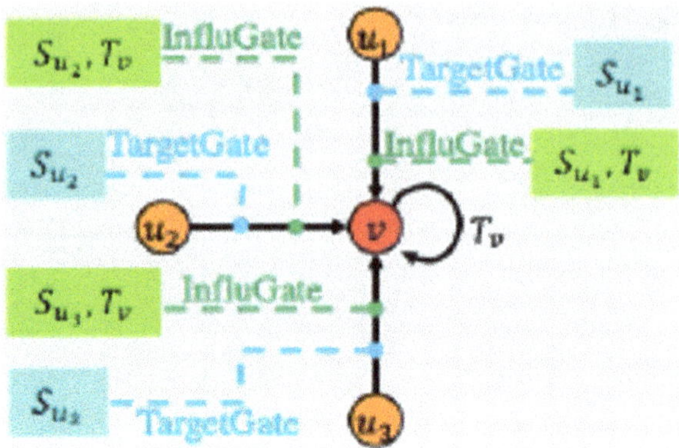

Fig. 11.4 Target GNN

- Transition: Changing the activation state X_u from 0 to 1 when u is selected as next seed.
- Reward: $r(u, S_t) = \Delta_v \sigma(S_t)$.
- Policy: The policy tells the agent how to pick next action.

With additional consideration on the effect of message passing in graph neural networks (GNNs) [493], ToupleGNN can be seen as a deep RL (DRL) framework. The DRL-based framework gives an effective and efficient solution not only for the influence maximization but also for the group influence maximization [135] and the combating cyberbullying [424]. Actually, this DRL framework has the potential to give a fine-designed solution for many more nonsubmodular optimization problems. We list a few in the following.

- **Address Challenges from Viral Marketing**: One of important applications of social networks is viral marketing. Research shows that people tend to accept recommendations from people they know and trust, rather than from other advertisement channels such as TV and commercial website [351]. This observation motivates intense research about popularity boosting in viral marketing area. The IM problem can be seen as a simplest model of viral marketing. In the real world, many facts have to be considered which generate many innovative research problems and challenge issues. A few examples are listed in Exercises. They may be solved with variations of ToupleGNN framework with challenges on parameter selection and action sequence termination rule.
- **Content Summarization**: In content summarization, the aim is to find a few sentences which can cover the most important information about a set of documents collected for certain purpose [112, 219, 470]. Summarization is a very important subject in natural language processing (NLP), which, recently, has been revolutionized by deep learning, to receive the development of models which enable to learn complex representations of languages data. To learn the pattern and relationship in language data, deep learning models for NLP have to employ large amounts of data to train deep neural networks. Those neural networks have ability to handle

variable-length input sequence with high-quality performance, and hence receive a lot of applications. With those techniques, Wang et al. have made research efforts on long-document summarization [419] and short-document summarization [420], respectively.

On the other hand, in [78], Dong et al. proposed a two stage algorithm by using social network techniques. In the first stage, they establish a network structure between documents. Then, partition the network into communities. From each community, extract subtopics. In the second stage, they formulate summarization from different subtopics as a combinatorial optimization problem of minimizing sentence distance and maximizing topic diversity. While the submodularity is proved for the problem in the second stage and hence a greedy algorithm is able to provide an approximation of good quality.

- **Corpora Extraction**: A corpora is a collection of linguistic data, such as cyber-security disclosures documents [198, 242, 268] and articles on web [71], which is always analyzed in various ways to establish patterns of grammar and vocabulary usage. Based on certain technology, ML (Machine Learning) algorithm is able to understand the semantics of the human language, that is, describe its own thinking like a human. Since larger size corpora always covers larger usage of general language, the accuracy of the ML algorithms gets improved as the size of corpora is growing. Currently, some existing corpora already grow into multiple billion of words. However, a large size of corpora may increase redundancy and noise and the complexity of computation. In fact, the computational complexity of ML algorithms always depends on the size of their search space which is usually proportional to the size of the corpora. Now, we face a controversial situation. On one hand, the larger size of the corpora benefits the accuracy of ML algorithm. On the other hand, the larger size of the corpora would decrease the efficiency of ML algorithm. How do we deal with the controversial? The solution is properly to extract a subset from the larger corpora so that instead of whole corpora, use the subset of the corpora for training ML algorithm, but not reduce the accuracy too much. That is, the efficiency receives a great improvement while the accuracy is kept within certain range. This is an attractive innovative research problem, which is possibly solved in the following way:

  - Construct a graph G on corpora based on a WMD (Word Mover's Distance), which gives a novel distance moving from words to words [224].
  - Partition G into communities C_i.
  - Assign a number k_i to C_i, proportional to size of C_i, with total number $\sum k_i = k$ equal to our expected size of subset of corpora.
  - For each community C_i, find fine-designed solution (k_i seeds) for the IM problem by DRL-based GRL method.
  - Put all obtained k seeds to form the subset of corpora.

- **Misinformation**: Recently, the social network plays an important pivotal role for communication. However, it is also susceptible to misinformation (or rumor) because of its rapid information dissemination capabilities. In [424], Wang et al.

present a framework leveraging deep reinforcement learning to combat cyberbullying in social networks. In this work, network embedding is performed in dynamic neural networks and the double deep Q-network (DDQN) is also involved. Experiments demonstrate that this approach outperforms other methods on realistic data. Meanwhile, in [135], Ghosh et al. addressed some issues in dealing with nonsubmodular optimizations. Actually, group influence maximization is a typical nonsubmodular optimization problem on decision making. To give a fine-designed solution, they introduce a robust framework, which integrates node embeddings from multiple graph neural networks, thereby utilizing diverse information for effective network analysis.

Misinformation clarification (or rumor blocking) is an important research subject in social networks. There exist many research issues regarding this subject. They contain challenges stemming from uncertainties and nonsubmodular optimizations. Efforts can be made on those challenges by designing new DRL-based model and developing new GRL methodologies based on above-mentioned approach of combating cyberbullying.

11.5 Federated Learning

Federated learning is a distributed machine learning. Its advantage is to offer model training without direct exposure of raw data. This advantage has made the federated training integrated into edge computing. However, this application of federated learning gets influenced by security vulnerabilities which are introduced by limited resource on edge devices. Meanwhile, blockchain provides a technology to bolster security, i.e., provides a possible tool to solve this security problem. But, it is still a challenge task to make practical development of blockchain on resource-constrained edge devices. To address this challenge, in [246], Li et al. introduce a blockchain-empowered heterogeneous multi-aggregator federated learning architecture. In this architecture, a novel light-weight Byzantine consensus mechanism is designed, to enable secure, fast model aggregation and synchronization. To dive into the heterogeneity problem in the architecture, we let aggregators associate with varied number of connected trainers with Non-IID data distributions and diverse training speed. This architecture also contains a multi-agent deep reinforcement learning optimization algorithm for helping aggregators to make decision on the training strategy, which would improve efficiency. Experiments demonstrate that this architecture achieves better models, outperforms others in baseline on realistic data.

Above research experience may help us to construct new secure federated learning systems for solving problems on graph data analysis in distributed environments, such as the following tasks.

- **Distributed Decision Making**: Knowledge discovered from online social networks can be used for decision making and strategy selection. Business expansion, market price determination, and political election are three examples. Due to large scale of social networks and privacy issues, secure distributed system or secure federated learning architecture can be employed in playing. Let us explain them, respectively.

A company may consider to expand its business if the company receive very positive feedback from market. However, where its expansion should be located? This may need to discovery more knowledge with help of effector detection. Especially, when there exist more than one companies in a market, a company selecting strategy has to consider reactions from others. This means that it is an important task to find a proper information from various locations and analyze them. Clearly, secure federated learning model is a good choice to take over this task since data transferring time can be saved and processing can be more efficient.

- **Marketing Price**: Determination of marketing price is an interesting and very complicated problem. It is usually based on the study of customers' behavior. There exist a lot of publications on this topic. Let us mention one of them to see a little clear picture. In [381], Tang and Yuan formulate a problem of assortment optimization and pricing. Usually, customers are located in different cities with different cultures and different economics situation. Actually, market price in different locations may not need to be unified, however, difference should not be very big since a bigger difference may reduce buying motivation of customers living in the location where the market price is high. In fact, nowadays, communication is developed very well. People can discovered the difference in very short time, especially, in group communications of internet. In the situation described as above, secure federated learning is better tool for companies to make decision based on analyze data in various locations.

- **Election**: The election is an excellent application for data mining in social networks with secure federated learning. The information resulting from analysis of network data can be used for the election result prediction and the strategy selection. For example, in 2012 political election for mayor of London, the election result was successfully predicted with analysis on social networks. As analysis posted on Facebook and Twitter, one found 7% more positive sentiment towards Mr. Johnson than Mr. Livingstone and hence predicted 54% of vote for Mr. Johnson. Google Insights, tracking web trend, found that of the total number of web searches for both candidates, 60% were for "Boris Johnson." Above facts leaded to a prediction that Mr. Johnson would win the election and actually he finally did. Mr Johnson's relatively large lead in Google searches this time around reflected not only voters' intentions but also a well-run digital campaign, so that the digital campaign becomes an important winning strategy in later elections. To get prediction more accurate, enlarge data collection area and make distributed analysis are necessary. They are suggesting to employ secure federated learning systems.
  As suggested above, new secure federated learning systems are expected. They may be constructed by improving and extending the existing systems to new applications, including those mentioned as above, and more.

- **Distributed Rumor Blocking**: In many existing works on rumor blocking, it is assumed that a single authority exists, who can generate a positive cascade to minimize the spread of rumor. This assumption is certainly unrealistic. In [400], Tong et al. study a more realistic scenario and assume that multiple agents generate multiple positive cascades with a peer-to-peer independent cascade information diffusion model. To make them work well, they put multiple agents in a non-cooperative game, so that under certain rules of non-cooperative game, each agent is able to make decision with goal to maximize own private utility function, but can achieve a good level of social utility, i.e., a good global performance. For this scenario of distributed rumor blocking, non-cooperative game model has a clear point which can be improved. In fact, those agents have the same goal, that is, block rumor. Hence, they tend to cooperate. Could the secure federated learning systems help for improvement? It is hopeful.

Finally, let us make a Remark on Reinforcement Learning Optimization. In every federated learning architecture, local servers are incentivized to enhance their training outcomes by local aggregation. The non-IID data distribution and CPU speed usually vary among trainers, which leads to a natural variation in model quality. However, in classic federated learning architectures, actual quality of the training is often neglected since the model based on the data volume of each trainer gets more attention and more weight. Moreover, they tend to wait for slowest trainer to complete its training before proceeding with aggregation. In [246], Li et al. addressed this issue, so that the model aggregation does not only take trainers' data amount as weight but also consider the performance of their actual trained model. Moreover, we design a deep reinforcement learning optimization algorithm, which can be used for trainers to select the most appropriate weights when perform local aggregation, targeting expedite model training. This optimization algorithm will be improved and attended to other distributed learning systems, especially to proposed secure federated learning systems as above. Recently, distributed optimization becomes more and more popular. Actually, in current new developments of technologies, concerned networks are often of large scales and data size becomes much large. Those changes make us have to consider distributed data storage and employ distributed algorithms. There exist a large number of research issues in distributed optimizations.

Exercises

1. Let a_1 and a_2 be two vectors in $\mathbb{R}^n$. Show that

$$u_1 = a_1 \text{ and } u_2 = a_2 - \frac{a_1^T u_1}{u_1^T u_1} u_1$$

   are orthogonal to each other.

2. Let a_1 and a_2 be two vectors in $\mathbb{R}^2$. Let S be the area of parallelogram spanned by a_1 and a_2. Show that

$$S^2 = (a_1^T a_1)(a_2^T a_2) - (a_1^T a_2)^2.$$

3. The softmax function is defined by

$$\text{softmax}_i(s; a) = \frac{e^{a^T s_i}}{\sum_{j=1}^{K} e^{a^T s_j}},$$

where a is a constant n-dimensional vector and s is a matrix of order $K \times n$. Set $n = 1$ and $p = \text{softmax}(s; a)$. Please show that

(a) $p_i/p_j = \exp(a^T(s_i - s_j))$.
(b) For any $i = 1, 2, .., K$, denote $s_i' = s_i + b$ for a constant vector b. Show that,
$$\text{softmax}(s'; a) = \text{softmax}(s; a).$$

4. For a probability vector $p = (p_i)$, the power law transformation is a function defined by
$$\text{Pow}(p; \alpha) = q \text{ with } q_i = \frac{p_i^{\alpha}}{\sum_j p_j^{\alpha}}.$$

Show that if s is a vector, then

$$\text{Pow}(\text{softmax}(s; 1); \alpha) = \text{softmax}(s; \alpha).$$

5. Suppose that ϵ_i are selected independently from a Gumbel distribution with location $\mu = 0$ and scale $\beta = 1/a$. Then is the expected choice probability if outcome categories are selected based on maximization of noise-perturbed scores $s_i + \epsilon_i$.
6. For $n = 1$, show the following:

(a) If $\lambda > 0$ is a constant, then $\text{softmax}(\lambda s'; a) = \text{softmax}(s; a)$ if and only if $a = 0$, $\lambda = 1$ or $s_i = s_j$ for all i and j.
(b) Show that
$$\text{softmax}(\lambda s; a/\lambda) = \text{softmax}(s; a).$$

(c) The softmax function need not be expressed in base e, but can equivalently be expressed in terms of any basis $b > 0$.

7. The log-sum-exp function is defined by

$$\text{lse}(z; \lambda) = \frac{1}{\lambda} \log(\sum_{i=1}^{n} \exp(\lambda z_i)), \quad \lambda > 0.$$

Show that

$$\text{softmax}(z; \lambda) = \nabla \text{lse}(z; \lambda).$$

8. Please prove that

$$\text{softma}_i(s; a) \geq \prod_{1 \leq j \leq K, j \neq i} \frac{1}{1 + e^{a^T(s_i - s_j)}}.$$

9. Any function f that takes an adjacency matrix A of a graph G as input should ideally satisfy one of the two following properties:

   - Permutation Invariance: $f(PAP^T) - f(A)$.
   - Permutation Equivariance: $f(PAP^T) = Pf(A)$.

   Here, P is a permutation matrix. Show that the shallow encoders defined by simply using node's ID as code is a permutation equivariant function.

10. Let a_1 and a_2 be two vectors in $\mathbb{R}^2$. Let $A = (a_1, a_2)$, i.e., a_1 and a_2 are two columns of A. Denote by S the area of parallelogram spanned by a_1 and a_2. Show that

$$S = |\det(A)|.$$

11. Let A be a square matrix of order n. Let $\lambda_1, \ldots, \lambda_n$ be all eigenvalues of A. Show that

$$\det(A) = \lambda_1 \cdots \lambda_n.$$

12. (Revenue Maximization [166]) The influence spread is the expected number of sold products. However, the firm may have more interest in revenue than influence spread. Usually, the revenue is the difference of a monotone submodular function and a modular function, which is submodular, but not monotone. For non-monotone submodular maximization, traditional approaches (e.g., greedy algorithm and local search) do not have a good performance. Please design a GRL method for solving this problem.

13. (Multiple Products [264, 480]) When a firm advertises more than one types of products in market with a limit budge, the revenue maximization would have a knapsack constraint, instead of size constraint. This constraint may require a different action sequence in the DRL-based GRL approach. Please give a specific design.

14. (Viral Marketing with Community Structure [168]) When marketing is carried out in a social network with a known community structure, we may wish to distribute seeds (free samples or discount coupons) fairly, that is, set up a upper bound for the number of seeds distributed in each community. In this case, the fairness would introduce a matroid constraint, which may induce a challenge issue on termination of action sequence in application of above DRL-based GRL method. Please give a specific design.

15. (Viral Marketing with Recommendations [164, 169]) When you purchase a product A in an online store, a recommendation could come out "One who buys A may also like to buy B." The recommendation is generated from analysis on historical data. Consider a directed social network $G = (V, E)$ with the IC model and k products in the market. Study the following problem: Consider a set of recommendations; each is in form "Who purchased products A_1, ..., A_h may also like to buy A_{h+1}". Given a budget B, find a set of customers for giving certain discount within the budget B, to maximize the expected total revenue. This problem is neither submodular nor supermodular. Therefore, it is a platform for the GRL method to play. Please design the GRL method for solving the problem.

16. (Marketing with Uncertain Customer Acceptable Price [381, 504]) Price is an important issue in the study of viral marketing. Zhu et al. [504] consider two stage process for product adoption. In the first stage, the customer gets explored to the product and may get influenced by features of the product. In the second stage, the customer decides to buy the product if price is acceptable. Each customer has his/her own acceptable price based on his/her evaluation on the product and his/her economic situation. Acceptable price is not easy to know exactly in marketing. However, from historical data, a distribution can be learned, so that it can be an important fact for seller to determine marketing price [381]. Can you design a GRL-based solution?

17. (Multi-feature Product [161]) A product may have multiple features considered by the customer for making a decision "buy or not". For example, except the quality, the price and the reputation (or brand image) may also be important features. Different persons have different levels of expertise on various features and hence may give different levels of influence on others. When consider multiple features of a product, different persons may also have different emphasis points. For example, a wealthy one may like a product with high reputation and a less wealthy one may prefer to have a product with high quality and less price. Therefore, in the study of viral marketing on multi-feature product, we have to employ the multi-feature diffusion model [161]. In this complicated case, machine learning has advantage. Please provide a machine learning solution.

18. (Block Rumor within a Community [103]) Every OSN has a community structure. For example, in an OSN, which is established from friendship, such as Facebook or LinkedIn, all persons with same interests in activities may form a community. Therefore, inside a community, the information about those activities spreads quite fast. Consider a community C. An edge is called a *bridge* to C if it crosses from C to another community, i.e., it contains exactly one endpoint in C. An endpoint u of a bridge to C is called a *bridge-end* for C if u is not in C. Suppose a rumor is found in community C. Since it is not easy to block the rumor spread inside community C, it is a naive idea to block the rumor at bridge-ends. In [103], it provided greedy approximation solutions for the problem of locating k protectors to maximize the number of bridge-ends getting protected. Now, please give a GRL-based solution.

19. (Influence-Based Community Detection [276]) Above work is under assumptions that (a1) a community structure is already known, and (a2) the cascade priority is uniform. However, the community structure is not so easy to know. Actually, the community detection is a quite complicated research problem. There are many research efforts made in the literature, which produce different approaches based on different principals. Especially, there is one which is based on influence so that in each community, members have bigger influence to each other [276], which may give a best service to our purpose. In Chap. 1, the influence-based community detection was studied. The problem is NP-hard and the objective function is nonsubmodular. Please employ the DRL-based GRL method to attack the problem.

20. (Rumor Blocking with Community Structure [167]) An interesting idea is to combine the rumor blocking and community detection and it may be simpler to solve them together. In fact, (b1) we only care about the community which contains the rumor source and (b2) since the objective function is the expected number of immunized bridge-ends, this gives us an opportunity to reduce the network size. To do so, we propose to solve a nonsubmodular maximization under a constraint on network size α, i.e., the number of network nodes is not exceed α. The objective function is the spread of immunization at bridge-ends under condition that all positive cascades pass through a subnetwork with at most α nodes. Let x_v be the indicator for a node v to be selected into the subnetwork. In the IC model, by the inclusive-exclusive principal, the objective function can be represented as

$$\Sigma_1 - \Sigma_2 + \Sigma_3 - \cdots,$$

where Σ_i is the sum of terms each of which is the union of i paths from protectors to bridge-ends and the coefficient is the expected number of bridge-ends that get immunized through those paths. Each Σ_i is monotone nondecreasing and supermodular and hence the objective function can be expressed as the difference of two monotone nondecreasing and supermodular functions

$$(\Sigma_1 + \Sigma_3 + \cdots) - (\Sigma_2 + \Sigma_4 + \cdots).$$

This is a variation of DS-decomposition, introduced in [248], i.e., the difference of two monotone submodular functions. Is this representation helpful for parameter selections in the GRL methods [19]?

21. (Rumor Source Detection [202]) In rumor blocking, knowing the location of rumor source is very helpful as we see in above discussion. However, it is hard to detect the location of rumor source. Actually, the rumor source is a special type of effector or information source. The information source detection is a special research topic in knowledge discovery. The rumor source and effector detection has been studied extensively. There exist various formulations. A lot of them can be reduced to nonsubmoduar optimization. Therefore, we may consider to make some efforts with the GRL approach.

22. (Rumor Blocking with Uncertain Source [495]) While the information source detection is hard to do, we would like to make efforts on rumor blocking with source uncertainty. In [495], we assume that rumor source is unknown. Consider a social network $G = (V, E)$ with IC model, Let R be a set of rumor sources and S a set of locations for protectors. Let $\sigma_r(S, R)$ denote the expected number of nodes that are activated by rumor source set R when S is selected to be allocated by protectors. Since R is not exactly known before we select S, we have to consider the worst case. Therefore, we minimize $f(S) = \max_R \sigma_r(R, S)$. It is proved in [495] that $f(S)$ is neither submodular nor supermodular. This problem can also be formulated into an integer programming as follows:

$$
\begin{aligned}
\min \quad & z \\
\text{subject to} \quad & \sigma_r(R, S) \leq z \quad \text{for all possible } R \\
& R \cap S = \emptyset,\, R \subseteq V,\, S \subseteq V.
\end{aligned}
$$

This formulation enable us to employ existing deep learning solutions [404]. Could you realize it?

23. (Rumor Blocking with Partial Information on Source [369]) In the real-world, it is rare completely to know the location of rumor source, and it is also rare completely not to know the location of rumor source. The most popular case is to have partial information about the location of rumor source. In [369], it is assumed to know a set of suspicious nodes. They consider three cases about suspicious nodes.

   • The probability of each suspicious node is known exactly [369].
   • The probability of each suspicious node is known to lie in an interval [369].
   • It is known that there are at most r rumor sources.

Shi et al. [369] formulated three robust optimization problems. Actually, in the literature, there are several different objective functions in optimization problems about robust rumor blocking. Different objective functions may give problems with different computational complexities. Therefore, much more robust optimization problems can be formulated on this subject. However, we found that with the GRL approach, we may be able to deal them with a uniform method with different selections on features or parameters. Please find a GRL-based solution.

24. (Clarify Composed Misinformation [130, 161, 254]) An object gets contaminated, possibly caused by not a single influence [130, 254]. The composed contamination have been studied extensively in the area of bio-informatics [130, 254]. However, on social influence, the composition is considered only recently [161]. This is due to the problem formulation. It is often a nonsubmodular optimization, which is not easy to find a good solution by traditional methods and the DRL-based GRL method is a favorite one to be employed. In the following, let us give a little explanation where and how those nonsubmodular optimization problems are formulated.

For composed misinformation, we will consider a more complicated background in the real-world. An object gets contaminated due to misinformation from possibly two or more factors. Each factor may diffuse individually, but not cause individual contamination. For example, the reputation of a product may have several good features, such as quality and price. The misinformation from competitors may mislead the consumers on different features in order to low down the reputation of the product. Therefore, to formulate such behavior, we have to use a model of composed influence. Meanwhile, for protectors, such a model is also required to formulate their positive cascades. In [161], we use a multi-layer social network to formulate the problem and showed that the rumor blocking in this model is nonsubmodular. Please design a GRL-based solution.

25. (Rumor Blocking with Multiple Agents [393]) Actually, in composed misinformation clarification, the cascade priority and the feature priority also play important roles to determine the submodularity of the spread of immunization. In fact, in [393], we indicated that the spread of immunization is submodular if the following all hold.

(b1) All positive cascades have priority higher than all negative.

(b2) All negative cascades have priority higher than all positive.

(b3) The cascade priority is strongly uniform, i.e., at every node, the priority ordering is the same.

However, if above three conditions are not all satisfied, then we have to face a nonsubmodular maximization to deal with the spread of immunization.

Especially, when (b3) is not satisfied, it may be more meaningful to convert the multi-layer model into lattice-point model, where the nonsubmodular optimization on lattice-point comes out. This would bring a new challenge to design the GRL solution. Could you do it?

Historical Notes

In recent years, as the development of machine learning gets more attentions, many research problems in computational social networks are studied with machine learning approaches. The graph representation learning plays an important role.

In the literature, many efforts have been made in graph representation learning for popularity predictions. Chen et al. [41] give a nice survey. The following is a brief summarization for method classifications made in [41].

Embedding-Based Methods: The embedding-based method is also called the sequence-based method. It is embedding graph information into a vector space by random walks and by maximizing the probability of observing neighborhood of an embedding node.

In a graph, a random walk is a sequence of nodes sampled randomly from the graph, which starts from a node to travel to a neighbor with certain probability, and continues the process. At each step, it may return to the starting node with some probability if the random walk has a restart strategy. There are two important types of random walks, DeepWalk [328] and node2vec [154], which have extensions LINE [379] and strct2vec [346], respectively.

Embedding-based methods are often used for popularity prediction in social networks with IC model. In those methods, each node is encoded into parameterized vector and the propagation probabilities between nodes are computed by learned node representations. The following is a list of examples:

- In [27], nodes are embedded into Euclidean space and the propagation probability between two nodes is given by the Euclidean distance between their representations.
- In [26], nodes are embedded in a continuous latent space. The contamination score of a node activated by the source node is computed through a diffusion kernel function in the latent space.
- In [429], LIS model is proposed, which is similar to embedded IC model. This model can be used for directly learn the user-specific influence and susceptibility, and capture context-dependent factors.
- In [111], a similarity between local influence and global user is considered when generate node extension.
- In [324], use a multi-task neural network to learn the influence probability.
- In [360], generate node sequence according to biased random walk.

Deep Learning Methods: Approaches in this class can be classified into the following groups:

- CNN-based models: We may find their applications in pattern recognition and image processing because CNNs have good performance in finding pattern in images. They also have applications in graph representation [209, 295, 454] although not many.
- GCN- and GAT-based models: GCN and GAT operate similarly. Therefore, they may be classified into the same group [67, 111, 239, 337, 418, 474].
- GNN-based models: GNN was first proposed in [354] for dealing with non-Euclidean data. There are many follow-up variations [35, 55, 173, 174, 192, 442].
- RNN-based models: RNN has a lot of applications to deal with sequential data [92, 443]. But it suffers from short-term memory. To solve this issue, two special subgroups are proposed, GRU-based model [240, 241, 328] and LSTM-based models [266, 340]. Moreover, transformer-based models [193, 407] are proposed for optimization of handling sequential data.
- RL-based models: They have received applications in influence maximization [40, 288], multi-scale diffusion prediction [455], and micro-scopic diffusion [456].

Solutions for Selected Exercises

Solutions are given here for the first four to five problems in exercises of each chapter.

Problem 1 in Chap. 1

($\Rightarrow$) Since f is submodular, we have

$$f(A \cup \{x\}) + f(B) \geq f((A \cup \{x\}) \cup B) + f((A \cup \{x\}) \cap B).$$

Since $A \subset B$ and $x \notin B$, we have

$$f(A \cup \{x\}) + f(B) \geq f(B \cup \{x\}) + f(A),$$

that is,

$$f(A \cup \{x\}) - f(A) \geq f(B \cup \{x\}) - f(B).$$

($\leftarrow$) Let $C = A \setminus B = \{y_1, \ldots, y_k\}$. Denote $C_i = \{y_1, \ldots, y_i\}$. Then we have

$$
\begin{aligned}
\Delta_C f(A \cap B) =& \Delta_{y_1} f(A \cap B) + \Delta_{y_2} f((A \cap B) \cup C_1) \\
& + \cdots + \Delta_{y_k} f((A \cap B) \cup C_{k+1}) \\
\geq& \Delta_{y_1} f(B) + \Delta_{y_2} f(B \cup C_1) \\
& + \cdots + \Delta_{y_k} f(B \cup C_{k+1}) \\
\geq& \Delta_C f(B).
\end{aligned}
$$

Therefore,

$$f((A \cap B) \cup C) - f(A \cap B) \geq f(B \cup C) - f(B),$$

W. Wu et al., *Computational Aspects of Social Networks*, Springer Optimization and Its Applications 234, https://doi.org/10.1007/978-3-032-14833-9

that is,

$$f(A) - f(A \cap B) \geq f(A \cup B) - f(B).$$

Problem 2 in Chap. 1

It sufficient to show that f is monotone nondecreasing if and only if for A and B of X with $A \subset B$, and any $x \in B \setminus A$,

$$\Delta_x f(A) \geq \Delta_x f(B). \tag{A.1}$$

Note that for $x \in B \setminus A$, $\Delta_x f(B) = 0$. Thus, inequality (A.1 is equivalent to

$$f(A \cup \{x\}) \geq f(A)$$

which must be true if f is monotone nondecreasing. Conversely, suppose $B \setminus A = \{x_1, \ldots, x_k\}$. Then

$$f(B) \geq f(A \cup \{x_1, \ldots, x_{k-1} \geq \cdots \geq f(A).$$

Problem 3 in Chap. 1

For any two sets A and B,

$$|A| + |B| = |A \cup B| = |A \cap B|.$$

In fact, an element x of $A \cup B$ is counted twice when count separately in A and B if and only if $x \in A \cap B$.

Problem 4 in Chap. 1

Let A and B be two subsets with $A \subset B$, and $x \notin B$. Denote $\eta(A) = \min(c, f(A)$. Since $c > 0$, we have

$$f(A) \leq f(B) \Rightarrow \eta(A) = \min(c, f(A)) \leq \eta(B) = \min(c, f(B)).$$

To show the submodularity of $\eta(\cdot)$, consider three cases.

Case 1. $f(A \cup \{x\}) > c$. In this case, we have

$$\begin{aligned}
\eta(A \cup \{x\}) - \eta(A) &= c - \eta(A) \\
&\geq c - \eta(B) \\
&= \eta(B \cup \{x\}) - \eta(B).
\end{aligned}$$

Case 2. $f(A \cup \{x\}) \leq c$ and $f(B) \leq c$. In this case, we have

$$\begin{aligned}
\eta(A \cup \{x\}) - \eta(A) &= f(A \cup \{x\}) - f(A) \\
&\geq f(B \cup \{x\}) - f(B) \\
&\geq \eta(B \cup \{x\}) - \eta(B).
\end{aligned}$$

Case 3. $f(A \cup \{x\}) \leq c$ and $f(B) > c$. In this case, we have

$$\begin{aligned}
\eta(A \cup \{x\}) - \eta(A) &= f(A \cup \{x\}) - f(A) \\
&\geq f(B \cup \{x\}) - f(B) \\
&\geq 0 = c - c = \eta(B \cup \{x\}) - \eta(B).
\end{aligned}$$

Furthermore, note that

$$f(\emptyset) = 0 \Rightarrow \eta(\emptyset) = \min(c, f(\emptyset)) = 0.$$

Therefore, if f is a polymatroid function, so is *eta*.

Problem 5 in Chap. 1

Assume $A \setminus B = \{x_1, x_2, \ldots, x_k\}$. Then we have

$$\begin{aligned}
f(A \cup B) - f(B) &= [f(B \cup \{x_1, \ldots, x_k\})] - [f(B \cup \{x_1, \ldots, x_{k-1}\})] \\
&\quad + \cdot + [f(B \cup \{x_1\}) - f(B)] \\
&\leq \sum_{i=1}^{k} [f(B \cup \{x_i\}) - f(B)].
\end{aligned}$$

Problem 1 in Chap. 2

Define

$$f_v(A) = \begin{cases} 1 & \text{if } |A| \geq |N_{in}(v)|/2 \\ 0 & \text{otherwise} \end{cases}$$

where A is the set of all active in-neighbors of v.

Problem 2 in Chap. 2

($\Leftarrow$) Suppose that every node v has in-degree at most two. Choose triggering set $T_v = N_{in}(v)$. Then v becomes active if and only if there exists an active in-neighbor of v if and only if $|A| \geq |N_{in}(v)|/2$ where A is the set of active in-neighbors of v.

($\Rightarrow$) For contradiction, suppose that there exists a node v with in-degree at least three. For any triggering set T_v, consider two cases.

Case 1. $T_v \neq \emptyset$. Choose a node $u \in T_v$. Let u be active and other in-neighbors inactive. Then according to the rule of triggering model, v should be active. However, according to the rule of positive influence, v should be inactive, a contradiction.

Case 2. $T_v = \emptyset$. Let all v's in-neighbors be active. Then according to the rule of triggering, v should be inactive. However, according to the rule of positive influence, v should be active, a contradiction.

Problem 3 in Chap. 2

(a) Define

$$f_v(S) = \begin{cases} 1 & \text{if } S \cap T_v \neq \emptyset \\ 0 & \text{if } A \cap T_v = \emptyset. \end{cases}$$

Then f_v is monotone nondecreasing and submodular.

(b) Let $\Pr[T_v, v \in V]$ denote the probability that at each node v, choose T_v as triggering set. Then

$$\sigma(S) = \sum \Pr[T_v, v \in V]\sigma(S \mid T_v, v \in V)$$

where $\sigma(S \mid T_v, v \in V)$ is the influence spread under condition that for each node v, triggering set T_v is fixed.

(c) By Theorem 2.4.3, $\sigma(S)$ is monotone nondecreasing and submodular.

Problem 4 in Chap. 2

Let $f_v(S)$ be the current threshold function at node v. Define

$$f_v^{new}(S) = \min(1, 2f_v(S)).$$

Then $f_v^{new}(S) \in [0, 1]$ and $f_v(S) \geq \theta$ for $\theta \in [0, 0.5]$ if and only if $f_v^{new}(S) \geq 2\theta$ for $2\theta \in [0, 1]$.

Problem 1 in Chap. 3

(a) Let $h(x) = f(x) - g(x)$. If $f(x) \neq g(x)$, then $h(x)$ is a non-zero polynomial, and its maximum degree is at most r. By the Fundamental Theorem of Algebra, a non-zero polynomial of degree r can have at most r distinct roots. Therefore, in the set $\{1, 2, \ldots, n\}$, there are at most r integers i such that $h(i) = 0$ (which is equivalent to $f(i) = g(i)$). When $f(x) \neq g(x)$, the probability that the randomized algorithm incorrectly returns "yes" is

$$P(\text{error}) = \frac{\text{Number of } i \text{ satisfying } f(i) = g(i)}{\text{Total number of possible } i} \leq \frac{r}{n}.$$

If $f(x) = g(x)$, then $f(i) = g(i)$ holds for all i, so the algorithm will definitely return "yes," and the error probability in this case is 0. In conclusion, the error probability of the algorithm is at most $\frac{r}{n}$.

(b) We could design a *repeated sampling algorithm* as follows:

- Choose a positive integer k (the number of repetitions).
- Independently, uniformly, and randomly select k integers $i_1, i_2, \ldots, i_k$ from the set $\{1, 2, \ldots, n\}$.
- For each i_j ($j = 1, 2, \ldots, k$), check whether $f(i_j) = g(i_j)$: – if there exists any i_j such that $f(i_j) \neq g(i_j)$, return "no"; – if $f(i_j) = g(i_j)$ holds for all i_j, return "yes."

Analysis of the Error Probability: When $f(x) \neq g(x)$, the probability of an erroneous result (i.e., $f(i_j) = g(i_j)$) in a single sampling is at most $\frac{r}{n}$. Since the k samplings are independent, the probability that all k samplings produce erroneous results is

$$P(\text{error}) \leq \left(\frac{r}{n}\right)^k.$$

By increasing the value of k, the error probability can be reduced to any desired small level (for example, choosing $k = \log_{n/r} \frac{1}{\epsilon}$ ensures that the error probability is less than ϵ). When $f(x) = g(x)$, the algorithm still definitely returns "yes," so the error probability remains 0.

Problem 2 in Chap. 3

The "Number of ways" column for each case has been filled in the following table.

Case	Identical balls?	Identical balls?	Balls can be empty?	Number of ways
1	Yes	No	Yes	$\binom{n+m-1}{m-1}$
2	Yes	No	No	$\binom{n-1}{m-1}$
3	No	Yes	No	$S(n, m)$
4	No	Yes	Yes	$\sum_{j=1}^{m} S(n, j)$
5	No	No	Yes	m^n
6	No	No	No	$m! \cdot S(n, m)$
7	Yes	Yes	Yes	$G(n, m)$
8	Yes	Yes	No	$G(n - m, m)$

In the table $S(n, m) = \frac{1}{m!} \sum_{k=0}^{m} (-1)^k \binom{m}{k} (m - k)^n$ is the Stirling number and $G(n, m)$ is the Gaussian binomial coefficients or partition function.

Case 1: Identical balls, Distinct bins, Bins can be empty

Claim: Number of ways $= \binom{n+m-1}{m-1}$

Proof This is a classic "stars and bars" problem. We model the distribution as placing n identical balls (stars: $\star$) into m distinct bins, separated by $m - 1$ dividers (bars: $|$). Total objects to arrange: n stars $+ m - 1$ bars $= n + m - 1$ objects. We need to choose positions for the $m - 1$ bars among these $n + m - 1$ positions. The number of ways is therefore $\binom{n+m-1}{m-1}$.

Case 2: Identical balls, Distinct bins, Bins cannot be empty

Claim: Number of ways $= \binom{n-1}{m-1}$

Proof Since bins cannot be empty, we first place 1 ball in each of the m bins (using m balls). This leaves $n - m$ balls to distribute freely (with bins now allowed to be empty). By Case 1, distributing $n - m$ identical balls into m distinct bins (with empty bins allowed) has $\binom{(n - m) + m - 1}{m - 1} = \binom{n - 1}{m - 1}$ ways.

Case 3: Distinct balls, Identical bins, Bins cannot be empty

Claim: Number of ways $= S(n, m)$ (Stirling number)

Proof Stirling number $S(n, m)$ can be defined recursively as follows:

$$S(n, m) = S(n - 1, m - 1) + m \cdot S(n - 1, m)$$

with base cases $S(n, 1) = 1$, $S(n, n) = 1$, and $S(n, m) = 0$ if $m > n$.

For the first ball, there is 1 way to place it in a bin. For the n-th ball: either place it in a new bin (contributing $S(n - 1, m - 1)$) or add it to one of the existing m bins (contributing $m \cdot S(n - 1, m)$). This recursion counts all ways to partition n distinct objects into m non-empty, indistinct subsets (bins), matching our problem.

Case 4: Distinct balls, Identical bins, Bins can be empty

Claim: Number of ways $= \sum_{j=1}^{m} S(n, j)$

Proof If bins can be empty, we allow distributions using $1, 2, \ldots$, up to m non-empty bins (since excess empty bins are indistinct and irrelevant). By definition, $S(n, j)$ counts ways to use exactly j non-empty bins. Summing over all possible $j \leq m$ gives the total number of distributions.

Case 5: Distinct balls, Distinct bins, Bins can be empty

Claim: Number of ways $= m^n$

Proof Each distinct ball can go into any of the m distinct bins. For the first ball, there are m choices; for the second ball, m choices; and so on, up to the n-th ball. By the multiplication principle, the total number of ways is $m \times m \times \cdots \times m = m^n$.

Case 6: Distinct balls, Distinct bins, Bins cannot be empty

Claim: Number of ways $= m! \cdot S(n, m)$

Proof First, partition n distinct balls into m non-empty groups (using $S(n, m)$ ways, as in Case 3). Since bins are distinct, we assign each group to a unique bin: there are $m!$ ways to permute the m groups among m bins. By the multiplication principle, the total number of ways is $m! \cdot S(n, m)$.

Case 7: Identical balls, Identical bins, Bins can be empty

Claim: Number of ways $= G(n, m)$ (integer partition function)

Proof This counts the number of integer partitions of n into at most m parts, where order does not matter (since bins are identical) and parts can be zero (since bins can be empty).

Formally, $G(n, m)$ is the number of solutions to

$$n_1 + n_2 + \cdots + n_m = n \quad \text{where } n_1 \geq n_2 \geq \cdots \geq n_m \geq 0.$$

Case 8: Identical balls, Identical bins, Bins cannot be empty
Claim: Number of ways $= G(n - m, m)$

Proof Since bins cannot be empty, first place 1 ball in each bin (using m balls), leaving $n - m$ balls to distribute. Now we need to partition $n - m$ identical balls into m identical bins (which can now be empty), which is exactly $G(n - m, m)$ by Case 7. Equivalently, this counts partitions of n into exactly m positive parts (since each bin has at least 1 ball).

Problem 3 in Chap. 3

(a) Expected Number of Balls

We model the problem as a sequence of independent trials, where each trial corresponds to placing one ball into a randomly chosen bin (with probability $\frac{1}{m}$). The process can be decomposed using the linearity of expectation. Let T be the total number of balls needed to fill all m bins. Let T_k be the number of additional balls needed to fill a new bin, given that $k - 1$ bins are already filled ($k = 1, 2, \ldots, m$). Then

$$T = T_1 + T_2 + \cdots + T_m.$$

For each T_k, when $k - 1$ bins are filled, there are $m - (k - 1)$ empty bins. The probability of placing a ball into a new (empty) bin is $p_k = \frac{m - k + 1}{m}$. So, T_k follows a geometric distribution with success probability p_k, and thus $\mathbb{E}[T_k] = \frac{1}{p_k}$. Hence $\mathbb{E}[T_k] = \frac{m}{m - k + 1}$. By linearity of expectation:

$$\mathbb{E}[T] = \sum_{k=1}^{m} \mathbb{E}[T_k] = \sum_{k=1}^{m} \frac{m}{m - k + 1}.$$

Reindexing the sum (let $j = m - k + 1$) gives

$$\mathbb{E}[T] = m \sum_{j=1}^{m} \frac{1}{j} = m \cdot H_m$$

where $H_m = 1 + \frac{1}{2} + \cdots + \frac{1}{m}$ is the m-th harmonic number. For large m, the harmonic number approximates as follows:

$$H_m \approx \ln m + \gamma + \frac{1}{2m} - \cdots$$

where $\gamma \approx 0.5772$ is the Euler-Mascheroni constant. Neglecting small terms:

$$\mathbb{E}[T] \approx m \left(\ln m + \gamma\right).$$

(b) Probability Estimation

We calculate $P(n)$, the probability that all m bins contain at least one ball after n balls are placed. Using the principle of inclusion-exclusion. Let A_i be the event that the i-th bin is empty after n balls. We need $P\left(\bigcap_{i=1}^{m} A_i^c\right)$, the probability that no bin is empty.

By inclusion-exclusion:

$$P\left(\bigcap_{i=1}^{m} A_i^c\right) = \sum_{k=0}^{m}(-1)^k \binom{m}{k} P\left(\bigcap_{i=1}^{k} A_i\right)$$

For $\bigcap_{i=1}^{k} A_i$, each ball must go into one of the remaining $m - k$ bins. The probability for one ball is $\frac{m-k}{m}$, so for n independent balls:

$$P\left(\bigcap_{i=1}^{k} A_i\right) = \left(\frac{m - k}{m}\right)^n = \left(1 - \frac{k}{m}\right)^n.$$

Substituting back, the probability that all bins are non-empty is

$$P(n) = \sum_{k=0}^{m}(-1)^k \binom{m}{k}\left(1 - \frac{k}{m}\right)^n.$$

For large n and m, use the approximation $\left(1 - \frac{k}{m}\right)^n \approx e^{-kn/m}$,

$$P(n) \approx \sum_{k=0}^{m}(-1)^k \binom{m}{k} e^{-kn/m}.$$

For $n = m \ln m + cm$ (where c is a constant), this simplifies further. The dominant term in the sum is $k = 0$ (contribution 1) and $k = 1$ (contribution $-me^{-(\ln m+c)} = -e^{-c}$). Higher-order terms are negligible, giving

$$P(n) \approx e^{-e^{-c}}.$$

Problem 4 in Chap. 3

(a) Expected Number of Collisions

Let us count collisions by considering pairs of balls. The total number of pairs of balls is $\binom{n}{2} = \frac{n(n-1)}{2}$. For any specific pair of balls, the probability they collide

$$P(\text{collision for a pair}) = \frac{1}{m}$$

(since the first ball can go to any bin, and the second must match that bin). Define $X_{i,j}$ as an indicator variable where $X_{i,j} = 1$ if the i-th and j-th balls collide, and 0 otherwise. Then the total number of collisions C is

$$C = \sum_{1 \le i < j \le n} X_{i,j}.$$

By linearity of expectation:

$$\mathbb{E}[C] = \sum_{1 \le i < j \le n} \mathbb{E}[X_{i,j}] = \binom{n}{2} \cdot \frac{1}{m} = \frac{n(n-1)}{2\,m}.$$

(b) Expected Number of Empty Bins

Let Y_i be an indicator variable where $Y_i = 1$ if the i-th bin is empty, and 0 otherwise. The total number of empty bins E is

$$E = \sum_{i=1}^{m} Y_i.$$

For a specific bin, the probability it remains empty after n balls is

$$P(Y_i = 1) = \left(1 - \frac{1}{m}\right)^n$$

(since each ball independently avoids the bin with probability $1 - \frac{1}{m}$). By linearity of expectation:

$$\mathbb{E}[E] = \sum_{i=1}^{m} \mathbb{E}[Y_i] = m \cdot \left(1 - \frac{1}{m}\right)^n.$$

(3) Probabilities for a Specified Bin

Consider a particular bin (e.g., the first bin). Each ball is placed into this bin with probability $p = \frac{1}{m}$, independently of other balls. This follows a binomial distribution with parameters n (trials) and $p = \frac{1}{m}$ (success probability). The probability of exactly k successes is

$$P(\text{exactly } k \text{ balls}) = \binom{n}{k} p^k (1-p)^{n-k} = \binom{n}{k} \left(\frac{1}{m}\right)^k \left(1 - \frac{1}{m}\right)^{n-k}.$$

Problem 1 in Chap. 4

Let A^* be an optimal solution. Denote $opt = f(A^*)$. Let us fix all random process until A_i is obtained. Then we have

$$\mathbb{E}[\Delta_{u_i} f(A_{i-1})] = \frac{1}{k} \cdot \sum_{u \in M_i} \Delta_u f(A_{i-1})$$

$$\geq \frac{1}{k} \cdot \sum_{u \in A^* \setminus A_i} \Delta_u f(A_{i-1})$$

$$\geq \frac{1}{k}(f(A^* \cup A_{i-1}) - f(A_{i-1}))$$

$$\geq \frac{1}{k}(opt - f(A_{i-1}))$$

where the first inequality is due to greedy choice of u_i, the second inequality is due to the submodularity of f, and the third inequality is due to the monotone nondecreasing property of f.

We now release the randomness of A_i and A_{i-1}. Then we have

$$\mathbb{E}[f(A_i)] - \mathbb{E}[f(A_{i-1})] \geq \frac{1}{k} \cdot (opt - \mathbb{E}[f(A_{i-1})]).$$

Denote $a_i = opt - \mathbb{E}[f(A_i)]$. Then we have

$$a_{i-1} - a_i \geq \frac{1}{k} \cdot a_{i-1}.$$

Thus,

$$a_i \leq (1 - \frac{1}{k})a_{i-1} \leq e^{-1/k} a_{i-1} \leq \cdots \leq e^{-1} a_0$$

where we use the inequality $1 + x \le e^x$. Note that $a_o = opt - \mathbb{E}[f(A_0)] = opt$. Finally, we obtain

$$\mathbb{E}[f(A_k)] \ge (1 - 1/e) \cdot opt.$$

Problem 2 in Chap. 4

Consider Algorithm 50. Let A_g be its output. Let $E = \{x_1, x_2, \ldots, x_n\}$ in the ordering $f(x_1) \ge f(x_2) \ge \cdots \ge f(x_n)$. Denote $E_i = \{x_1, \ldots, x_i\}$. Then we have

$$f(A_g) = f(x_1)|E_1 \cap A_g| + \sum_{i=2}^{n} f(x_i)(|E_i \cap A_g| - |E_{i-1} \cap A_g|)$$

$$= \sum_{i=1}^{n-1} |E_i \cap A_g|(f(x_i) - f(x_{i+1})) + |E_n \cap A_g|f(x_n).$$

Algorithm 50 Greedy Algorithm for Problem 2 in Chapter 4

Input: An independent system $(E, \mathcal{I})$ and a function $f : E \to \mathbb{R}_{\ge 0}$.
Output: An independent set A_g.
1: Sort all elements of E in the ordering $f(x_1) \ge f(x_2) \ge \cdots \ge f(x_n)$;
2: Initially, $A \leftarrow \emptyset$;
3: **for** $i \leftarrow 1$ to n **do**
4:     **if** $A \cup \{x_i\} \in \mathcal{I}$ **then**
5:         $A \leftarrow A \cup \{x_i\}$
6:     **end if**
7: **end for**
8: **return** $A_g = A$.

Let A^* be the maximum solution. Similarly, we have

$$f(A^*) = \sum_{i=1}^{n-1} |E_i \cap A^*|(f(x_i) - f(x_{i+1})) + |E_n \cap A^*|f(x_n).$$

Therefore,

$$\frac{f(A_g)}{f(A^*)} \ge \min_{1 \le i \le n} \frac{|A_g \cap E_i|}{|A^* \cap E_i|}.$$

Clearly, $|A^* \cap E_i| \le v(E_i)$. We claim that $|A_g \cap E_i| \ge u(E_i)$. To prove our claim, it is sufficient to show that $A_g \cap E_i$ is a maximal independent subset of E_i. For contradiction, suppose that $A_g \cap E_i$ is a maximal independent subset of E_i. Then there exists an element $x_j \in E_i \setminus A_g$ such that $(A_g \cap E_i) \cup \{x_j\}$ is independent. Note that in the computation of Algorithm 50, $I \cup \{x_j|\}$ is a subset of $(A_g \cap E_i) \cup \{x_j\}$. Hence, $I \cup \{x_j\}$ is independent. Thus, $x_j \in A_g$, a contradiction. Therefore, our claim is proved.

Finally, we obtain

$$\frac{f(A_g)}{f(A^*)} \geq \min_{1 \leq i \leq n} \frac{v(E_i)}{u(E_i)} \geq \min_{F \subseteq E} \frac{v(F)}{u(F)}.$$

Problem 3 in Chap. 4

Consider an independent system $(E, \mathcal{I})$. For any subset A of E, define

$$\mathcal{I}_A = \{B \subseteq E \mid A \not\subseteq B\}.$$

We claim that $(E, \mathcal{I}_A)$ is a matroid.

To prove our claim, consider any $F \subseteq E$ and two cases.

Case 1. $A \not\subseteq F$. In this case, F has unique maximal independent set which is itself. Hence, $u(F) = v(F)$.

Case 2. $A \subseteq F$. In this case, every maximal independent set contained in F is in form $F \setminus \{x\}$ for an element $x \in A$. Therefore, $u(F) = v(F) = |F| - 1$.

Now, $\mathcal{I} = \cap_{A \notin \mathcal{I}} \mathcal{I}_A$.

Problem 4 in Chap. 4

We prove that if independent system $(E, \mathcal{I})$ is the intersection of h matroids $(E, \mathcal{I}_i)$ for $1 \leq i \leq h$, then for any subset F of E,

$$v(F)/u(F) \leq h.$$

Then, the proof is completed by using the result in Problem 2.

Let $F \subseteq E$. Consider two maximal independent sets I and J of F with respect to $(E, \mathcal{I})$. Let $I_i \supseteq I$ be a maximal independent subset of $I \cup J$ with respect to $(E, \mathcal{I}_i)$.

Note that for any $e \in J \setminus I$, if $e \in \cap_{i=1}^{h}(I_i \setminus I)$, then $I \cup \{e\} \in \cap_{i=1}^{h} I_i = I$, contradicting the maximality of I. Hence, e appears in at most $h - 1$ $(I_i \setminus I)$'s. Thus,

$$\sum_{i=1}^{h} |I_i| - h|I| = \sum_{i=1}^{h} |I_i \setminus I|$$
$$\leq (h - 1)|J \setminus I|$$
$$\leq (h - 1)|J|.$$

Now, let $J_i \supseteq J$ be a maximal independent subset of $I \cup J$ with respect to $(E, \mathcal{I}_i)$. Since $(E, \mathcal{I}_i)$ is a matroid, we have $|I_i| = |J_i||$ for every $0 \leq i \leq h$. Therefore,

$$|J| \leq \left(\sum_{i=1}^{h} |J_i| - h|J| \right) + |J|$$

$$= \left(\sum_{i=1}^{h} |I_i| - h|J| \right) + |J|$$

$$\leq h|I|.$$

Problem 1 in Chap. 5

Assume that f is defined on subsets of V. Define

$$m_f(X) = f(\emptyset) + \sum_{j \in X} \Delta_j f(V \setminus \{j\})$$

and $p_f = f - m_f$. Then m_f is modular. Hence p_f is submodular. Moreover, $p_f(\emptyset) = 0$ and for any $v \in V \setminus X$,

$$p_f(X \cup \{v\}) - p_f(X) = f(X \cup \{v\}) - f(X) - \Delta_v f(V \setminus \{v\}) > 0.$$

Thus, p_f is monotone nondecreasing. Hence, p_f is a polymatroid function.

Problem 2 in Chap. 5

Assume that m is defined on subsets of V. Define $m'(X) = m(X) - m(\emptyset)$. Let

$$V^+ = \{v \in V \mid m'(\{v\}) \geq 0\}$$

and

$$V^- = \{v \in V \mid m'(\{v\}) < 0\}.$$

Define

$$m^+(X) = m'(X \cap V^+)$$

and

$$m^-(X) = m'(X \cap V^-).$$

Then m^+ and m^- are monotone nondecreasing and $m = m^+ - m^-$.

Problem 3 in Chap. 5

The statement can be disproved by the following counterexample.

Define $f(X) = \min(\max(|X|, 1), 3) - 1$ on subsets of V. Clearly, f is monotone nondecreasing. Assume $f(X) = h(X) + g(X)$ where $h(X)$ is monotone nondecreasing and submodular, and $g(X)$ is monotone nondecreasing and supermodular.

Let A and B be subsets of V such that $A \subset B, |A| = 1$ and $|B| = 2$. Let $v \in V \setminus B$. Then we have

$$\Delta_v f(\emptyset) = 0, \ \Delta_v(A) = 1, \ \Delta_v f(B) = 0.$$

Thus,

$$\Delta_v f(\emptyset) + \Delta_v f(B) < \Delta_v f(A).$$

On the other hand, since h is monotone nondecreasing and submodular, we have $\Delta_v h(\emptyset) \geq \Delta_v h(A)$ and $\Delta_v h(B) \geq 0$. Thus,

$$\Delta_v h(\emptyset) + \Delta_v h(B) \geq \Delta_v h(A).$$

Moreover, since g is monotone nondecreasing and supermodular, we have $\Delta_v g(\emptyset) \geq 0$ and $\Delta_v g(B) \geq \Delta_v g(A)$. Thus,

$$\Delta_v g(\emptyset) + \Delta_v g(B) \geq \Delta_v(A).$$

Therefore,

$$\Delta_v f(\emptyset) + \Delta_v f(B) \geq \Delta_v f(A),$$

a contradiction.

Problem 4 in Chap. 5

Let $\alpha' \leq \alpha(f)$. Suppose that α' can be computed in polynomial time. Define $g(X) = \sqrt{|X|}$. Then

$$\alpha(g) = \min_{A \subset B \subseteq V \setminus \{x\}} \{\sqrt{|A| + 1} - \sqrt{|A|} - \sqrt{|B| + 1} + \sqrt{|B|}\}$$

$$= \min_{A \subset B \subseteq V \setminus \{x\}} \{\sqrt{|A| + 1} - \sqrt{|A|} - \sqrt{|A| + 2} + \sqrt{|A| + 1}$$

$$= 2\sqrt{n + 1} - sqrtn - \sqrt{n + 2} > 0.$$

Define

$$h(X) = f(X) + \frac{|\alpha'|}{\alpha(g)}.$$

Then $\alpha(h) \geq 0$ and

$$f = h - \frac{|\alpha'|}{\alpha(g)}.$$

Problem 5 in Chap. 5

(a) Consider the following linear program (LP) and its dual LP.

$$\text{(Primal LP)} \quad \max \quad z^T x$$
$$\text{s.t.} \quad x(S) \leq f(S), \forall S \subset V$$
$$x(V) = f(V).$$

$$\text{(Dual LP)} \quad \min \quad \sum_{S \subseteq V} y_S f(S)$$
$$\text{s.t.} \quad \sum_{S \subseteq V} y_S e_S = z$$
$$y_S \geq 0, \forall S \subseteq V$$

where e_S is a vector whose ith component is 1 at $i \in S$, 0 at $i \notin S$. Suppose $z_1 \geq z_2 \geq \cdots \geq z_n$. Define $S_i = \{1, 2, \ldots, i\}$. Let

$$x_i^* = f(S_i) - f(S_{i-1}), \forall i = 1, 2, \ldots, n.$$

Since f is submodular, for any $S = \{i_1 \geq i_2 \geq \cdots \geq i_k\} \subseteq V$,

$$x^*(S) = \sum_{i \in S} x_i^* = \sum_{1 \leq j \leq k} (f(S_{i_j}) - f(S_{i_j - 1})$$
$$\leq \sum_{1 \leq j \leq k} (f(\{i_1, \ldots, i_j\}) - f(\{i_1, \ldots, i_{j-1}\}))$$
$$\leq f(S) - f(\emptyset) = f(S).$$

Thus, x^* is a feasible solution of the primal LP.
Let

$$y_S^* = \begin{cases} z_i - z_{i+1} & \text{if } S = S_i \text{ for some } i, \\ z_n & \text{if } S = V, \\ 0 & \text{otherwise.} \end{cases}$$

Then y^* is a feasible solution of the dual LP.

Moreover, we have

$$z^T x^* = \sum_{i \in V} z_i (f(S_i) - f(S_{i-1}))$$

$$= \sum_{i=1}^{n-1} (z_i - z_{i+1}) f(S_i) + z_n f(S_n)$$

$$= \sum S \subseteq V \, y_S^* f(S) = \hat{f}(z).$$

Therefore, x^* is a primal optimal solution. Hence, $\hat{f}(z) = \max_{x \in P_f} z^T x$.
(b) Suppose $z_1 \geq z_2 \geq \cdots \geq z_n$. Denote $S_i = \{1, 2, \ldots, i\}$. Then

$$\hat{f}(z) = \sum_{i=1}^{n} z_i x_i^* = \sum_{i=1}^{n} z_i (f(S_i) - f(S_{i-1})).$$

Thus,

$$\sum_{i=1}^{n} z_i ((f(S_i) + g(S_i)) - (f(S_{i-1}) + g(S_{i-1})))$$

$$= \sum_{i=1}^{n} z_i (f(S_) - f(S_{i-1})) + \sum_{i=1}^{n} z_i (g(S_i) - g(S_{i-1}))$$

$$= \hat{f}(z) + \hat{g}(z)$$

and

$$\sum_{i=1}^{n} z_i (\lambda f(S_) - \lambda f(S_{i-1})) = \lambda \sum_{i=1}^{n} z_i (f(S_) - f(S_{i-1})) = \lambda \hat{f}(z).$$

(c) ($\Rightarrow$) Let x^* be an optimal solution for

$$\max_{x \in P_f} (\alpha z_1 + (1 - \alpha) z_2)^T x.$$

Then

$$\hat{f}(\alpha z_1 + (1 - \alpha) z_2) = (\alpha z_1 + (1 - \alpha) z_2)^T x^*$$

$$= \alpha z_1^T x^* + (1 - \alpha) z_2^T x^*$$

$$\leq \alpha \hat{f}(z_1) + (1 - \alpha) \hat{f}(z_2).$$

($\Leftarrow$) Since f is convex, we have that for $S, T \subseteq V$,

$$\frac{1}{2} \cdot \hat{f}(e_S) + \frac{1}{2} \cdot \hat{f}(e_T) \geq \hat{f}(\frac{1}{2}(e_S + e_T)) = \frac{1}{2} \hat{f}(e_S + e_T).$$

Note that $\hat{h}(e_S) = f(S)$, $\hat{f}(e_T) = f(T)$, and

$$(e_S + e_T)_i = \begin{cases} 2 & \text{if } i \in S \cap T \\ 1 & \text{if } i \in S \setminus T \text{ or } i \in T \setminus S \\ 0 & \text{otherwise,} \end{cases}$$

that is, $\hat{f}(e_S + e_T) = f(S \cap T) + f(S \cup T)$. Therefore, $f(S) + f(T) \geq f(S \cup T) + f(S \cap T)$.

Answer to the Question

The proof is not correct because Lovasz extension may not be differentiable for some set functions.

Problem 1 in Chap. 6

Let $T = T_1 \subset T_2 \subset \cdots \subset T_k = S$ where $T_i \setminus T_{i-1} = \{a_i\}$. For each $2 \leq i \leq k$,

$$f(T_i) - f(T_{i-1}) \geq f(S) - f(S \setminus \{a_i\}) \geq 0.$$

Hence $f(T) \leq f(S)$.

Next, let $S = T_1 \subset T_2 \subset \cdots \subset T_k = S$ where $T_i \setminus T_{i-1} = \{a_i\}$. For each $2 \leq i \leq k$,

$$f(T_i) - f(T_{i-1}) \leq f(S \cup \{a_i\}) - f(S) \leq 0.$$

Thus, $f(T) \leq f(S)$.

Problem 2 in Chap. 6

Let C be a maximum solution. By result in Problem 1, we have

$$\begin{aligned} 2f(S) + f(X \setminus S) &\geq f(S \cup C) + f(S \cap C) + f(X \setminus S) \\ &\geq f(X) + f(C \setminus S) + f(S \cap C) \\ &\geq f(X) + f(C) + f(\emptyset) \\ &\geq f(C) = opt. \end{aligned}$$

Therefore, $\max(f(S), f(X \setminus S)) \geq \frac{1}{3} \cdot opt$.

Problem 3 in Chap. 6

We will use the following property of matroid: For $S, C \in \mathcal{I}$ with $|C| = |S|$, there exists a one-to-one mapping $\pi : C \setminus S \to S \setminus C$ such that $(S \setminus \{\pi(e)\}) \cup \{e\} \in \mathcal{I}$ for any $e \in C \setminus S$.

The proof is divided into three cases.

Case 1. $|S| = |C|$. In this case, there exists a bijection $\pi : C \setminus S \setminus C$ such that $(S \setminus \{\pi(e)\}) \cup \{e\} \in \mathcal{I}$ for any $e \in C \setminus S$. Since S is locally maximal, we have

$$f(S) \geq f((S \setminus \{\pi(e)\}) \cup \{e\}) \text{ for any } e \in C \setminus S.$$

Hence,

$$f(S \cup \{e\}) - f(S) \leq f(S \setminus \{\pi(e)\}) \cup \{e\}) - f(S \setminus \{\pi(e)\})$$
$$\leq f(S) - f(S \setminus \{\pi(e)\}).$$

Therefore,

$$f(S \cup C) - f(S) \leq \sum_{e \in C \setminus S} [f(S \cup \{e\} - f(S)]$$
$$\leq \sum_{e \in C \setminus S} [f(S) - f(S \setminus \{\pi(e)\})]$$
$$= \sum_{e' \in S \setminus C} [f(S) - f(S \setminus \{e'\})]$$
$$\leq f(S) - f(S \cap C).$$

Case 2. $|S| > |C|$. In this case, there exists a one-to-one (not on to) mapping $\pi : C \setminus S \setminus C$ such that $(S \setminus \{\pi(e)\}) \cup \{e\} \in \mathcal{I}$ for any $e \in C \setminus S$. Since S is locally maximal, we have

$$f(S) \geq f((S \setminus \{\pi(e)\}) \cup \{e\}) \text{ for any } e \in C \setminus S.$$

Hence, for every $e \in C \setminus S$,

$$f(S \cup \{e\}) - f(S) \leq f(S \setminus \{\pi(e)\}) \cup \{e\}) - f(S \setminus \{\pi(e)\})$$
$$\leq f(S) - f(S \setminus \{\pi(e)\}).$$

Thus,

$$f(S \cup C) - f(S) \leq \sum_{e' \in \pi(C \setminus S)} [f(S) - f(S \setminus \{r'\})].$$

Moreover, since deletion operation cannot improve S, we have

$$0 \leq \sum_{e' \in (S \setminus C) \setminus \pi(C \setminus S)} [f(S) - f(S \setminus \{r'\})].$$

Therefore,

$$f(S \cup C) - f(S) = \sum_{e' \in S \setminus C} [f(S) - f(S \setminus \{e'\})]$$
$$\leq f(S) - f(S \cap C).$$

Case 3. $|C| > |S|$. In this case, we choose a subset D of C such that $|D| = |S|$ and $S \cap D = S \cap C$. Then there exists a bijection $\pi : D \setminus S \to S \setminus C$ such that $(S \setminus \{\pi(e)\}) \cup \{e\} \in \mathcal{I}$ for any $e \in D \setminus S$. Since S is locally maximal, we have

$$f(S) \geq f((S \setminus \{\pi(e)\}) \cup \{e\}) \text{ for any } e \in D \setminus S.$$

Hence, for every $e \in D \setminus S$,

$$f(S \cup \{e\}) - f(S) \leq f(S \setminus \{\pi(e)\}) \cup \{e\}) - f(S \setminus \{\pi(e)\})$$
$$\leq f(S) - f(S \setminus \{\pi(e)\}).$$

Moreover, for $e \in C \setminus D$, since deletion cannot improve S, we have

$$f(S \cup \{e\}) - f(S) \leq 0.$$

Thus,

$$f(S \cup C) - f(S) \leq \sum_{e \in D \setminus S} [f(S \cup \{e\} - f(S)]$$
$$\leq \sum_{e \in D \setminus S} [f(S) - f(S \setminus \{\pi(e)\})]$$
$$= \sum_{e' \in S \setminus C} [f(S) - f(S \setminus \{e'\})]$$
$$\leq f(S) - f(S \cap C).$$

Problem 4 in Chap. 6

Let C be an optimal solution for

$$\max\{f(S) \mid S \in \mathcal{I}\}.$$

This, $opt = f(C)$. Denote $C' = C \cap (X \setminus S_1)$. By the result in Problem 3, we have

$$2 \cdot f(S_1) \geq f(S_1 \cup C) + f(S_1 \cap C),$$
$$2 \cdot f(S_2) \geq f(S_2 \cup C') + f(S_2 \cap C').$$

Therefore,

$$4 \cdot \max(f(S_1), f(S_2))$$
$$\geq f(S_1 \cup C) + f(S_1 \cap C) + f(S_2 \cup C') + f(S_2 \cap C')$$
$$\geq f(S_1 \cup S_2 \cup C) + f(C') + f(S_1 \cap C) + f(S_2 \cap C')$$
$$\geq f(S_1 \cup S_2 \cup C) + f(C) + f(\emptyset) + f(S_2 \cap C')$$
$$\geq f(C) = opt.$$

Problem 1 in Chap. 7

($\Longrightarrow$) For contradiction, suppose there exists $z \in A \setminus \{x\}$ such that $s(x) \geq s(z) + dist(z, x)$. For any $y \in V$,

$$r_x(y) = s(x) + dist(x, y) \geq s(z) + dist(z, x) + dist(x, y)$$
$$\geq s(z) + dist(z, y) = r_z(y).$$

Hence,
$$r_A(y) = \min(r_x(y), r_{A \setminus \{x\}}(y)) = r_{A \setminus \{x\}}(y).$$

Thus, for any $y_1, y_2 \in V$,

$$r_A(y_1) - r_A(y_2) = r_{A \setminus \{x\}}(y_1) - r_{A \setminus \{x\}}(y_2),$$

that is, A and $A \setminus \{x\}$ are not separable, a contradiction.
($\Longleftarrow$) Since $s(x) < r_{A \setminus \{x\}}(x)$, we have

$$r_A(x) = \min(r_x(x), r_{A \setminus \{x\}}(x))$$
$$= \min(s(x), r_{A \setminus \{x\}}(x))$$
$$= s(x) < r_{A \setminus \{x\}}(x).$$

Choose $z \in A \setminus \{x\}$ such that $s(z) = \min_{y \in A \setminus \{x\}} s(y)$. Then

$$s(z) = r_z(z) \geq r_{A \setminus \{x\}}(z)$$
$$= \min_{y \in A \setminus \{x\}} (s(y) + dist(y, z))$$
$$\geq \min_{y \in A \setminus \{x\}} s(y) = s(z).$$

Then

$$r_{A \setminus \{x\}}(z) = s(z) = r_A(z).$$

Therefore,

$$r_A(x) - r_A(z) < r_{A \setminus \{x\}}(x) - r_{A \setminus \{x\}}(z).$$

That is, A and $A \setminus \{x\}$ are separated by x and z.

Problem 2 in Chap. 7

Suppose there exist two distinct maximal node subsets A and B such that u is detectable in A and also in B. Then we have

$$s(u) < r_{A \setminus \{u\}}(u) < s(v) + dist(v, u) \forall v \in A \setminus \{u\}$$

and

$$s(u) < r_{B \setminus \{u\}}(u) < s(v) + dist(v, u) \forall v \in B \setminus \{u\}.$$

Therefore,

$$s(u) < s(v) + dist(v, u) \forall v \in (A \cup B) \setminus \{u\}.$$

Thus, u is detectable in $A \cup B$, contradicting the maximality of A and B.

Problem 3 in Chap. 7

For contradiction, suppose there are two distinct detectable sets A and B in $\mathcal{F}$ such that they cannot be separated by any two nodes x and y. Hence,

$$r_A(x) - r_A(y) = r_B(x) - r_B(y).$$

Select $x = \text{argmax}\{s(z) \mid z \in A\}$ and $y = \text{argmax}\{s(z) \mid z \in B\}$. Then

$$r_A(x) = \min(r_x(x), r_{A\setminus\{x\}}(x))$$
$$= \min(s(x), r_{A\setminus\{x\}}(x))$$
$$= s(x).$$

Similarly, $r_B(y) = s(y)$. Therefore,

$$s(x) + s(y) = r_A(y) + r_B(x). \qquad\qquad (A.2)$$

Note that

$$r_A(y) = \min_{z \in A}\{s(z) + dist(z, y)\} \geq \min_{z \in A} s(z) = s(x).$$

If $y \notin A$, then $dist(z, y) > 0$ for any $z \in A$ and hence

$$r_A(y) > s(x).$$

Similarly, we have $r_B(x) \geq s(y)$ and if $y \notin A$, then $r_A(y) > s(x)$. By (A.2), we have to have $y \in A$ and $x \in B$. Therefore, $x, y \in A \cap B$ and $s(x) = s(y)$.

Consider $y' = \text{argmin}\{s(z) \mid z \in A \triangle B\}$ where $A \triangle B = (A \setminus B) \cup (B \setminus A)$. Without loss of generality, assume $y' \in A \setminus B$. Then $r_A(y') = s(y')$. Note that $r_A(x) = s(x) = r_B(x)$ and

$$r_A(x) - r_A(y') = r_B(x) - r_B(y').$$

Thus, $r_B(y') = s(y')$. Since $y' \notin B$, we have

$$r_{B\setminus A}(y') = \min_{z \in B\setminus A}\{s(z) + dist(z, y')$$
$$> \min_{z \in B\setminus A} s(z)$$
$$\geq \min_{z \in A\triangle B} s(z) = s(y') = r_B(y').$$

Therefore,
$$r_B(y') = \min(r_{A\cap B}(y'), r_{B\setminus A}(y')) = r_{A\cap B}(y').$$

Since $A \cap B \subseteq A \setminus \{y'\}$, we have

$$s(y') = r_B(y') = r_{A\cap B}(y') \geq r_{A\setminus\{y'\}}(y'),$$

contradicting the fact that A is detectable and Problem 1.

Problem 4 in Chap. 7

Let A be a detectable set and $B \subset A$. Then for any $x \in B$,

$$s(x) < r_A(x) \leq r_B(x).$$

Therefore, B is detectable by Problem 1.

Problem 5 in Chap. 7

Note that $\sigma(S \mid R)$ can be expressed as

$$\sigma(S \mid R) = \sum_{H \subseteq G} \Pr[H] \sigma^H(S] mid R)$$

where $\Pr[H]$ is the probability that all alive edges form subgraph H and $\sigma^H(S \mid R)$ is the number of nodes that accept the positive cascade of S under the existence of the rumor cascade of R. Therefore, it is sufficient to consider the deterministic diffusion model.

Let $\Sigma(S \mid R)$ denote the set of nodes that accept the positive cascade of S under the existence of the rumor cascade of R. Then for two sets A and B with $A \subset B$,

$$\Sigma(A \mid R) \subseteq \Sigma(B \mid R).$$

Consider $x \notin B$. It is easy to see

$$\Sigma(x \mid R) \setminus \Sigma(A \mid R) \supseteq \Sigma(x \mid R) \setminus \Sigma(B \mid R).$$

Hence,

$$\begin{aligned}
\sigma(A \cup \{x\} \mid R) - \sigma(A \mid R) &= |\Sigma(x \mid R) \setminus \Sigma(A \mid R)| \\
&\geq |\Sigma(x \mid R) \setminus \Sigma(B \mid R)| \\
&= \sigma(B \cup \{x\} \mid R) - \sigma(B \mid R).
\end{aligned}$$

Therefore, $\sigma(S \mid R)$ is submodular with respect to S. Finally, observe that the priority selection at each node does not have impact in above argument.

Problem 1 in Chap. 8

Please note the following facts:

- The social utility function γ does not appear in the definition of Nash Equilibrium.
- If define $\gamma'(a) = \sum_{i=1}^{k} \alpha_i(a)$, then $(\gamma', \cup_{i=1}^{k})$ would be a basic valid utility system.

These two facts indicate that putting "basic" in condition does not make sense.

Problem 2 in Chap. 8

It is hard to employ the non-cooperative game as a model for this rumor blocking problem. In fact, it is hard to define the social utility to meet the significance in the real world.

Problem 3 in Chap. 8

In the non-cooperate model of distributed rumor blocking, if nodes are allowed to select different priority, then the social utility may be not submodular. Consider a social network as shown in Fig. A.1; every edge has influence probability one. Suppose that there are two cascades C_1 and C_2 distributed from agents 1 and 2, respectively. Suppose that node v_3 selects priority $P(C_1) > P(C_r) > P(C2)$ and node v_4 selects priority $P(C_2) > P(C_r) > P(C_1)$. The rumor source is allocated at node v_6.

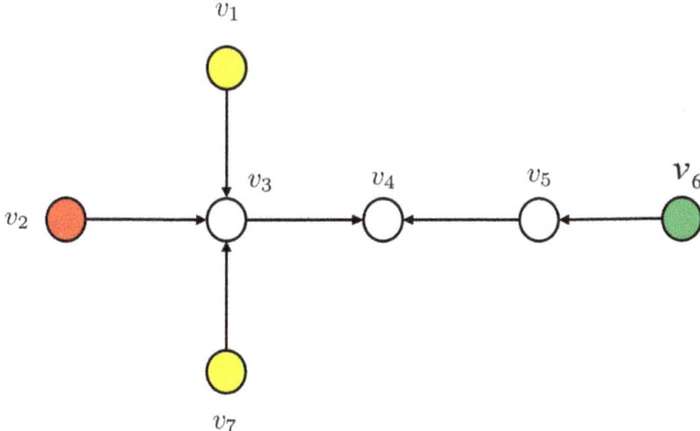

Fig. A.1 A counterexample

Then, we have $\gamma(\emptyset, \{v_2\}) = 5$, $\gamma(\{v_1\}, \{v_2\}) = \gamma(\{v_7\}, \{v_2\}) = \gamma(\{v_1, v_7\}, \{v_2\}) = 4$. Thus,

$$\gamma(\{v_1\}, \{v_2\}) + \gamma(\{v_7\}, \{v_2\}) < \gamma(\emptyset, \{v_2\}) + \gamma(\{v_1, v_7\}, \{v_2\}).$$

Therefore, the social utility γ is not submodular.

Problem 4 in Chap. 8

In the non-cooperative game model of competitive influence, private utility $sigma_i$ may not be submodular with respect to S_j for $j \neq i$. For a counterexample, consider two players 1 and 2 and a social network as shown in Fig. A.2. Every edge has influence probability one. Then, we have $\sigma_2(\emptyset, \{v_2\}) = 5$, $\sigma_2(\{v_1\}, \{v_2\}) = \sigma_2(\{v7\}, \{v_2\}) = \sigma_2(\{v_1, v_7\}, \{v_2\}) = 2$. Hence,

$$\sigma_2(\{v_1\}, \{v_2\}) + \sigma_2(\{v_7\}, \{v_2\}) < \sigma_2(\emptyset, \{v_2\}) + \sigma_2(\{v_1, v_7\}, \{v_2\}).$$

Therefore, σ_2 is not submodular with respect to S_1.

Problem 5 in Chap. 8

The answer is no. In fact, does not matter what priority choice each node makes the social utility $\sigma(S_1, \ldots, S_p)$ is submodular with respect to $\cup_{i=1}^{p} S_i$, and the private

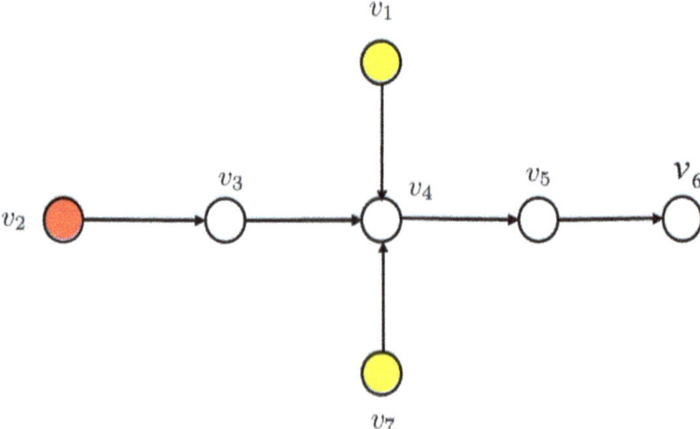

Fig. A.2 A counterexample

utility $\sigma_i(S_1, \ldots, S_p$ is submodular with respect to S_i. In fact, the priority does not affect the influence of $\cup_{i=1}^p S_i$. Therefore, the social utility is invariant when the priority is changed.

To study the submodularity of private utility σ_i, it is sufficient to work on the social network with deterministic model (i.e., the influence probability of every edge is one) since the private utility σ_i in the IC model can be expressed as a linear combination of the private utility σ_i in finitely many subgraphs of G with the deterministic model.

Now, assume that the social network is with the deterministic model and all S_j for $j \neq i$ are fixed. Let $\Gamma(S_i)$ denote the set of nodes which receive the influence of S_i. Then for any two node subsets A and B, we have

$$\Gamma(A \cup B) = \Gamma(A) \cup \Gamma(B)$$

$$\Gamma(A \cap B) \subseteq \Gamma(A) \cap \Gamma(B).$$

Hence,
$$|\Gamma(A)| + |\Gamma(B)| \geq |\Gamma(A \cup B)| + |\Gamma(A \cap B)|.$$

Note that

$$\sigma_i(S_1, \ldots, S_p)|_{S_i=A} = |\Gamma(A)|$$
$$\sigma_i(S_1, \ldots, S_p)|_{S_i=B} = |\Gamma(B)|$$
$$\sigma_i(S_1, \ldots, S_p)|_{S_i=A \cup B} = |\Gamma(A \cup B)|$$
$$\sigma_i(S_1, \ldots, S_p)|_{S_i=A \cap B} = |\Gamma(A \cap B)|.$$

Therefore,

$$\sigma_i(S_1, \ldots, S_p)|_{S_i=A} + \sigma_i(S_1, \ldots, S_p)|_{S_i=B} \geq \sigma_i(S_1, \ldots, S_p)|_{S_i=A \cup B} + \sigma_i(S_1, \ldots, S_p)|_{S_i=A \cap B}.$$

Hence, σ_i is submodular with respect to S_i.

Problem 1 in Chap. 9

Recall that a realization is a function $\phi : V \to O = \{\text{active, inactive}\}$. When ϕ is fixed, the state of each edge (u, v) can be fixed by setting

$$(u, v) \text{ is } \begin{cases} \text{alive} & \text{if } \phi(u) = \phi(v) = \text{ active} \\ \text{dead}. \end{cases}$$

Let $G(\phi)$ be the subgraph induced by alive edges of G. Then the influence is reduced to that in $G(\phi)$ with deterministic diffusion model. Let $I(A)$ denote the set of influenced nodes of A through graph $G(\phi)$. Consider two node subsets A and B with $A \subset B$. Consider also a node $x \in V \setminus A$. Then

$$I(\{x\}) - I(B) \subseteq I(\{x\}) - I(A).$$

Therefore,

$$U(A \cup \{x\}, \phi) - U(A, \phi) = \left| \sum_{u \in I(\{x\}) \setminus I(A)} w(u) \right|$$

$$\geq \left| \sum_{w \in I(\{x\}) \setminus I(B)} w(u) \right|$$

$$= U(B \cup \{x\}, \phi) - U(B, \phi).$$

Problem 2 in Chap. 9

Since realization ϕ is unchanged, we may treat ϕ as fixed. Then the information diffusion process can be considered to be carried out in the alive graph $G(\phi)$ with deterministic model, where an edge (u, v) is alive if and only if $\phi(u) = \phi(v) = $ active and $G(\phi)$ is the subgraph induced by all alive edges. In such a diffusion process, the set of nodes activated by $dom(\psi)$ equals the union of sets of nodes activated by node x over $dom(\psi)$.

Problem 3 in Chap. 9

Let $\sigma_{m,full}(S_1, \ldots, S_k)$ denote the maximum influence spread when all $|\cup_{i=1}^{k} S_i|$ seeds are deployed in the network through k stages: At the ith stage, seeds in S_i will be deployed in the network. Let $S = \cup_{i=1}^{k} S_i$. Then

$$\sigma_{m,full}(S, \emptyset, \ldots, \emptyset) \geq \sigma(S_1, \ldots, S_k) \geq \sigma_{m,full}(\emptyset, \ldots, \emptyset, S).$$

Note that

$$\sigma_m(S) = \sigma_{m,full}(S, \emptyset, \ldots, \emptyset) = \sigma_{m,full}(\emptyset, \ldots, \emptyset, S).$$

This completes the proof.

Problem 4 in Chap. 9

Note that $\Pr[\phi] = \Pr[$ node 1 is active$]$. Moreover, under the IC model,

$$\Pr[\text{ node 1 is active}] = 1 - (1 - p_{21})(1 - p_{31}),$$

and under the LT model,

$$\Pr[\text{ node 1 is active}] = p_{21} + p_{31}.$$

Therefore, under the IC model,

$$\Pr[\phi] = p_{21} + p_{31} - p_{21}p_{31},$$

and under the LT model,

$$\Pr[\phi] = p_{21} + p_{31}.$$

Problem 1 in Chap. 10

This is because the following two robust LPs are equivalent:

$$\begin{aligned}
\min_{x} \max_{c \in U_c} \quad & c^T x \\
\text{subject to} \quad & a_i^T x \le b_i, \forall a_i \in U_{a_i}, \forall b_i \in U_{b_i}, i = 1, \ldots, m.
\end{aligned}$$

$$\begin{aligned}
\min \quad & \alpha \\
\text{subject to} \quad & c^T x \le \alpha, \forall c \in U_c \\
& a_i^T x \le b_i, \forall a_i \in U_{a_i}, \forall b_i \in U_{b_i}, i = 1, \ldots, m.
\end{aligned}$$

Problem 2 in Chap. 10

The uncertainty of b_i can be removed because the worst-case scenario is achieved at $b_i = b_i^-$.

Problem 3 in Chap. 10

Consider an n-clique $G = (V, E)$ and a parameter space $\Theta = \times_{e \in E}[0, 1]$. Consider any seed set S with $|S| = k$. Choose a parameter vector $\theta = (p_e)_{e=(u,v) \in E}$ where

$$p_e = \begin{cases} 0 \text{ if } u \in S \text{ or } v \in S \\ 1 \text{ otherwise.} \end{cases}$$

Then $\sigma_\theta(S) = k$. However, $\sigma_\theta(S_\theta^*) = n$ if $k \geq 2$. (Let S_θ^* contain a node in S and another node not in S.) Therefore,

$$g(\Theta, S) = \frac{k}{n}.$$

Problem 4 in Chap. 10

Consider a graph G consisting of two disjoint $(n/2)$-cliques, A and B. Denote $p = 2/n$ and let $\epsilon = 1/(10n)$ be a small positive constant. For every edge e, set parameter space $[l_e, r_e]$ where $l_e = p - \epsilon$ and $r_e = p + \epsilon$. Choose parameter vector θ such that for every edge e,

$$p_e = \begin{cases} l_e \text{ if } e \in A \\ r_e \text{ if } e \in B. \end{cases}$$

By Erdös-Rényi theorem, there exists a realization (i.e., subgraph induced by all alive edges) such that every connected component in A has size $O(\log n)$ and B has a connected component of size at least cn for a constant $c > 0$. Therefore, suppose $S = \{v\}$, we have

$$\sigma_\theta(S) = \begin{cases} O(\log n) \text{ if } v \in A, \\ cn \qquad\quad \text{ if } v \text{ lies in the big connected component.} \end{cases}$$

Therefore

$$g(\Theta, S) = O(\frac{\log n}{n})$$

and

$$r_e - l_e = 2\epsilon = O(1/n).$$

Problem 1 in Chap. 11

$$
\begin{aligned}
u_1^T u_2 &= u_1^T \left(a_2 - \frac{u_1^T a_2}{u_1^T u_1} u_1 \right) \\
&= u_1^T a_2 - \frac{u_1^T a_2}{u_1^T u_1} u_1^T u_1 \\
&\quad - u_1^T a_2 - u_1^T a_2 \\
&= 0.
\end{aligned}
$$

Problem 2 in Chap. 11

Let u_2 be defined in Problem 1. Then u_2 and a_1 are orthogonal and

$$
s = \|a_2\| \cdot \|u_2\|.
$$

Note that

$$
\begin{aligned}
\|u_2\|^2 &= u_2^T u_2 \\
&= a_2^T a_2 - \frac{(a_1^T a_2)^2}{a_1^T a_1}.
\end{aligned}
$$

Thus

$$
\begin{aligned}
S^2 &= \|a_1\|^2 \cdot \|u_2\|^2 \\
&= (a_1^T a_1)(u_2^T u_2) \\
&= (a_1^T a_1)(a_2^T a_2) - (a_1^T a_2)^2.
\end{aligned}
$$

Problem 3 in Chap. 11

(a)

$$
\begin{aligned}
\frac{p_i}{p_j} &= \frac{\exp(a^T s_i)}{\sum_k \exp(a^T s_k)} \cdot \frac{\sum_k \exp(a^T s_k)}{\exp(a^T s_j)} \\
&= \frac{\exp(a^T s_i)}{\exp(a^T s_j)} \\
&= \exp(a^T (s_i - s_j)).
\end{aligned}
$$

(b) The ith component of softmax$(s'; a)$ is

$$
\begin{aligned}
p_i' &= \frac{\exp(a^T s_i')}{\sum_k \exp(a^T s_k')} \\
&= \frac{\exp(a^T (s + b))}{\sum_k \exp(a^T (s + b))} \\
&= \frac{\exp(a^T s)}{\sum_k \exp(a^T s)} \\
&= p_i,
\end{aligned}
$$

where p_i is the ith component of softmax$(s; a)$. Therefore,

$$
\mathrm{softmax}(s'; a) = \mathrm{softmax}(s; a).
$$

Problem 4 in Chap. 11

Let $q = \mathrm{Pow})\mathrm{softma}(s; 1); \alpha)$. Then

$$
q_i = \frac{\left(\frac{\exp s_i}{\sum_k \exp s_k}\right)^\alpha}{\sum_j \left(\frac{\exp s_j}{\sum_k \exp s_k}\right)^\alpha} = \frac{(\exp s_i)^\alpha}{\sum_j (\exp s_j)^\alpha} = \frac{\exp(\alpha s_i)}{\sum_j \exp(\alpha s_j)}.
$$

References

1. A. A. Ageev and M. I. Sviridenko: Pipage rounding: A new method of constructing algorithms with proven performance guarantee. *Journal of Combinatorial Optimization*, 8(3):307–328 (2004).
2. C.C. Aggarwal, P. Yu: Online analysis of community evolution in data streams. In *Proceedings of the SIAM international conference on data mining* (SDM 2005), pp 56–67.
3. R. Albert, H. Jeong, and A. Barabasi, Error and attack tolerance of complex networks, *Nature*, vol. 406, 2000.
4. H. Almeida, D. Guedes, W. Meira Jr, and M. J. Zaki: Is there a best quality metric for graph clusters? In *Machine Learning and Knowledge Discovery in Databases*. Springer, 2011, pp. 44–59.
5. N. Alon, I. Gamzu, M. Tennenholtz: Optimizing budget allocation among channels and influencers. In: *Proceedings of the 21st International Conference on World Wide Web*, 2012, pp. 381–388.
6. S.P. Anderson, A. De Palma, J.-F. Thisse: *Discrete Choice Theory of Product Differentiation*, MIT press, 1992.
7. C. Aslay, L. V. Lakshmanan, W. Lu, and X. Xiao: Influence maximization in online social networks. In *Proc. 7th ACM Int. Conf. Web Search Data Mining*, Feb. 2018, pp. 775–776.
8. S. Asur, S. Parthasarathy, D. Ucar: An event-based framework for characterizing the evolutionary behavior of interaction graphs. *ACM Trans Knowl Discov Data (TKDD)* 3(4):16 (2009).
9. L. Backstrom, D. Huttenlocher, J. Kleinberg, X. Lan: Group formation in large social networks: membership, growth, and evolution. In *Proceedings of the 12th ACM SIGKDD international conference on knowledge discovery and data mining*, ACM, pp 44–54 (2006).
10. Ashwinkumar Badanidiyuru and Jan Vondrák: Fast algorithms for maximizing submodular functions. In *Proceedings of the 25th Annual ACM-SIAM Symposium on Discrete Algorithms*, pages 1497–1514, 2014.
11. A. Barabasi, R. Albert, and H. Jeong: Scale-free characteristics of random networks: the topology of the world-wide web, *Physica A*, vol. 281, 2000.
12. A. L. Barabasi, H. Jeong, Z. Nda, E. Ravasz, A. Schubert, and T. Vicsek: Evolution of the social network of scientific collaborations. *Physica A: Statistical Mechanics and its Applications*, vol. 311, 2002.

W. Wu et al., *Computational Aspects of Social Networks*, Springer Optimization and Its Applications 234, https://doi.org/10.1007/978-3-032-14833-9

13. E. Berger: Dynamic Monopolies of Constant Size. *Journal of Combinatorial Theory Series B* 83: 191–200 (2001).

14. Shishir Bharathi, David Kempe, Mahyar Salek: Competitive influence maximization in social networks. In *WINE 2007*: 306–311. (Their conjecture is in another version of paper https://david-kempe.com/publications/viral-competitive.pdf.)

15. Yuanjun Bi, Weili Wu, Li Wang: Community Expansion in Social Network. In *DASFAA (1) 2013*: 41–55.

16. Yuanjun Bi, Weili Wu, Ailian Wang, Lidan Fan: Community Expansion Model Based on Charged System Theory. In *COCOON 2013*: 780–790.

17. Yuanjun Bi, Weili Wu, Yuqing Zhu, Lidan Fan, Ailian Wang: A nature-inspired influence propagation model for the community expansion problem. *J. Comb. Optim.* 28(3): 513–528 (2014).

18. Andrew An Bian, Joachim M. Buhmann, Andreas Krause,Sebastian Tschiatschek: Guarantees for Greedy Maximization of Non-submodular Functions with Applications, arXiv:1703.02100v3 [cs.DM] 13 June 2017.

19. Jeff A. Bilmes: Submodularity In Machine Learning and Artificial Intelligence. CoRR arXiv:2202.00132 (2022).

20. V. D. Blondel, J.-L. Guillaume, R. Lambiotte, and E. Lefebvre: Fast unfolding of communities in large networks. *J. Stat. Mech., Theory Exp.*, vol. 10, p. P10008, Oct. 2008.

21. A. Blum, M. Hajiaghayi, K. Ligett, A. Roth: Regret minimization and the price of total anarchy. In: *Proc. 39th ACM Symp. on Theory of Computing*, pp. 373–382 (2008).

22. Christian Borgs, Michael Brautbar, Jennifer T. Chayes, Brendan Lucier: Maximizing Social Influence in Nearly Optimal Time. In *SODA* 2014: 946–957.

23. Christian Borgs, Jennifer T. Chayes, Adrian Marple, Shang-Hua Teng: An Axiomatic Approach to Community Detection. In *ITCS* 2016: 135–146.

24. A. Borodin, Y. Filmus, J. Oren: Threshold Models for Competitive Influence in Social Networks. In *Proceedings of the Internet and Network Economics—6th International Workshop*, WINE 2010, Stanford, CA, USA, 13–17 December 2010; pp. 539–550.

25. Michaela Borzechowski: The complexity class Polynomial Local Search (PLS) and PLS-complete problems. Thesis in Freie Universität Berlin, https://www.mi.fu-berlin.de/inf/groups/ag-ti/theses/download/Borzechowski16.pdf.

26. Simon Bourigault, Cedric Lagnier, Sylvain Lamprier, Ludovic Denoyer, and Patrick Gallinari. Learning social network embeddings for predicting information diffusion. In *Proceedings of the 7th ACM international conference on Web search and data mining*, pages 393–402, 2014.

27. Simon Bourigault, Sylvain Lamprier, and Patrick Gallinari. Representation learning for information diffusion through social networks: an embedded cascade model. In *Proceedings of the Ninth ACM international conference on Web Search and Data Mining*, pages 573–582, 2016.

28. A. Bozorgi, S. Samet, J. Kwisthout, T. Wareham: Community-based influence maximization in social networks under a competitive linear threshold model. *Knowl.-Based Syst.*, 134: 149–158 (2017).

29. U. Brandes, D. Delling, M. Gaertler, R. Gorke, M. Hoefer, Z. Nikoloski, and D. Wagner: On modularity clustering. *IEEE Transactions on Knowledge and Data Engineering (TKDE)*, 20(2): 172–188 (2008).

30. J. Brown, P. Reinegen: Social ties and word-of-mouth referral behavior. *J Consum Res*, 14:350–362 (1987).

31. J. Bruna, W. Zaremba, and Y. Szlam, A.and LeCun: Spectral networks and locally connected networks on graphs. In *ICLR*, 2014.

32. Niv Buchbinder, Moran Feldmany, Joseph (Seffi) Naorz, Roy Schwartz: Submodular Maximization with cardinality constraints, In *Proceedings, 25th annual ACM-SIAM symposium on Discrete algorithms*, pp. 1433–1452, 2014.

33. C. Budak, D. Agrawal, A.E. Abbadi: Limiting the spread of misinformation in social networks. In *20th Intl. World Wide Web Conference*, pp. 665–674, 2011.

34. G. Calinescu, C. Chekuri, M. Pál, and J. Vondrák. Maximizing a monotone submodular function subject to a matroid constraint. *SIAM Journal on Computing (SICOMP)*, 40(6):1740–1766 (2011)

35. Qi Cao, Huawei Shen, Jinhua Gao, Bingzheng Wei, and Xueqi Cheng: Popularity prediction on social platforms with coupled graph neural networks. In *Proceedings of the 13th International Conference on Web Search and Data Mining*, pages 70–78, 2020.
36. T. Carnes, C. Nagarajan, S. Wild, A. van Zuylen: A. Maximizing Influence in a Competitive Social Network: A Follower's Perspective. In *Proceedings of the Ninth International Conference on Electronic Commerce*, Minneapolis, MN, USA, 19–22 August 2007; pp. 351–360.
37. D. Chakrabarti, R. Kumar, A. Tomkins: Evolutionary clustering. In *Proceedings of the 12th ACM SIGKDD international conference on knowledge discovery and data mining*, (ACM, 2006) pp 554–560.
38. C. Chekuri, J. Vondrák, and R. Zenklusen: Dependent randomized rounding via exchange properties of combinatorial structures. In *Proc., IEEE Symposium on Foundations of Computer Science (FOCS)*, pages 575–584, 2010.
39. Chandra Chekuri, Jan Vondrák, and Rico Zenklusen: Submodular function maximization via the multilinear relaxation and contention resolution schemes. *SIAM J. Comput.*, 43(6):1831–1879 (2014).
40. Haipeng Chen, Wei Qiu, Han-Ching Ou, Bo An, and Milind Tambe: Contingency aware influence maximization: A reinforcement learning approach. In *Uncertainty in Artificial Intelligence*, pages 1535–1545. PMLR, 2021.
41. Tiantian Chen, Jianxiong Guo, Weili Wu: Graph representation learning for popularity prediction problem: A survey. *Discret. Math. Algorithms Appl.* 14(7): 2230003:1–2230003:29 (2022).
42. Tiantian Chen, Siwen Yan, Jianxiong Guo, Weili Wu: ToupleGDD: A Fine-Designed Solution of Influence Maximization by Deep Reinforcement Learning. *IEEE Trans. Comput. Soc. Syst.* 11(2): 2210–2221 (2024).
43. Tiantian Chen, Bin Liu, Wenjing Liu, Qizhi Fang, Jing Yuan, Weili Wu: A random algorithm for profit maximization in online social networks. *Theor. Comput. Sci.* 803: 36–47 (2020).
44. W. Chen, L.V.S. Lakshmanan, C. Castillo: Information and Influence Propagation in Social Networks. *Synthesis Lectures on Data Management*. Morgan & Claypool Publishers, San Rafael (2013).
45. Wei Chen, Tian Lin, Zihan Tan, Mingfei Zhao, Xuren Zhou: Robus influence maximization. *KDD'16*, San Francisco, CA, USA, 2016.
46. W. Chen, W. Lu, and N. Zhang: Time-critical influence maximization in social networks with time-delayed diffusion process. In *Twenty-Sixth AAAI Conference on Artificial Intelligence*, 2012.
47. Wei Chen, Binghui Peng, Grant Schoenebeck, Biaoshuai Tao: Adaptive Greedy versus Non-Adaptive Greedy for Influence Maximization. *AAAI 2020*: 590–597 .
48. Wei Chen and Binghui Peng. On adaptivity gaps of influence maximization under the independent cascade model with full adoption feedback. In *Proceedings of the 30th International Symposium on Algorithms and Computation (ISAAC)*, 2018.
49. Wei Chen, Binghui Peng, Grant Schoenebeck, Biaoshuai Tao: Adaptive Greedy versus Non-adaptive Greedy for Influence Maximization. *J. Artif. Intell. Res.* 74: 303–351 (2022).
50. Wei Chen, Chi Wang, Yajun Wang: Scalable influence maximization for prevalent viral marketing in large-scale social networks. *KDD* 2010: 1029–1038.
51. W. Chen, Y. Wang, and S. Yang: Efficient influence maximization in social networks, in *KDD'09*, (New York, NY, USA), pp. 199-208, ACM, 2009.
52. W. Chen, Y. Wang, and Y. Yuan: Combinatorial multi-armed bandit and its extension to probabilistically triggered arms. CoRR, arXiv:1407.8339, 2014.
53. Wei Chen, Yifei Yuan, Li Zhang: Scalable Influence Maximization in Social Networks under the Linear Threshold Model. In *ICDM* 2010: 88–97
54. Xin Chen, Qingqin Nong, Yan Feng, Yongchang Cao, Suning Gong, Qizhi Fang, Ker-I Ko: Centralized and decentralized rumor blocking problems. *J. Comb. Optim.* 34(1): 314–329 (2017).
55. Xueqin Chen, Fan Zhou, Kunpeng Zhang, Goce Trajcevski, Ting Zhong, and Fengli Zhang: Information diffusion prediction via recurrent cascades convolution. In *2019 IEEE 35th International Conference on Data Engineering (ICDE)*, pages 770–781. IEEE, 2019.

56. Xujin Chen, Xiao-Dong Hu, Changjun Wang: Approximation for the minimum cost doubly resolving set problem. *Theor. Comput. Sci.* 609: 526–543 (2016).

57. Z. Chen, K. Zhu, and L. Ying: Detecting multiple information sources in networks under the SIR model, *IEEE Trans. Netw. Sci. Eng.*, 3 (1): 17–31 (2016).

58. Jie Cheng, Yangyang Zhang, Qiang Ye, Hongwei Du: High-precision shortest distance estimation for large-scale social networks. In *INFOCOM* 2016: 1–9.

59. F. R. K. Chung and L. Lu. Concentration inequalities and martingale inequalities: A survey. *Internet Mathematics*, 3(1):79–127 (2006).

60. A. Clauset, M. E. J. Newman, and C. Moore: Finding community structure in very large networks, *Phys. Rev. E, Stat. Phys. Plasmas Fluids Relat. Interdiscip. Top.*, vol. 70, no. 6, p. 066111, 2004.

61. A. Clauset, C. Moore, M.E.J. Newman Hierarchical structure and the prediction of missing links in networks. *Nature*, 453: 98–101 (2008).

62. A. Clauset, C. R. Shalizi, and M. E. J. Newman: Power-law distributions in empirical data, *SIAM Reviews*, 2007.

63. E. Cohen, D. Delling, T. Pajor, and R. F. Werneck: Sketch-based influence maximization and computation: Scaling up with guarantees. In *Proc. 23rd ACM Int. Conf. Conf. Inf. Knowl. Manage.*, Nov. 2014, pp. 629–638.

64. Michele Conforti and Gérard Cornuéjols: Submodular set functions, matroids and the greedy algorithm: Tight worst-case bounds and some generalizations of the Rado-Edmonds theorem. *Discrete Applied Mathematics* 7 (3): 251–274 (1984).

65. M. Coscia, G. Rossetti, F. Giannotti, and D. Pedreschi: Uncovering hierarchical and overlapping communities with a local-first approach. *ACM Trans. Knowl. Discovery Data*, 9 (1): p. 6 (2014).

66. M. Coscia, G. Rossetti, F. Giannotti, and D. Pedreschi: DEMON: A local-first discovery method for overlapping communities. In *roc. KDD*, pp. 615–623, 2012.

67. Hsing-Huan Chung, Hen-Hsen Huang, and Hsin-Hsi Chen: Predicting future participants of information propagation trees. In *IEEE/WIC/ACM International Conference on Web Intelligence*, pages 321–325, 2019.

68. W.H. Cunningham: Decomposition of submodular functions. *Combinatorica* 3(1): 53–68 (1983).

69. P. Dagum, R. Karp, M. Luby, and S. Ross: An optimal algorithm for Monte Carlo estimation. *SIAM J. Comput.*, 29: 1484–1496 (2000).

70. Abhimanyu Das and David Kempe: Submodular meets spectral: Greedy algorithms for subset selection, sparse approximation and dictionary selection. In *ICML*, pp. 1057–1064, 2011.

71. S. Das, M. Chen: Yahoo! for Amazon: opinion extraction from small talk on the web. *Management Science*, 53: 1375–1388 (2007).

72. H. Dai, B. Dai, and L. Song: Discriminative embeddings of latent variable models for structured data. In *ICML*, 2016.

73. Dongping Deng, Hongwei Du, Xiaohua Jia, Qiang Ye: Minimum-Cost Information Dissemination in Social Networks. In *WASA* 2015: 83–93.

74. Z. Ding, X. Zhang, D. Sun, and B. Luo: Overlapping community detection based on network decomposition, In *Sci. Rep.*, vol. 6, Apr. 2016, Art. no. 24115.

75. Thang N. Dinh, My T. Thai: Community detection in scake-free networks: approximation algorithms for maximizing modularity. *IEEE Journal on Selected Area in Communications* 31 (6): 997–1006 (2013).

76. T. N. Dinh, Y. Xuan, and M. T. Thai: Towards social-aware routing in dynamic communication networks, in *IEEE IPCCC*, 2009, pp. 161–168.

77. P. Domingos, M. Richardson: Mining the network value of customers. In: *KDD*, pp 57–66 (2001).

78. Luobing Dong, Meghana N. Satpute, Weili Wu, Ding-Zhu Du: Two-Phase Multidocument Summarization Through Content-Attention-Based Subtopic Detection. *IEEE Trans. Comput. Soc. Syst.* 8(6): 1379–1392 (2021).

79. Ding-Zhu Du, Ker-I Ko: *Theory of Computational Complexity* (2nd Ed.), (John-Wiley, 2014).

80. Ding-Zhu Du, Ker-I Ko, Xiaodong Hu: *Design and Analysis of Approximation Algorithms*, (Springer, 2012).

81. Ding-Zhu Du, Panos M. Pardalos, Xiaodong Hu, Weili Wu: *Introduction to Combinatorial Optimization*, (Springer, 2022).

82. Ding-Zhu Du, P.M. Pardalos, and Weili Wu: *Mathematical Theory of Optimization*, (Kluwer Academic Publishers, 2001).

83. Ding-Zhu Du and Xiang-Sun Zhang: A convergence theorem of Rosen's gradient projection method. *Mathematical Programming*, 36: 135–144 (1986).

84. Ding-Zhu Du and Xiang-Sun Zhang: Global convergence of Rosen's gradient projection method. *Mathematical Programming*, 44: 357–366 (1989).

85. Hongmin W. Du, Yingfan L. Du, Zhao Zhang: Adaptive Influence Maximization: Adaptability via Nonadaptability, *INFORMS Journal on Computing*, 36(5): 1190–1200 (2024).

86. Hongmin W. Du, Xiang Li, Guanghua Wang: New approximations for monotone submodular maximization with knapsack constraint. *J. Comb. Optim.* 48(4): 28 (2024).

87. Hongwei Du, Caiwei Yuan, He Yuan, Shanshan Wei, Wen Xu: Identify Connected Positive Influence Dominating Set in Social Networks Using Two-Hop Coverage. *IEEE Trans. Comput. Soc. Syst.* 6(5): 956–967 (2019).

88. Hongwei Du, Zhao Zhang, Zhenhua Duan, Cong Tian, Ding-Zhu Du: Full view camera sensor coverage and group set coverage. In *Proceedings of 15th EAI International Wireless Internet Conference*, Dallas, USA, 2022.

89. N. Du, Y. Liang, M.F. Balcan, L. Song: Budgeted influence maximization for multiple products arXiv:1312.2164.

90. N. Du, L. Song, M. Gomez-Rodriguez, and H. Zha: Scalable influence estimation in continuous-time diffusion networks. In *Advances in neural information processing systems*, pp. 3147–3155, 2013.

91. N. Du, B. Wu, X. Pei, B. Wang, and L. Xu: Community detection in large-scale social networks. In *in Proceedings of the 9th WebKDD and 1st SNA-KDD 2007 workshop on Web mining and social network analysis*. ACM, 2007, pp. 16–25.

92. Nan Du, Hanjun Dai, Rakshit Trivedi, Utkarsh Upadhyay, Manuel Gomez-Rodriguez, and Le Song: Recurrent marked temporal point processes: Embedding event history to vector. In *Proceedings of the 22nd ACM SIGKDD international conference on knowledge discovery and data mining*, pages 1555–1564, 2016.

93. J. Duch and A. Arenas: Community detection in complex networks using extremal optimization, *Physical review E*, vol. 72, no. 2, p. 027104, 2005.

94. R. I. M. Dunbar: Do online social media cut through the constraints that limit the size of offline social networks? *Roy. Soc. Open Sci.*, vol. 3, no. 1, p. 150292, 2016, https://doi.org/10.1098/rsos.150292.

95. M. Edelson, T. Sharot, R. J. Dolan, and Y. Dudai: Following the crowd: Brain substrates of long-term memory conformity, *Sci.*, 333 (6038): 108–111 (2011).

96. Jack Edmonds: Matroids and the greedy algorithm. *Mathematical programming*, 1 (1): 127–136 (1971).

97. J. Elman. Finding structure in time. *Cog. Sci.*, 14(2):179–211 (1990).

98. J. Fakcharoenphol, S. Rao, and K. Talwar: A tight bound on approxi- mating arbitrary metrics by tree metrics. In *Proceedings of STOC'03*, pages 448-455, New York, NY, USA, 2003.

99. M. Faloutsos, P. Faloutsos, and C. Faloutsos: On power-law relationships of the internet topology, *ser. SIGCOMM '99*. New York, NY, USA: ACM, 1999, pp. 251–262.

100. Lidan Fan, Weili Wu, Kai Xing, Wonjun Lee: Precautionary rumor containment via trustworthy people in social networks. *Discret. Math. Algorithms Appl.* 8(1): 1650004:1–1650004:18 (2016).

101. Lidan Fan, Weili Wu, Xuming Zhai, Kai Xing, Wonjun Lee, Ding-Zhu Du: Maximizing rumor containment in social networks with constrained time. *Soc. Netw. Anal. Min.* 4(1): 214 (2014).

102. Lidan Fan, Zaixin Lu, Weili Wu, Yuanjun Bi, Ailian Wang, Bhavani M. Thuraisingham: An individual-based model of information diffusion combining friends' influence. *J. Comb. Optim.* 28(3): 529–539 (2014).

103. Lidan Fan, Zaixin Lu, Weili Wu, Bhavani M. Thuraisingham, Huan Ma, Yuanjun Bi: Least Cost Rumor Blocking in Social Networks. In *ICDCS* 2013: 540–549.
104. U. Feige, V. Mirrokni, and J. Vondráak: Maximizing nonmonotone submodular functions. In *Proceedings of the IEEE Foundations of Computer Science*, 2007, pp. 461–471.
105. Uriel Feige and Rani Izsak: Welfare maximization and the supermodular degree. In *ITCS*, pages 247–256, 2013.
106. Michal Feldman, David Kempe, Brendan Lucier, Renato Paes Leme: Pricing public goods for private sale. *EC 2013*: 417–434.
107. M. Feldman, J. S. Naor, and R. Schwartz: Nonmonotone submodular maximization via a structural continuous greedy algorithm. In *ICALP 2011*, 2011.
108. M. Feldman, J. Naor, R. Schwartz: A unified continuous greedy algorithm for submodular maximization. In *IEEE FOCS 2011*, pp. 570–579.
109. M. Feldman and R. Izsak: Constrained monotone function maximization and the supermodular degree. In *ACM-SIAM SODA*, 2014.
110. Moran Feldman, Rani Izsak: Building a Good Team: Secretary Problems and the Supermodular Degree. In *SODA* 2017: 1651–1670.
111. Shanshan Feng, Gao Cong, Arijit Khan, Xiucheng Li, Yong Liu, and Yeow Meng Chee. Inf2vec: Latent representation model for social influence embedding. In *2018 IEEE 34th International Conference on Data Engineering (ICDE)*, pages 941–952. IEEE, 2018.
112. R. Ferreira, L.S. Cabral, F. Freitas, R.D. Lins, G.F. Silva, S.J. Simske, L. Favaro: A multi-document summarization system based on statistics and linguistic treatment. *Expert Syst. Appl.*, 41 (13): 5780–5787 (2014).
113. M. L. Fisher, G. L. Nemhauser, and L. A. Wolsey: An analysis of approximations for maximizing submodular set functions—II. Springer Berlin Heidelberg, Berlin, Heidelberg, 73–87.
114. G. W. Flake, S. Lawrence, and C. L. Giles: Efficient identification of web communities. In *Proceedings of the sixth International Conference on Knowledge Discovery and Data Mining. ACM*, 2000, pp. 150–160.
115. L. Ford and D. R. Fulkerson: *Flows in networks*, Princeton, Princeton University Press, 1962, vol. 1962.
116. Santo Fortunato: Comprehensive review: Community detection in graphs, *Physics Reports*, 486: 75–174 (2010).
117. S. Fortunato and M. Barthélemy: Resolution limit in community detection, *Proc. Nat. Acad. Sci. USA*, vol. 104, no. 1, pp. 36–41, 2007.
118. S. Fortunato and C. Castellano: Community structure in graphs, *Encyclopedia of Complexity and Systems Science*, 2008.
119. K. Fujii and S. Sakaue: Beyond adaptive submodularity: Approximation guarantees of greedy policy with adaptive submodularity ratio. In *Proc. Int. Conf. Mach. Learn.*, 2019, pp. 2042–2051.
120. S. Fujishige: *Submodular functions and optimization*, volume 58. Elsevier Science, 2005.
121. Takuro Fukunaga: Adaptive Algorithm for Finding Dominating Sets in Uncertain, *IEEE/ACM Transactions on Networking*, 28 (1): 387–398 (2020).
122. Ling Gai, Hongwei Du, Lidong Wu, Junlei Zhu, Yuehua Bu: Blocking Rumor by Cut. *J. Comb. Optim.* 36(2): 392–399 (2018).
123. G. Gallo, M. D. Grigoriadis, and R. E. Tarjan: A fast parametric maximum flow algorithm and applications. *SIAM Journal on Computing*, 18 (1): 30–55 (1989).
124. Chuangen Gao, Hai Du, Weili Wu, Hua Wang: Viral marketing of online game by DS decomposition in social networks. *Theor. Comput. Sci.* 803: 10–21 (2020).
125. Chuangen Gao, Shuyang Gu, Ruiqi Yang, Jiguo Yu, Weili Wu, Dachuan Xu: Interaction-aware influence maximization and iterated sandwich method. *Theor. Comput. Sci.* 821: 23–33 (2020).
126. Chuangen Gao, Shuyang Gu, Jiguo Yu, Hai Du, Weili Wu: Adaptive seeding for profit maximization in social networks. *J. Glob. Optim.* 82(2): 413–432 (2022).
127. R. Gandhi, S. Khuller, S. Parthasarathy, and A. Srinivasan. Dependent rounding and its applications to approximation algorithms. *Journal of the ACM*, 53(3):324–360 (2006).

128. Chuangen Gao, Shuyang Gu, Ruiqi Yang, Hongwei Du, Smita Ghosh, Hua Wang: Robust Profit Maximization with Double Sandwich Algorithms in Social Networks. In *ICDCS* 2019: 1539–1548.

129. Chuangen Gao, Shuyang Gu, Ruiqi Yang, Jiguo Yu, Weili Wu, Dachuan Xu: Interaction-aware influence maximization and iterated sandwich method. *Theor. Comput. Sci.* 821: 23-33 (2020).

130. Hong Gao, Frank K. Hwang, My T. Thai, Weili Wu, Taieb Znati: Construction of d(H)-disjunct matrix for group testing in hypergraphs. *J. Comb. Optim.* 12(3): 297–301 (2006).

131. Xiaofeng Gao, Zhenhao Cao, Sha Li, Bin Yao, Guihai Chen, and Shaojie Tang. Taxonomy and evaluation for microblog popularity prediction. *ACM Transactions on Knowledge Discovery from Data (TKDD)*, 13(2):1–40 (2019).

132. Priyanshi Garg, Weili Wu: Social network analysis and applications: A review of the broad research aspects of social network structure. *Discret. Math. Algorithms Appl.* 14(6): 2230001:1–2230001:31 (2022).

133. S. O. Gharan and J. Vondrák: Submodular maximization by simulated annealing, in *SODA 2011*, 2011.

134. Priyanshi Garg, Weili Wu: Social network analysis and applications: A review of the broad research aspects of social network structure. *Discret. Math. Algorithms Appl.* 14(6): 2230001:1–2230001:31 (2022).

135. Smita Ghosh, Tiantian Chen, Weili Wu: Enhanced Group Influence Maximization in Social Networks Using Deep Reinforcement Learning, *IEEE Transactions on Computational Social Systems*, 12(2): 573–585 (2025).

136. R. Ghosh, K. Lerman: Community detection using a measure of global influence, *Advances in Social Network Mining and Analysis*, LNCS 5498 (2008) 20–35.

137. M. Girvan and M. E. J. Newman: Community structure in social and biological networks, *Proc. Nat. Acad. Sci. USA*, 99 (12): 7821–7826 (2002).

138. J. Goldenberg, B. Libai, E. Muller: Talk of the network: a complex systems look at the underlying process of word-of-mouth. *Mark Lett* 12:211–223 (2001).

139. J. Goldenberg, B. Libai, E. Muller: Using complex systems analysis to advance marketing theory development: modeling heterogeneity effects on new product growth through stochastic cellular automata. *Acad Mark Sci Rev* 2001:1 (2001).

140. O. Goldschmidt and D. S. Hochbaum: A polynomial algorithm for the k-cut problem for fixed k, *Mathematics of operations research*, 19 (1): 24–37 (1994).

141. Daniel Golovin, Andreas Krause: Adaptive Submodularity: Theory and Applications in Active Learning and Stochastic Optimization. *J. Artif. Intell. Res.* 42: 427–486 (2011).

142. Daniel Golovin and Andreas Krause. Adaptive submodularity:theory and applications in active learning and stochastic optimization. *Journal of Artificial Intelligence Research*, 42:427–486 (2011) arXiv version (arxiv.org/abs/1003.3967).

143. Manuel Gomez-Rodriguez, Jure Leskovec, and Andreas Krause. Inferring networks of diffusion and influence. *ACM Transactions on Knowledge Discovery from Data (TKDD)*, 5(4):1–37 (2012).

144. Suning Gong, Qingqin Nong, Han Xiao, Ding-Zhu Du: Uncertainty in Study of Social Networks: Robust Optimization and Machine Learning. *Asia Pac. J. Oper. Res.* 40(1): 2340006:1–2340006:16 (2023).

145. Alkis Gotovos, Amin Karbasi, and Andreas Krause. Non-monotone adaptive submodular maximization. In *IJCAI*, 2015.

146. Chengcheng Gou, Huawei Shen, Pan Du, Dayong Wu, Yue Liu, and Xueqi Cheng: Learning sequential features for cascade outbreak prediction. *Knowledge and Information Systems*, 57(3):721–739 (2018).

147. A. Goyal, F. Bonchi, and L. V. S. Lakshmanan: Learning influence probabilities in social networks. In *WSDM*, pages 241–250, 2010.

148. A. Goyal, F. Bonchi, and L. V. S. Lakshmanan: A data-based approach to social influence maximization. *PVLDB*, 5(1):73–84 (2011).

149. A. Goyal, W. Lu, and L. V. S. Lakshmanan: SIMPATH: An efficient algorithm for influence maximization under the linear threshold model. In *Proc. IEEE 11th Int. Conf. Data Mining*, Dec. 2011, pp. 211–220.

150. A. Goyal, W. Lu, and L. Lakshmanan: Celf++: optimizing the greedy algorithm for influence maximization in social networks. In *Proceedings of the 20th international conference companion on World wide web*, pp. 47–48, ACM, 2011.

151. S. Goyal, M. Kearns: Competitive contagion in networks. In *Proc. 43rd ACM Symp. on Theory of Computing*, pp. 759–774 (2012).

152. M. Granovetter: Threshold models of collective behavior. *Am J Sociol*, 83(6):1420–1443 (1978).

153. M. Grötschel, L. Lovász, and A. Schrijver: *Geometric Algorithms and Combinatorial Optimization*. Springer-Verlag, 2nd edition, 1988.

154. Aditya Grover and Jure Leskovec. node2vec: Scalable feature learning for networks. In *Proceedings of the 22nd ACM SIGKDD international conference on Knowledge discovery and data mining*, pages 855–864, 2016.

155. Shuyang Gu, Chuangen Gao, Jun Huang, Weili Wu: Profit maximization in social networks and non-monotone DR-submodular maximization. *Theor. Comput. Sci.* 957: 113847 (2023).

156. S. Gu, H. Du, M.T. Thai, D.-Z. Du: Friending. In: Du D-Z, Pardalas PM, Zhang Z (eds) *Nonlinear combinatorial optimization*. Springer, Berlin, pp 265–272, 2019.

157. Shuyang Gu, Chuangen Gao, Ruiqi Yang, Weili Wu, Hua Wang, Dachuan Xu: A general method of active friending in different diffusion models in social networks. *Soc. Netw. Anal. Min.* 10(1): 41 (2020).

158. R. Guimerà, M. Sales Pardo Missing and spurious interactions and the reconstruction of complex networks. lem Proc. Natl. Acad. Sci. USA, 106: 22073 (2009).

159. Jianxiong Guo, Tiantian Chen, Weili Wu: Budgeted Coupon Advertisement Problem: Algorithm and Robust Analysis. *IEEE Trans. Netw. Sci. Eng.* 7(3): 1966–1976 (2020).

160. Jianxiong Guo, Tiantian Chen, Weili Wu: Continuous Activity Maximization in Online Social Networks. *IEEE Trans. Netw. Sci. Eng.* 7(4): 2775–2786 (2020).

161. Jianxiong Guo, Tiantian Chen, Weili Wu: A Multi-Feature Diffusion Model: Rumor Blocking in Social Networks. *IEEE/ACM Trans. Netw.* 29(1): 386–397 (2021).

162. Jianxiong Guo, Yi Li, Weili Wu: Targeted Protection Maximization in Social Networks. *IEEE Trans. Netw. Sci. Eng.* 7(3): 1645–1655 (2020).

163. Jianxiong Guo, Qiufen Ni, Weili Wu, Ding-Zhu Du: Composite Community-Aware Diversified Influence Maximization With Efficient Approximation. *IEEE/ACM Trans. Netw.* 32(2): 1584–1599 (2024).

164. Jianxiong Guo, Weili Wu: A Novel Scene of Viral Marketing for Complementary Products. *IEEE Trans. Comput. Soc. Syst.* 6(4): 797–808 (2019).

165. Jianxiong Guo, Weili Wu: Discount advertisement in social platform: algorithm and robust analysis. *Soc. Netw. Anal. Min.* 10(1): 57 (2020).

166. Jianxiong Guo, Weili Wu: A k-Hop Collaborate Game Model: Adaptive Strategy to Maximize Total Revenue. *IEEE Trans. Comput. Soc. Syst.* 7(4): 1058–1068 (2020).

167. Jianxiong Guo, Weili Wu: A k-Hop Collaborate Game Model: Extended to Community Budgets and Adaptive Nonsubmodularity. *IEEE Trans. Syst. Man Cybern. Syst.* 52(9): 5567–5578 (2022).

168. Jianxiong Guo, Weili Wu: Influence Maximization: Seeding Based on Community Structure. *ACM Trans. Knowl. Discov. Data* 14(6): 66:1–66:22 (2020).

169. Jianxiong Guo, Weili Wu: Viral Marketing for Complementary Products, in Du, Pardalos, Zhang (eds) *Nonlinear Combinatorial Optimization*, (Springer, 2019) pp. 309–315.

170. Jianxiong Guo, Weili Wu: Adaptive Influence Maximization: If Influential Node Unwilling to Be the Seed. *ACM Trans. Knowl. Discov. Data* 15(5): 84:1–84:23 (2021).

171. A. Gupta, V. Nagarajan, and R. Ravi: Approximation algorithms for optimal decision trees and adaptive TSP problems, *Math. Oper. Res.*, 42 (3): 876–896 (2017).

172. N. Guttmann-Beck and R. Hassin: Approximation algorithms for minimum k-cut, *Algorithmica*, 27(2): 198–207 (2000).

173. William L. Hamilton: *Graph Representation Learning*. Synthesis Lectures on Artificial Intelligence and Machine Learning, Vol. 14, No. 3 , Pages 1–159 (2020).

174. W.L. Hamilton, R. Ying, and J. Leskovec: Inductive representation learning on large graphs. In *NeurIPS*, 2017.

175. William L. Hamilton, Rex Ying, and Jure Leskovec. Representation learning on graphs: Methods and applications. *IEEE Data Eng. Bull.*, 40(3):52–74 (2017).

176. Kai Han, Keke Huang, Xiaokui Xiao, Jing Tang, Aixin Sun, and Xueyan Tang. Efficient algorithms for adaptive influence maximization. *PVLDB*, 11(9):1029–1040 (2018).

177. S. Harenberg et al., Community detection in large-scale networks: A survey and empirical evaluation. *WIREs Comput. Stat.*, 6 (6): 426–439 (2014).

178. D. Halliday, R. Resnick, J. Walker: *Fundamentals of physics extended*. Wiley.com, 2010.

179. Meng Han, Mingyuan Yan, Zhipeng Cai, Yingshu Li, Xingquan Cai, Jiguo Yu: Influence maximization by probing partial communities in dynamic online social networks. *Trans. Emerg. Telecommun. Technol.* 28(4): (2017).

180. M.A. Hasan, V. Chaoji, S. Salem, M. Zaki: Link prediction using supervised learning. In *Proc. of SDM 06 Workshop on Link Analysis, Counterterrorism and Security* (2006).

181. H. Van Hasselt, A. Guez, and D. Silver: Deep reinforcement learning with double Q-learning. In *Proc. AAAI Conf. Artif. Intell.*, vol. 30, no. 1, 2016, pp. 2094–2100.

182. A. Hassidim, Y. Singer: Robust guarantees of stochastic greedy algorithms. In *Proceedings of the 34 th International Conference on Machine Learning*, Sydney, Australia, PMLR 70, 2017.

183. D. Hausmann and B. Korte: K-greedy algorithms for independence systems, *Oper. Res. Ser. A-B*, 22 (1): 219–228 (1978).

184. D. Hausmann, B. Korte, and T. Jenkyns: Worst case analysis of greedy type algorithms for independence systems, *Math. Prog. Study*, 12: 120–131 (1980).

185. Qiang He, Hongwei Du, Ziwei Liang: Positive Influence Maximization in Signed Networks Within a Limited Time. *IEEE Trans. Comput. Soc. Syst.* 10(5): 2624–2635 (2023).

186. Xinran He, David Kempe: Price of Anarchy for the N-Player Competitive Cascade Game with Submodular Activation Functions. In *WINE 2013*: 232–248.

187. Xinran He, David Kempe: Robust Influence Maximization. In *KDD 2016*: 885–894.

188. X. He, G. Song, W. Chen, Q. Jiang: Influence blocking maximization in social networks under the competitive linear threshold model. In: *Proceedings of the SDM*, pp. 463–474 (2012).

189. W.J. Hopp, X. Xu, Product line selection and pricing with modularity in design, Manuf. Serv. *Oper. Manage.* 7 (3): 172–187 (2005).

190. Y. Hu et al., *Phys. Rev. E* 78(2): 026121 (2008), https://doi.org/10.1103/PhysRevE.78.026121.

191. Huimin Huang, Zaiqiao Meng, Hong Shen: Competitive and complementary influence maximization in social network: A follower's perspective Knowl. *Based Syst.*, 213 (2021) (2021), Article 106600, https://doi.org/10.1016/j.knosys.2020.106600.

192. Zhenhua Huang, Zhenyu Wang, and Rui Zhang: Cascade2vec: Learning dynamic cascade representation by recurrent graph neural networks. *IEEE Access*, 7:144800–144812 (2019).

193. Zhenhua Huang, Zhenyu Wang, Rui Zhang, Yangyang Zhao, and Fadong Zheng: Learning bidirectional social influence in information cascades using graph sequence attention networks. In *Companion Proceedings of the Web Conference 2020*, pages 19–21, 2020.

194. P. Hui, J. Crowcroft, and E. Yoneki: Bubble rap: Social-based forwarding in delay-tolerant networks, *IEEE Trans. Mob. Com.*, 10 (11): 1576–1589 (2011).

195. Hui-Ju Hung, Hong-Han Shuai, De-Nian Yang, Liang-Hao Huang, Wang-Chien Lee, Jian Pei, Ming-Syan Chen: When Social Influence Meets Item Inference. *KDD 2016*: 915–924.

196. R. Iyer and J. Bilmes: Algorithms for Approximate Minimization of the Difference between Submodular Functions, In *Proc. UAI* (2012).

197. Rani Izsak: Working Together: Committee Selection and the Supermodular Degree. *AAMAS 2017*: 1578–1580.

198. N. Jegadeesh, D. Wu: Word Power: A New Approach for Content Analysis. *Journal of Financial Economics* 110 (2013): 712–729.

199. Maximilian Jenders, Gjergji Kasneci, and Felix Naumann. Analyzing and predicting viral tweets. In *Proceedings of the 22nd international conference on world wide web*, pages 657–664, 2013.

200. T. Jenkyns: The efficacy of the greedy algorithm, *Cong. Num.*, 17: 341–350 (1976).

201. Z. Jiang. J. Liu, and S. Wang: Traveling salesman problems with pagerank distance on complex networks reveal community structure. *Phys. A, Stat. Mech. Appl.*, 463: 293–302 (2016).

202. Rong Jin, Priyanshi Garg, Weili Wu, Qiufen Ni, Rosanna E. Guadagno: A Greedy Monitoring Station Selection for Rumor Source Detection in Online Social Networks. *IEEE Trans. Comput. Soc. Syst.* 11(2): 2644–2655 (2024).

203. Rong Jin, Weili Wu: Schemes of propagation models and source estimators for rumor source detection in online social networks: A short survey of a decade of research. *Discret. Math. Algorithms Appl.* 13(4): 2130002:1–2130002:12 (2021).

204. David S. Johnson, Christos H. Papadimitriou, and Mihalis Yannakakis: How easy is local search? *Journal of computer and system sciences* 37(1): 79–100 (1988).

205. K. Jung, W. Heo, and W. Chen: Irie: Scalable and robust influence maximization in social networks. In *Data Mining (ICDM), 2012 IEEE 12th International Conference on*, pp. 918–923, IEEE, 2012.

206. R. Kannan, S. Vempala, and A. Vetta: On clusterings: Good, bad and spectral, *Journal of the ACM (JACM)*, 51 (3): 497–515 (2004).

207. D. R. Karger and C. Stein: A new approach to the minimum cut problem, *Journal of the ACM (JACM)*, 43 (4): 601–640 (1996).

208. G. Karypis and V. Kumar: A fast and high quality multilevel scheme for partitioning irregular graphs, *SIAM Journal on scientific Computing*, 20 (1): 359–392 (1998).

209. Zekarias T Kefato, Nasrullah Sheikh, Leila Bahri, Amira Soliman, Alberto Montresor, and Sarunas Girdzijauskas: Cas2vec: Network-agnostic cascade prediction in online social networks. In *2018 Fifth International Conference on Social Networks Analysis, Management and Security (SNAMS)*, pages 72–79. IEEE, 2018.

210. David Kempe, Jon M. Kleinberg, Éva Tardos: Influential Nodes in a Diffusion Model for Social Networks. *ICALP* 2005: 1127–1138.

211. D. Kempe, J.M. Kleinberg, and E. Tardos. Maximizing the spread of influence through a social network. *KDD'2003*: pp. 137–146.

212. E. B. Khalil, B. Dilkina, and L. Song: Scalable diffusion-aware optimization of network topology. In *Proc. 20th ACM SIGKDD Int. Conf. Knowl. Discovery Data Mining*, 2014, pp. 1226–1235.

213. S. Khuller, A. Moss, and J. S. Naor, The budgeted maximum coverage problem, *Information Processing Letters*, 70 (1): 39–45 (1999)

214. J. Kim, S.-K. Kim, and H. Yu: Scalable and parallelizable processing of influence maximization for large-scale social networks. In *ICDE*, pages 266–277, 2013.

215. Y. Kim and J. Srivastava: Impact of social influence in e-commerce decision making. In *Proceedings of the ninth international conference on Electronic commerce*. ACM, pp. 293–302, 2007.

216. M. Kimura, K. Saito, and H. Motoda: Minimizing the spread of contamination by blocking links in a network. In *Proc. 23rd AAAI Conf. Artif. Intell.*, 2008, vol. 8, pp. 1175–1180.

217. T.N. Kipf and M. Welling: Semi-supervised classification with graph convolutional networks. In *ICLR*, 2016.

218. M. Kivelä, A. Arenas, M. Barthelemy, J. P. Gleeson, Y. Moreno, and M. A. Porter: Multilayer networks, *J. Complex Netw.*, 2 (3): 203–271 (2014), https://doi.org/10.1093/comnet/cnu016.

219. Y. Ko and J. Seo: An effective sentence-extraction technique using contextual information and statistical approaches for text summarization. *Pattern Recognit. Lett.*, 29 (9): 1366–1371 (2008).

220. Shoubin Kong, Qiaozhu Mei, Ling Feng, Fei Ye, and Zhe Zhao: Predicting bursts and popularity of hashtags in real-time. In *Proceedings of the 37th international ACM SIGIR conference on Research & development in information retrieval*, pages 927–930, 2014.

221. B. Korte and D. Hausmann: An analysis of the greedy heuristic for independence systems, *Annals of Discrete Math.*, 2: 65–74 (1978).
222. Andreas Krause, Daniel Golovin, Sarah J. Converse: Sequential Decision Making in Computational Sustainability via Adaptive Submodularity. *AI Mag.* 35(2): 8–18 (2014).
223. S. Kumar and N. Shah: False information on Web and social media: A survey, 2018, arXiv:1804.08559. [Online]. Available: http://arxiv.org/abs/1804.08559.
224. Matt J. Kusner, Yu Sun, Nicholas I. Kolkin, and Kilian Q. Weinberger: From word embeddings to document distance, In *Proceedings of the 32nd International Conference on Machine Learning (ICML 2015)*, pages 957–955, 2015.
225. K. Kutzkov, A. Bifet, F. Bonchi, and A. Gionis: Strip: stream learning of influence probabilities. In *KDD*, pages 275–283, 2013.
226. Sylvain Lamprier: A recurrent neural cascade-based model for continuous-time diffusion. In *International Conference on Machine Learning*, pages 3632–3641. PMLR, 2019.
227. A. Lancichinetti, S. Fortunato, and F. Radicchi: Benchmark graphs for testing community detection algorithms. *Physical review E*, 78 (4): 046110 (2008).
228. A. Lancichinetti, M. Kivel, J. Saramki, S. Fortunato: Characterizing the community structure of complex networks, *PLoS One* 5 (8): e11976 (2010).
229. T. Lappas, E. Terzi, D. Gunopulos, and H. Mannila : Finding effectors in social networks. In *Proc. 16th ACM SIGKDD Int. Conf. Knowl. Discovery Data Mining (KDD)*, 2010, pp. 1059–1068.
230. A. Lázár, D. Ábel, and T. Vicsek: Modularity measure of networks with overlapping communities, *EPL (Europhys. Lett.)*, vol. 90, no. 1, p. 18001, 2010.
231. J. Lee, V. Mirrokni, V. Nagarajan, and M. Sviridenko: Nonmonotone submodular maximization under matroid and knapsack constraints. In *Proceedings of the ACM on Theory of Computing*, 2009, pp. 323–332.
232. E. Leicht, P. Holme, and M.E.J Newman: Vertex similarity in networks. *Phys. Rev. E*, 73(2): 026120 (2006).
233. Kristina Lerman and Tad Hogg. Using a model of social dynamics to predict popularity of news. In *Proceedings of the 19th international conference on World wide web*, pages 621–630, 2010.
234. J. Leskovec, L. A. Adamic and B. A. Huberman: The dynamics of viral marketing. *ACM Transactions on the Web (TWEB)*, 1(1): 5, (2007).
235. J. Leskovec, L. Backstrom, R. Kumar, A. Tomkins: Microscopic evolution of social networks. In *Proceedings of the 14th ACM SIGKDD international conference on knowledge discovery and data mining*, (ACM, 2008) pp 462–470.
236. J. Leskovec, L. Backstrom, and J. Kleinberg: Meme-tracking and the dynamics of the news cycle. In *Proc. of the 15th SIGKDD*. ACM, 2009, pp. 497–506.
237. J. Leskovec, K. J. Lang, and M. Mahoney, Empirical comparison of algorithms for network community detection. In *Proc. 19th Int. Conf. World Wide Web*, 2010, pp. 631–640.
238. J. Leskovec, A. Krause, C. Guestrin, C. Faloutsos, J. VanBriesen, and N. Glance: Cost-effective outbreak detection in networks. In *ACM KDD'07*, (New York, NY, USA), pp. 420–429, ACM, 2007.
239. Carson K Leung, Alfredo Cuzzocrea, Jiaxing Jason Mai, Deyu Deng, and Fan Jiang: Personalized deepinf: enhanced social influence prediction with deep learning and transfer learning. In *2019 IEEE international conference on big data (big data)*, pages 2871–2880. IEEE, 2019.
240. Cheng Li, Xiaoxiao Guo, and Qiaozhu Mei: Joint modeling of text and networks for cascade prediction. In *Proceedings of the International AAAI Conference on Web and Social Media*, vol. 12, pages 640–643, 2018.
241. Cheng Li, Jiaqi Ma, Xiaoxiao Guo, and Qiaozhu Mei: Deepcas: An end-to-end predictor of information cascades. In *Proceedings of the 26th international conference on World Wide Web*, pages 577–586, 2017.
242. F. Li: The information content of forward-looking statements in corporate filings - a naive Bayesian machine learning approach. *Journal of Accounting Research* 48: 1049–1102 (2010).

243. Ji Li, Zhipeng Cai, Mingyuan Yan, Yingshu Li: Using crowdsourced data in location-based social networks to explore influence maximization. In *INFOCOM* 2016: 1–9.
244. Shengdong Li, Chuanwen Luo, Yuqing Zhu, Weili Wu: Bold driver and static restart fused adaptive momentum for visual question answering. *Knowl. Inf. Syst.* 65(2): 921–943 (2023).
245. S. Li, Y. Zhu, D. Li, D. Kim, and H. Huang: Rumor restriction in online social networks. In *Proc. of the 32nd IPCCC*. IEEE, 2013, pp. 1–10
246. Xiao Li, Weili Wu: A Blockchain-Empowered Multiaggregator Federated Learning Architecture in Edge Computing With Deep Reinforcement Learning Optimization. *IEEE Transactions on Computational Social Systems*, 12(2): 645–657 (2025).
247. Xiang Li, Hongbin George Du: A short proof for stronger version of DS decomposition in set function optimization. *J. Comb. Optim.* 40(4): 901–906 (2020).
248. Xiang Li, Hongbin George Du, Panos M. Pardalos: A variation of DS decomposition in set function optimization. *J. Comb. Optim.* 40(1): 36–44 (2020).
249. Yandi Li, Haobo Gao, Yunxuan Gao, Jianxiong Guo, Weili Wu: A Survey on Influence Maximization: From an ML-Based Combinatorial Optimization. *ACM Trans. Knowl. Discov. Data* 17(9): 133:1–133:50 (2023).
250. Y. Li, J. Fan, Y. Wang, and K. L. Tan: Influence maximization on social graphs: A survey, *IEEE Trans. Knowl. Data Eng.*, 30 (10): 1852–1872 (2018).
251. Y. Li, J. Liu, and C. Liu: A comparative analysis of evolutionary and memetic algorithms for community detection from signed social networks. *Soft Comput.*, 18 (2): 329–348 (2014).
252. Y. Li, W. Chen, Y. Wang, and Z.-L. Zhang: Influence diffusion dynamics and influence maximization in social networks with friend and foe relationships. In *WSDM*, pages 657–666, 2013.
253. Y. Li, D. Zhang, and K.-L. Tan: Real-time targeted influence maximization for online advertisements. In *Proceedings of the VLDB Endowment*, vol. 8, no. 10, 2015.
254. Yingshu Li, My T. Thai, Zhen Liu, Weili Wu: Protein-protein interaction and group testing in bipartite graphs. *IJBRA* 1(4): 414–419 (2005).
255. Zhihang Li, Hongwei Du, Xiang Li: Topic-Aware Information Coverage Maximization in Social Networks. *IEEE Trans. Comput. Soc. Syst.* 11(2): 1722–1732 (2024).
256. Z. Li and J. Liu: A multi-agent genetic algorithm for community detection in complex networks, *Phys. A, Stat. Mech. Appl.*, vol. 449, pp. 336–347, May 2016.
257. Z. Li, S. Zhang, R.-S. Wang, X.-S. Zhang, and L. Chen: Quantitative function for community detection, *Physical review E*, vol. 77, no. 3, p. 036109, 2008.
258. Ziwei Liang, Qiang He, Hongwei Du, Wen Xu: Targeted influence maximization in competitive social networks. *Inf. Sci.* 619: 390–405 (2023).
259. Ziwei Liang, He Yuan, Hongwei Du: Two-stage pricing strategy with price discount in online social networks. *Theor. Comput. Sci.* 882: 1–14 (2021).
260. David Liben-Nowell, Jon M. Kleinberg: The link-prediction problem for social networks. *J. Assoc. Inf. Sci. Technol.* 58(7): 1019–1031 (2007).
261. S. Lim, J. Hao, Z. Lu, X. Zhang, and Z. Zhang: Approximating the k-minimum distance rumor source detection in online social networks, in *Proc. 27th Int. Conf. Comput. Commun. Netw. (ICCCN)*, Jul. 2018, pp. 1–9.
262. Bin Liu, Xiao Li, Huijuan Wang, Qizhi Fang, Junyu Dong, Weili Wu: Profit Maximization problem with Coupons in social networks. *Theor. Comput. Sci.* 803: 22–35 (2020).
263. Bin Liu, Xiao Li, Huijuan Wang, Qizhi Fang, Junyu Dong, Weili Wu: Profit Maximization problem with Coupons in social networks. *Theor. Comput. Sci.* 803: 22–35 (2020)
264. Bin Liu, Yuxia Yan, Qizhi Fang, Junyu Dong, Weili Wu, Huijuan Wang: Maximizing profit of multiple adoptions in social networks with a martingale approach. *J. Comb. Optim.* 38(1): 1–20 (2019).
265. C. Liu, J. Liu, and Z. Jiang: A multiobjective evolutionary algorithm based on similarity for community detection from signed social networks, *IEEE Trans. Cybern.*, 44 (12): 2274–2287 (2014).
266. Chaochao Liu, Wenjun Wang, Pengfei Jiao, Xue Chen, and Yueheng Sun: Cascade modeling with multihead self-attention. In *2020 International Joint Conference on Neural Networks (IJCNN)*, pages 1–8. IEEE, 2020.

267. Jiancong Liu, Zhiheng You, Ziwei Liang, Hongwei Du: User-driven competitive influence maximization in social networks. *Theor. Comput. Sci.* 1018: 114813 (2024).

268. T. Loughran, B. McDonald: Textual analysis in accounting and finance: A survey. *Journal of Accounting Research*, 54 (4): 1187–1230 (2016).

269. L. Lü and T. Zhou. Link prediction in complex networks: A survey. *Physica A*, 390(6):1150–1170 (2011).

270. W. Lu, F. Bonchi, A. Goyal, and L. V. S. Lakshmanan: The bang for the buck: fair competitive viral marketing from the host perspective. In *KDD*, pages 928–936, 2013.

271. Wei Lu, Wei Chen, Laks V.S. Lakshmanan: From competition to complementarity: comparative influence diffusion and maximization, In *Proc. the VLDB Endowsment*, Vol 9, No. 2: 60-71 (2015).

272. Zaixin Lu, Weili Wu, Weidong Chen, Jiaofei Zhong, Yuanjun Bi, Zheng Gao: The maximum community partition problem in networks, *Discrete Mathematics, Algorithms and Applications* 5(4): (2013).

273. Zaixin Lu, Wei Zhang, Weili Wu, Bin Fu, Ding-Zhu Du: Approximation and Inapproximation for the Influence Maximization Problem in Social Networks under Deterministic Linear Threshold Model. In *ICDCS Workshops* 2011: 160–165.

274. Zaixin Lu, Zhao Zhang, Weili Wu: Solution of Bharathi-Kempe-Salek conjecture on influence maximization on in-arborescence, *Journal of Combinatorial Optimization*, 33(2): 803–808 (2017).

275. Zaixin Lu, Wei Zhang, Weili Wu, Bin Fu, Ding-Zhu Du: Approximation and inapproximation for the influence maximization problem in social networks under deterministic linear threshold Model, In *ICDCS Workshops 2011*: pp. 160–165.

276. Zaixin Lu Yuqing Zhu, Wei Li, Weili Wu Xiuzhen Cheng: Influence-based community partition for social networks, *Computational Social Networks* 1: 1–18 (2014).

277. W. Luo and W. P. Tay: Finding an infection source under the SIS model. In *Proc. IEEE Int. Conf. Acoust., Speech Signal Process.*, May 2013, pp. 2930–2934.

278. W. Luo and W. P. Tay: Estimating infection sources in a network with incomplete observations. In *Proc. IEEE Global Conf. Signal Inf. Process.*, Dec. 2013, pp. 301–304.

279. Huan Ma, Zaixin Lu, Deying Li, Yuqing Zhu, Lidan Fan, Weili Wu: Mining hidden links in social networks to achieve equilibrium. *Theor. Comput. Sci.* 556: 13–24 (2014).

280. Huan Ma, Zaixin Lu, Lidan Fan, Weili Wu, Deying Li, Yuqing Zhu: A Nash Equilibrium Based Algorithm for Mining Hidden Links in Social Networks. In *COCOA 2013*: 141–152.

281. H. Ma, H. Yang, M. Lyu, and I. King: Mining social networks using heat diffusion processes for marketing candidates selection. In *Proceeding of the 17th ACM conference on Information and knowledge management.* pp. 233–242, 2008.

282. M. Macy: Chains of Cooperation: Threshold Effects in Collective Action. *American Sociological Review* 56 (1991).

283. M. Macy, R. Willer: From Factors to Actors: Computational Sociology and Agent-Based Modeling. *Ann. Rev. Soc.* 2002.

284. T. Maehara, K. Murota: A framework of discrete DC programming by discrete convex analysis. *Math. Program. Ser. A* 152: 435–466 (2015).

285. T. Maehara, N. Marumo, K. Murota: Continuous relaxation for discrete DC programming. In *Proceedings of the 3rd International Conference on Modelling, Computation and Optimization in Information Systems and Management Sciences*, pp. 181–190 (2015).

286. F. D. Malliaros and M. Vazirgiannis: Clustering and community detection in directed networks: A survey. *Physics Reports*, vol. 533, no. 4, pp. 95–142, 2013.

287. Sahil Manchanda, Akash Mittal, Anuj Dhawan, Sourav Medya, Sayan Ranu, and Ambuj Singh: Gcomb: Learning budget-constrained combinatorial algorithms over billion-sized graphs. *Advances in Neural Information Processing Systems*, 33: 20000–20011 (2020).

288. S. Mancoridis, B. S. Mitchell, C. Rorres, Y.-F. Chen, and E. R. Gansner: Using automatic clustering to produce high-level system organizations of source code. In *IWPC*, vol. 98. Citeseer, 1998, pp. 45–2.

289. Dongyu Mao, Weili Wu, Ding-Zhu Du: Co-Activity Maximization in Online Social Networks. *IEEE Trans. Comput. Soc. Syst.* 11(1): 66–75 (2024).

290. Miller McPherson, Lynn Smith-Lovin, and James M Cook. Birds of a feather: Homophily in social networks. *Annual review of sociology*, 27(1):415–444 (2001).

291. Nimrod Megiddo and Christos H Papadimitriou: On total functions, existence theorems and computational complexity. *Theoretical Computer Science* 81 (2): 317–324 (1991).

292. C. Merkwirth and T. Lengauer: Automatic generation of complementary descriptors with molecular graph networks. *J. Chem. Inf. Model*, 45(5):1159–1168 (2005).

293. Wil Michiels, Emile Aarts, and Jan Korst: *Theoretical aspects of local search.* Springer Science & Business Media, 2007.

294. Volodymyr Mnih, Koray Kavukcuoglu, David Silver, Andrei A Rusu, Joel Veness, Marc G Bellemare, Alex Graves, Martin Riedmiller, Andreas K Fidjeland, Georg Ostrovski, Stig Petersen, Charles Beattie, Amir Sadik, Ioannis Antonoglou, Helen King, Dharshan Kumaran, Daan Wierstra, Shane Legg, Demis Hassabis: Human-level control through deep reinforcement learning, *Nature*, vol. 518, pp. 529–533, 2015.

295. Soheila Molaei, Hadi Zare, and Hadi Veisi: Deep learning approach on information diffusion in heterogeneous networks. *Knowledge-Based Systems*, 189:105153 (2020).

296. S. Morris: Contagion. *Review of Economic Studies* 67(2000).

297. E. Mossel and S. Roch, On the submodularity of influenc in social networks. In *STOC 2007*: pp. 128–134.

298. Wolfgang Mulzer and Yannik Stein: Computational Aspects of the Colorful Carathéodory Theorem. In: arXiv preprint arXiv:1412-3347 (2014).

299. S. Najafi, I. Duenyas, S. Jasin, J. Uichanco: Multi-product dynamic pricing with limited inventories under cascade click model, 2019, Available at SSRN 3362921.

300. M. Narasimhan and J. Bilmes: A submodular-supermodular procedure with applications to discriminative structure learning, In *Proc. UAI* (2005).

301. George L. Nemhauser, Laurence A. Wolsey: Best Algorithms for Approximating the Maximum of a Submodular Set Function. *Math. Oper. Res.* 3(3): 177–188 (1978).

302. G.L. Nemhauser, L.A. Wolsey, and M.L. Fisher. An analysis of approximations for maximizing submodular set functions, *Mathematical Programming*, 14(1): 265–294 (1978).

303. M. E. J. Newman: Fast algorithm for detecting community structure in networks, *Phys. Rev. E, Stat. Phys. Plasmas Fluids Relat. Interdiscip. Top.*, vol. 69, no. 6, p. 066133, 2004.

304. M. E. J. Newman: Finding community structure in networks using the eigenvectors of matrices, *Phys. Rev. E, Stat. Phys. Plasmas Fluids Relat. Interdiscip. Top.*, vol. 74, no. 3, p. 036104, 2006.

305. M.E. Newman: Modularity and community structure in networks. In *Proceedings of National Academy of Sciences*, vol 103 no 23 (2006) pp. 8577–8582.

306. M. E. J. Newman and M. Girvan: Finding and evaluating community structure in networks, *Phys. Rev. E, Stat. Phys. Plasmas Fluids Relat. Interdiscip. Top.*, vol. 69, p. 026113, Feb. 2004.

307. Hung T. Nguyen, Thang N. Dinh, My T. Thai: Cost-aware Targeted Viral Marketing in billion-scale networks. In *INFOCOM 2016*: 1–9.

308. N. P. Nguyen, T. N. Dinh, Y. Xuan, and M. T. Thai: Adaptive algorithms for detecting community structure in dynamic social networks, In *IEEE INFOCOM*, April 2011, pp. 2282–2290.

309. H.T. Nguyen, M.T. Thai, T.N. Dinh A billion-scale approximation algorithm for maximizing benefit in viral marketing. *IEEE/ACM Transactions on Networking*, pp. 2419–2429 (2017).

310. H. T. Nguyen, M. T. Thai, and T. N. Dinh: Stop-and-stare: Optimal sampling algorithms for viral marketing in billion-scale networks. In *Proc. Int. Conf. Manage. Data*, Jul. 2016, pp. 695–C710.

311. D. T. Nguyen, H. Zhang, S. Das, M. T. Thai, and T. N. Dinh: Least cost influence in multiplex social networks: Model representation and analysis. In *IEEE 13th International Conference on Data Mining (ICDM) 2013*, pp. 567–576, IEEE, 2013.

312. H. Nguyen and R. Zheng: On budgeted influence maximization in social networks, *IEEE Journal on Selected Areas in Communications*, vol. 31, no. 6, pp. 1084–1094, 2013.

313. Qiufen Ni, Hongwei Du: On strict submodularity of social influence. *J. Comb. Optim.* 41(2): 348–356 (2021).

314. Qiufen Ni, Smita Ghosh, Chuanhe Huang, Weili Wu, Rong Jin: Discount allocation for cost minimization in online social networks. *J. Comb. Optim.* 41(1): 213–233 (2021).

315. Qiufen Ni, Jianxiong Guo, Chuanhe Huang, Weili Wu: Information coverage maximization for multiple products in social networks. *Theor. Comput. Sci.* 828-829: 32–41 (2020).

316. Qiufen Ni, Jianxiong Guo, Chuanhe Huang, Weili Wu: Community-based rumor blocking maximization in social networks: Algorithms and analysis. *Theor. Comput. Sci.* 840: 257–269 (2020).

317. Qiufen Ni, Jianxiong Guo, Weili Wu, Huan Wang: Influence-based Community Partition with Sandwich Method for Social Networks. *IEEE Trans. Comput. Soc. Syst.* 10(2): 819–830 (2023).

318. Qiufen Ni, Jianxiong Guo, Weili Wu, Huan Wang, Jigang Wu: Continuous Influence-Based Community Partition for Social Networks. *IEEE Trans. Netw. Sci. Eng.* 9(3): 1187–1197 (2022).

319. Nadia Niknami, Jie Wu: Competitive influence maximisation model with monetary incentive. *Int. J. Parallel Emergent Distributed Syst.* 37(6): 680–695 (2022).

320. A. Noack: Modularity clustering is force-directed layout. *Phys. Rev. E*, vol. 79, p. 026102, 2009.

321. N. Ohsaka, T. Akiba, Y. Yoshida, and K.-I. Kawarabayashi: Fast and accurate influence maximization on large networks with pruned Monte- Carlo simulations. In *Proc. AAAI*, Jun. 2014, pp. 138–C144.

322. J. B. Orlin: A faster strongly polynomial time algorithm for submodular function minimization. *Mathematical Programming*, 118: 237–251 (2009).

323. G. Palla, I. Derényi, I. Farkas, and T. Vicsek: Uncovering the overlapping community structure of complex networks in nature and society, *Nature*, vol. 435, no. 7043, pp. 814–818, 2005.

324. George Panagopoulos, Fragkiskos Malliaros, and Michalis Vazirgiannis. Multi-task learning for influence estimation and maximization. *IEEE Transactions on Knowledge and Data Engineering*, 2020.

325. B. Pásztor, L. Mottola, C. Mascolo, G. Picco, S. Ellwood, and D. Macdonald: Selective Reprogramming of Mobile Sensor Networks through Social Community Detection. In *EWSN*, vol. 5970, 2010, pp. 178–193.

326. D. Peleg: Local Majority Voting, Small Coalitions, and Controlling Monopolies in Graphs: A Review. *3rd Colloq. on Structural Information and Communication*, 1996.

327. Binghui Peng, Wei Chen: Adaptive Influence Maximization with Myopic Feedback. In *NeurIPS 2019*: 5575–5584.

328. Bryan Perozzi, Rami Al-Rfou, and Steven Skiena. Deepwalk: Online learning of social representations. In *Proceedings of the 20th ACM SIGKDD international conference on Knowledge discovery and data mining*, pages 701–710, 2014.

329. Canh V. Pham, Hieu V. Duong, Bao Q. Bui, My T. Thai: Budgeted Competitive Influence Maximization on Online Social Networks. In *CSoNet 2018*: 13–24.

330. Canh V. Pham, Hieu V. Duong, Bao Q. Bui, My T. Thai: Competitive Influence Maximization within Time and Budget Constraints in Online Social Networks: An Algorithmic Approach. *Appl. Sci.* 2019, 9, 2274; https://doi.org/10.3390/app9112274.

331. Canh V. Pham, Hieu V. Duong, My T. Thai: Importance Sample-Based Approximation Algorithm for Cost-Aware Targeted Viral Marketing. In *CSoNet 2019*: 120–132.

332. S. Phillips and J. Westbrook: Online load balancing and network flow. In *Proceedings of the twenty-fifth annual ACM symposium on Theory of computing*. ACM, 1993, pp. 402–411.

333. Y. Ping, Z. Cao, and H. Zhu: Sybil-aware least cost rumor blocking in social networks. In *Proc. of the 2014 IEEE Global Communications Conference*. IEEE, 2014, pp. 692–697.

334. P. Pons and M. Latapy: Computing communities in large networks using random walks. *J. Graph Algorithms Appl.*, vol. 10, no. 2, pp. 191–218, 2006.

335. M. A. Porter, J.-P. Onnela, and P. J. Mucha: Introductory review: Communities in networks, *Notices of the American Mathematical Society* 56: 1082 (2009)

336. V. Qazvinian, E. Rosengren, D. R. Radev, and Q. Mei: Rumor has it: Identifying misinformation in microblogs. In *Proc. of EMNLP. Association for Computational Linguistics*, 2011, pp. 1589–1599.

337. Jiezhong Qiu, Jian Tang, Hao Ma, Yuxiao Dong, Kuansan Wang, and Jie Tang: Deepinf: Social influence prediction with deep learning. In *Proceedings of the 24th ACM SIGKDD International Conference on Knowledge Discovery & Data Mining*, pages 2110–2119, 2018.

338. F. Radicchi, C. Castellano, F. Cecconi, V. Loreto, and D. Parisi: Defining and identifying communities in networks. In *Proc. of the Nat. Acad. Sci.* 101(9), 2658 (2004), https://doi.org/10.1073/pnas.0400054101.

339. U. N. Raghavan, R. Albert, and S. Kumara: Near linear time algorithm to detect community structures in large-scale networks. *Phys. Rev. E, Stat. Phys. Plasmas Fluids Relat. Interdiscip. Top.*, vol. 76, no. 3, p. 036106, 2007.

340. Mohammad Raihanul Islam, Sathappan Muthiah, Bijaya Adhikari, B Aditya Prakash, and Naren Ramakrishnan: Deepdiffuse: Predicting the'who'and'when'in cascades. In *2018 IEEE International Conference on Data Mining (ICDM)*, pages 1055–1060. IEEE, 2018.

341. Yingli Ran, Zhao Zhang, Hongwei Du, Yuqing Zhu: Approximation algorithm for partial positive influence problem in social network. *J. Comb. Optim.* 33(2): 791–802 (2017).

342. Guoyao Rao, Yongcai Wang, Wenping Chen, Deying Li, Weili Wu: Matching influence maximization in social networks. *Theor. Comput. Sci.* 857: 71–86 (2021).

343. Guoyao Rao, Yongcai Wang, Wenping Chen, Deying Li, Weili Wu: Union acceptable profit maximization in social networks. *Theor. Comput. Sci.* 917: 107–121 (2022).

344. Jiajia Rao, Hongwei Du, Xiaoting Yan, Chuang Liu: Detecting Overlapping Community in Social Networks Based on Fuzzy Membership Degree. In *CSoNet* 2016: 99–110.

345. J. Ratkiewicz, M. Conover, M. Meiss, B. Gonçalves, A. Flammini, and F. Menczer: Detecting and tracking political abuse in social media. In *Proc. of the 5th ICWSM*, 2011.

346. Leonardo FR Ribeiro, Pedro HP Saverese, and Daniel R Figueiredo. struc2vec: Learning node representations from structural identity. In *Proceedings of the 23rd ACM SIGKDD international conference on knowledge discovery and data mining*, pages 385–394, 2017.

347. M. Richardson, V. Domingos: Mining knowledge-sharing sites for viral marketing. In *KDD*, pp 61–70 (2002).

348. M. Rosvall and C. T. Bergstrom: Maps of random walks on complex networks reveal community structure, *Proc. Nat. Acad. Sci. USA*, vol. 105, no. 2, pp. 1118–1123, 2008.

349. T. Roughgarden: Intrinsic robustness of the price of anarchy. In: *Proc. 40th ACM Symp. on Theory of Computing*, pp. 513–522 (2009).

350. Kazumi Saito, Masahiro Kimura, Kouzou Ohara, Hiroshi Motoda: Learning Continuous-Time Information Diffusion Model for Social Behavioral Data Analysis, In *ACML 2009*, pp. 322–337.

351. A. Salehi-Abari and T. White: The impact of naive agents in heterogeneous trust-aware societies. Springer, vol. LNAI 5683, pp. 110–122, 2010.

352. Kanthi K. Sarpatwar, Baruch Schieber, Hadas Shachnai: Constrained submodular maximization via greedy local search. *Oper. Res. Lett.* 47(1): 1–6 (2019).

353. Guillaume Salha, Nikolaos Tziortziotis, and Michalis Vazirgiannis. Adaptive submodular influence maximization with myopic feedback. In *2018 IEEE/ACM International Conference on Advances in Social Networks Analysis and Mining (ASONAM)*, pages 455–462. IEEE, 2018.

354. F. Scarselli, M. Gori, A.C. Tsoi, M. Hagenbuchner, and G. Monfardini: The graph neural network model. *IEEE Trans. Neural Netw. Learn. Syst*, 20(1): 61–80 (2009).

355. A. A. Schäfer and M. Yannakakis: Simple local search problems that are hard to solve, *SIAM J. Comput.*, 20: 56–87 (1991).

356. T. Schelling: *Micromotives and Macrobehavior*. Norton, New York (1978).

357. A. Schrijver: A combinatorial algorithm minimizing submodular functions in strong polynomial time. *J. Combinatorial Theory (B)*, 80: 346–355 (2000).

358. L. Seeman and Y. Singer: Adaptive seeding in social networks. In *FOCS*, pages 459–468, 2013.

359. D. Shah and T. Zaman: Rumors in a network: Who's the culprit? *IEEE Trans. Inf. Theory*, 57 (8): 5163–5181 (2011).

360. Jiaxing Shang, Shuo Huang, Dingyang Zhang, Zixuan Peng, Dajiang Liu, Yong Li, and Lexi Xu: Rne2vec: information diffusion popularity prediction based on repost network embedding. *Computing*, 103(2):271–289 (2021).

361. H. Shen, X. Cheng, K. Cai, and M. Hu: Detect overlapping and hierarchical community structure in networks. *Phys. A, Stat. Mech. Appl.*, vol. 388, no. 8, pp. 1706–1712, 2009.

362. Huawei Shen, Dashun Wang, Chaoming Song, and Albert-Laszlo Barabasi. Modeling and predicting popularity dynamics via reinforced poisson processes. In *Proceedings of the AAAI Conference on Artificial Intelligence*, volume 28, pages 291–297, 2014.

363. Y. Shen, T. N. Dinh, H. Zhang, and M. T. Thai: Interest-matching information propagation in multiple online social networks. In *Proceedings of the 21st ACM International Conference on Information and Knowledge Management, CIKM 12*, (New York, NY, USA), pp. 1824–1828, ACM, 2012.

364. Yilin Shen, Nam P. Nguyen, My T. Thai: Exploiting the Robustness on Power-Law Networks. In *COCOON 2011*: 379–390.

365. Yilin Shen, Dung T. Nguyen, Ying Xuan, My T. Thai: New techniques for approximating optimal substructure problems in power-law graphs. *Theor. Comput. Sci.* 447: 107–119 (2012).

366. N. Shervashidze, P. Schweitzer, E. Leeuwen, K. Mehlhorn, and K. Borgwardt: Weisfeiler-lehman graph kernels. *JMLR*, 12:2539–2561, (2011).

367. J. Shi and J. Malik: Normalized cuts and image segmentation, *Pattern Analysis and Machine Intelligence, IEEE Transactions on*, 22 (8): 888–905 (2000).

368. T. Shi, S. Cheng, Z. Cai, and J. Li: Adaptive connected dominating set discovering algorithm in energy-harvest sensor networks. In *Proc. 35th Annu. IEEE Int. Conf. Comput. Commun., (INFOCOM)*, Apr. 2016, pp. 1–9

369. Qihao Shi, Wujian Yang, Can Wang, Mingli Song: Robust misinformation prevention with uncertainty on suspicious nodes. *Neurocomputing* 577: 127344 (2024).

370. Yaron Singer. Influence maximization through adaptive seeding. ACM SIGecom Exchanges, 15 (1): 32–59 (2016).

371. K. Steinhaeuser and N. V. Chawla: Identifying and evaluating community structure in complex networks. *Pattern Recognit. Lett.*, 31 (5): 413–421 (2010).

372. M. Stoer and F. Wagner: A simple min-cut algorithm. *J. ACM*, 44(4):585–591 (1997).

373. Lichao Sun, Weiran Huang, Philip S Yu, and Wei Chen. Multi-round influence maximization. In *Proceedings of the 24th ACM SIGKDD International Conference on Knowledge Discovery & Data Mining*, pages 2249–2258. ACM, 2018.

374. M. Sviridenko: A note on maximizing a submodular set function subject to knapsack constraint, *Operations Research Letters*, 32: 41–43 (2004)-

375. W.J. Sviridenko, M.J. Vondrk: Optimal approximation for submodular and supermodular optimization with bounded curvature. *Math. Oper. Res.* 42(2): 1197–1218 (2017).

376. Z. Svitkina and L. Fleischer, Submodular approximation: Sampling-based algorithms and lower bounds. *SIAM J. Comput.* 40: 1715–1737 (2011)

377. A. Tagarelli, A. Amelio, and F. Gullo: Ensemble-based community detection in multilayer networks, *Data Mining Knowl. Discovery*, vol. 31, no. 5, pp. 1506–C1543, 2017.

378. T. Takahashi and N. Igata: Rumor detection on twitter, in *Proc. of the 6th ICIS-ISIS*, 2012, pp. 452–457.

379. Jian Tang, Meng Qu, Mingzhe Wang, Ming Zhang, Jun Yan, and Qiaozhu Mei. Line: Large-scale information network embedding. In *Proceedings of the 24th international conference on world wide web*, pages 1067–1077, 2015.

380. Jing Tang, Keke Huang, Xiaokui Xiao, Laks V. S. Lakshmanan, Xueyan Tang, Aixin Sun, and Andrew Lim. Efficient approximation algorithms for adaptive seed minimization. In *Proceedings of the 2019 International Conference on Management of Data, SIGMOD Conference 2019*, Amsterdam, The Netherlands, June 30–July 5, 2019., pages 1096–1113, 2019.

381. Shaojie Tang, Jing Yuan: Product sequencing and pricing under cascade browse model. *Oper. Res. Lett.* 48(6): 687–692 (2020).

382. Shaojie Tang, Jing Yuan: Streaming adaptive submodular maximization. *Theor. Comput. Sci.* 944: 113644 (2023).

383. Shaojie Tang, Jing Yuan: Partial-monotone adaptive submodular maximization. *J. Comb. Optim.* 45(1): 35 (2023).

384. Wenyi Tang, Guangchun Luo, Yubao Wu, Ling Tian, Xu Zheng, Zhipeng Cai: A Second-Order Diffusion Model for Influence Maximization in Social Networks. *IEEE Trans. Comput. Soc. Syst.* 6(4): 702–714 (2019).

385. Youze Tang, Xiaokui Xiao, Yanchen Shi: Influence maximization: near-optimal time complexity meets practical efficiency. In *SIGMOD Conference* 2014: 75–86.

386. Youze Tang, Yanchen Shi, Xiaokui Xiao: Influence Maximization in Near-Linear Time: A Martingale Approach. In *SIGMOD Conference* 2015: 1539–1554.

387. Zhongzheng Tang, Jingwen Chen, Chenhao Wang, Tian Wang, Weijia Jia: Greedy+Max: An Efficient Approximation Algorithm for k-Submodular Knapsack Maximization. In *COCOA (1)* 2023: 287–299.

388. B. Taskar, M.F. Wong, P. Abbeel, D. Koller: *Link prediction in relational data Proceedings of Neural Information Processing Systems*, MIT Press, Cambridge, MA (2003), pp. 659–666.

389. T. Tokuyama and J. Nakano, Geometric algorithms for the minimum cost assignment problem. *Random Structures & Algorithms*, 6 (4): 393–406 (1995).

390. Guangmo Tong: StratLearner: Learning a Strategy for Misinformation Prevention in Social Networks. In *NeurIPS* 2020.

391. Guangmo Tong, Lei Cui, Weili Wu, Cong Liu, Ding-Zhu Du: Terminal-set-enhanced community detection in social networks. In *INFOCOM* 2016: 1–9.

392. Guangmo Amo Tong, Ding-Zhu Du: Beyond Uniform Reverse Sampling: A Hybrid Sampling Technique for Misinformation Prevention. In *INFOCOM* 2019: 1711–1719.

393. Guangmo Amo Tong, Ding-Zhu Du, Weili Wu: On Misinformation Containment in Online Social Networks. In *NeurIPS* 2018: 339–349.

394. Guangmo Amo Tong, Shasha Li, Weili Wu, Ding-Zhu Du: Effector Detection in Social Networks. *IEEE Trans. Comput. Soc. Syst.* 3(4): 151–163 (2016).

395. Guangmo Tong. Adaptive influence maximization under general feedback models. arXiv preprint arXiv:1902.00192, 2019.

396. Guangmo Tong, Ruiqi Wang: On Adaptive Influence Maximization Under General Feedback Models. *IEEE Trans. Emerg. Top. Comput.* 10(1): 463–475 (2022).

397. Guangmo Tong, Ruiqi Wang, Zheng Dong, Xiang Li: Time-Constrained Adaptive Influence Maximization. *IEEE Trans. Comput. Soc. Syst.* 8(1): 33–44 (2021). -

398. Guangmo Tong, Weili Wu, Panos M. Pardalos, Ding-Zhu Du: On positive-in fluence target-domination. *Optim. Lett.* 11(2): 419–427 (2017).

399. Guangmo Tong, Ruiqi Wang, Xiang Li, Weili Wu, Ding-Zhu Du: An Approximation Algorithm for Active Friending in Online Social Networks. In *ICDCS* 2019: 1264–1274.

400. Guangmo Tong, Weili Wu, Ding-Zhu Du: Distributed Rumor Blocking With Multiple Positive Cascades. *IEEE Trans. Comput. Soc. Syst.* 5(2): 468–480 (2018).

401. Guangmo Tong, Weili Wu, Ling Guo, Deying Li, Cong Liu, Bin Liu, Ding-Zhu Du: An Efficient Randomized Algorithm for Rumor Blocking in Online Social Networks. *IEEE Trans. Netw. Sci. Eng.* 7(2): 845–854 (2020).

402. Guangmo Tong, Weili Wu, Shaojie Tang, Ding-Zhu Du: Adaptive Influence Maximization in Dynamic Social Networks. *IEEE/ACM Trans. Netw.* 25(1): 112–125 (2017).

403. H. Tong, B. A. Prakash, T. Eliassi-Rad , M. Faloutsos, and C. Faloutsos: Gelling, and melting, large graphs by edge manipulation. In *Proc. 21st ACM Int. Conf. Inf. Knowl. Manage.*, 2012, pp. 245–254.

404. Niki Triantafyllou, Maria M. Papathanasiou: Deep learning enhanced mixed integer optimization: Learning to reduce model dimensionality. *Computers & Chemical Engineering* 187: 108725 (2024).

405. T. Valente: *Network Models of the Diffusion of Innovations*. Hampton Press, 1995.

406. L.G. Valiant: The complexity of enumeration and reliability problem. *SIAM J. Comput.* 8 (3): 410–421 (1979).

407. Ashish Vaswani, Noam Shazeer, Niki Parmar, Jakob Uszkoreit, Llion Jones, Aidan N Gomez, Lukasz Kaiser, and Illia Polosukhin: Attention is all you need. *Advances in neural information processing systems*, 30:5998–6008 (2017).

408. S.A. Vavasis: *Nonlinear Optimization: Complexity Issues*, Oxford Science, New York, 1991.

409. P. Velickovic, G. Cucurull, A. Casanova, A. Romero, P. Lio, and Y. Bengio: Graph attention networks. In *ICLR*, 2018.

410. A. Vetta: Nash equilibria in competitive societies, with applications to facility location, traffic routing and auctions. In: *Proceedings 43rd annual IEEE symposium on foundations of computer science.* (2002) pp 416–425.

411. B. Viswanath, A. Post, K. P. Gummadi, and A. Mislove: An analysis of social network-based sybil defenses. In *SIGCOMM '10. ACM*, 2010, pp. 363–374.

412. J. Vondrák: Symmetry and approximability of submodular maximization problems. In *FOCS '09*, 2009, pp. 651–670.

413. Paul Wagenseller III, Feng Wang, Weili Wu: Size Matters: A Comparative Analysis of Community Detection Algorithms. *IEEE Trans. Comput. Soc. Syst.* 5(4): 951–960 (2018).

414. Peng-Jun Wan, Ding-Zhu Du, Panos M. Pardalos, Weili Wu: Greedy approximations for minimum submodular cover with submodular cost. *Comput. Optim. Appl.* 45(2): 463–474 (2010).

415. Ailian Wang, Weili Wu, Lei Cui: On Bharathi-Kempe-Salek conjecture on influence maximization on in-arborescence, *Journal of Combinatorial Optimization*, 31(4): 1678–1684 (2016).

416. C. Wang, W. Chen, and Y. Wang: Scalable influence maximization for independent cascade model in large-scale social networks. *Data Min. Knowl. Discov.*, 25(3):545–576, 2012.

417. Feng Wang, David Hongwei Du, Erika Camacho, Kuai Xu, Wonjun Lee, Yan Shi, Shan Shan: On positive influence dominating sets in social networks. *Theor. Comput. Sci.* 412(3): 265–269 (2011).

418. Feng Wang, Jinhua She, Yasuhiro Ohyama, and Min Wu: Learning multiple network embeddings for social influence prediction. *IFAC-PapersOnLine*, 53(2):2868–2873 (2020).

419. Guanghua Wang, Priyanshi Garg, and Weili Wu: Segmented Summarization and Refinement: A Pipeline for Long-Document Analysis on Social Media. *Journal of Social Computing* 5 (2): 132–144 (2024).

420. Guanghua Wang and Weili Wu: Dynamic Spectral Clustering and Rewriting Models: A Novel Pipeline Approach for Efficient Multi-Document Summarization. Submitted for publication, 2024.

421. M. Wang, C. Wang, J. X. Yu, and J. Zhang, Community detection in social networks: An in-depth benchmarking study with a procedureoriented framework. In *Proc. 41st Int. Conf. VLDB*, 2015, pp. 998–1009.

422. Senzhang Wang, Xia Hu, Philip S Yu, and Zhoujun Li. Mmrate: inferring multi-aspect diffusion networks with multi-pattern cascades. In *Proceedings of the 20th ACM SIGKDD international conference on Knowledge discovery and data mining*, pages 1246–1255, 2014.

423. S. Wang, X. Zhao, Y. Chen, Z. Li, K. Zhang, and J. Xia: Negative influence minimizing by blocking nodes in social networks. In *Proc. Workshops 27th AAAI Conf. Artif. Intell.*, 2013, pp. 134–136.

424. Wenting Wang, Tiantian Chen, Weili Wu: Reinforcement Learning for Combating Cyberbullying in Online Social Networks. In *COCOA (2)*, 480–493 (2023).

425. X. Wang and J. Liu: A layer reduction based community detection algorithm on multiplex networks, *Phys. A, Stat. Mech. Appl.*, vol. 471, pp. 244–252, Apr. 2017.

426. X. Wang, Y. Zhang, W. Zhang, X. Lin: Dominated competitive influence maximization with time-critical and time-delayed diffusion in social networks. *J. Comput. Sci.*, 28: 318–327 (2018).

427. Y. Wang, G. Cong, G. Song, and K. Xie: Community-based greedy algorithm for mining top-k influential nodes in mobile social networks. In *KDD*, pages 1039–1048, 2010.

428. Y. Wang, Y. Sun, Z. Liu, S. Sarma, M. Bronstein, and J. Solomon: Dynamic graph CNN for learning on point clouds. *ACM TOG*, 38(5):1–12 (2019).

429. Yongqing Wang, Huawei Shen, Shenghua Liu, and Xueqi Cheng. Learning user-specific latent influence and susceptibility from information cascades. In *Twenty-Ninth AAAI Conference on Artificial Intelligence*, pages 477–484, 2015.

430. Zhefeng Wang, Yu Yang, Jian Pei, and Enhong Chen: Activity maximization by effective information diffusion in social networks, *IEEE Trans. Knowl. Data Eng.* 29(11): 2374–2387 (2017).

431. Justin Ward and Stanislav Živn'y. Maximizing k-submodular functions and beyond. *ACM Transactions on Algorithms*, 12(4):1–26 (2016).

432. D. Watts: A Simple Model of Global Cascades in Random Networks. *Proc. Natl. Acad. Sci.* 99: 5766–71 (2002).

433. R. S. Weiss and E. Jacobson: A method for the analysis of the structure of complex organizations, *American Sociological Review*, pp. 661–668, 1955.

434. B. Weisfeiler and A. Lehman. A reduction of a graph to a canonical form and an algebra arising during this reduction. *Nauchno-Technicheskaya Informatsia*, 2(9):12–16 (1968).

435. L.A. Wolsey: An analysis of the greedy algorithm for submodular set covering problem, *Combinatorica*, 2: 385–393 (1982).

436. L.A. Wolsey: Maximizing real-valued submodular functions: Primal and dual heuristics for location problems. *Mathematics of Operations Research*, 7: 410–425 (1982).

437. Chenchen Wu, Yishui Wang, Zaixin Lu, Panos M. Pardalos, Dachuan Xu, Zhao Zhang, Ding-Zhu Du: Solving the degree-concentrated fault-tolerant spanning subgraph problem by DC programming. *Math. Program.* 169(1): 255–275 (2018).

438. Shanchan Wu and Louiqa Raschid. Prediction in a microblog hybrid network using bonacich potential. In *Proceedings of the 7th ACM international conference on Web search and data mining*, pages 383–392, 2014.

439. Weili Wu, Hongwei Du, Huijuan Wang, Lidong Wu, Zhenhua Duan, Cong Tian: On general threshold and general cascade models of social influence, *J. Comb. Optim.* 35(1): 209–215 (2018).

440. Wili Wu, Zhao Zhang, Ding-Zhu Du, Set function optimizations, *Journal of the Operations Research Society of China* 7: 183–193 (2019).

441. Weili Wu, Zhao Zhang, Ding-Zhu Du: Global Approximation of Local Optimality: Nonsubmodular Optimization. *J. Oper. Res. Soc. China* (2023). https://doi.org/10.1007/s40305-023-00475-3.

442. Yao Wu, Hong Huang, and Hai Jin: Information diffusion prediction with personalized graph neural networks. In *International Conference on Knowledge Science, Engineering and Management*, pages 376–387. Springer, 2020.

443. Shuai Xiao, Junchi Yan, Xiaokang Yang, Hongyuan Zha, and Stephen Chu: Modeling the intensity function of point process via recurrent neural networks. In *Proceedings of the AAAI Conference on Artificial Intelligence*, volume 31, pages 1597–1603, 2017.

444. J. Xie, S. Kelley, and B. K. Szymanski: Overlapping community detection in networks: The state-of-the-art and comparative study, *ACM Comput. Surv.*, 45 (4): 43 (2013).

445. Wen Xu, Zaixin Lu, Weili Wu, Zhiming Chen: A novel approach to online social influence maximization. *Soc. Netw. Anal. Min.* 4(1): 153 (2014).

446. W. Xu and H. Chen: Scalable rumor source detection under independent Cascade model in online social networks. In *Proc. 11th Int. Conf. Mobile Ad-Hoc Sensor Netw. (MSN)*, Dec. 2015, pp. 236–242.

447. B. Yan and S. Gregory: Detecting community structure in networks using edge prediction methods, *Statist. Mech.*, vol. 2012, p. P09008, 2012.

448. Ruidong Yan, Hongwei Du, Yi Li, Wenping Chen, Yongcai Wang, Yuqing Zhu, Deying Li: Target users' activation probability maximization with different seed set constraints in social networks. *Theor. Comput. Sci.* 838: 111–125 (2020).

449. Ruidong Yan, Deying Li, Weili Wu, Ding-Zhu Du, Yongcai Wang: Minimizing Influence of Rumors by Blockers on Social Networks: Algorithms and Analysis. *IEEE Trans. Netw. Sci. Eng.* 7(3): 1067–1078 (2020).

450. Ruidong Yan, Yi Li, Deying Li, Weili Wu, Yongcai Wang: SSDBA: the stretch shrink distance based algorithm for link prediction in social networks. *Frontiers Comput. Sci.* 15(1): 151301 (2021).

451. Ruidong Yan, Yi Li, Weili Wu, Deying Li, Yongcai Wang: Rumor Blocking through Online Link Deletion on Social Networks. *ACM Trans. Knowl. Discov. Data* 13(2): 16:1–16:26 (2019).

452. R. Yan, Y. Zhu, D. Li, Z. Ye: Minimum cost seed set for threshold influence problem under competitive models. *World Wide Web* 2018.

453. Mihalis Yannakakis: Computational complexity. In *Local search in combinatorial optimization* (1988), pp. 19–55.

454. Cheng Yang, Maosong Sun, Haoran Liu, Shiyi Han, Zhiyuan Liu, and Huanbo Luan: Neural diffusion model for microscopic cascade prediction. *arXiv preprint* arXiv:1812.08933, 2018.

455. Cheng Yang, Jian Tang, Maosong Sun, Ganqu Cui, and Zhiyuan Liu: Multi-scale information diffusion prediction with reinforced recurrent networks. In *IJCAI*, pages 4033–4039, 2019.

456. Cheng Yang, Hao Wang, Jian Tang, Chuan Shi, Maosong Sun, Ganqu Cui, and Zhiyuan Liu. Full-scale information diffusion prediction with reinforced recurrent networks. *IEEE Transactions on Neural Networks and Learning Systems*, pages 1–13, 2021.

457. Dingda Yang, Xiangwen Liao, Huawei Shen, Xueqi Cheng, Guolong Chen: Relative influence maximization in competitive social networks. *Sci. China Inf. Sci.* 60(10): 108101:1–108101:3 (2017).

458. D.N. Yang, H.J. Hung, W.C. Lee, W. Chen: Maximizing acceptance probability for active friending in online social networks. *KDD* 2013: pp 713–721.

459. J. Yang and J. Leskovec: Defining and evaluating network communities based on ground-truth. In *Proc. ICDM*, Dec. 2012, pp. 745–754.

460. J. Yang, and J. Leskovec. SNAP: Network Datasets: DBLP Collaberation Network. Accessed: Aug. 2017. [Online] http://snap.stanford.edu/data/com-DBLP.html.

461. Mengmeng Yang, Kai Chen, Zhongchen Miao, and Xiaokang Yang. Cost-effective user monitoring for popularity prediction of online user-generated content. In *2014 IEEE International Conference on Data Mining Workshop*, pages 944–951. IEEE, 2014.

462. Wenguo Yang, Jianmin Ma, Yi Li, Ruidong Yan, Jing Yuan, Weili Wu, Deying Li: Marginal Gains to Maximize Content Spread in Social Networks. *IEEE Trans. Comput. Soc. Syst.* 6(3): 479–490 (2019).

463. Wenguo Yang, Jing Yuan, Weili Wu, Jianmin Ma, Ding-Zhu Du: Maximizing Activity Profit in Social Networks. *IEEE Trans. Comput. Soc. Syst.* 6(1): 117–126 (2019).

464. Z. Yang, R. Algesheimer, and C. J. Tessone, A comparative analysis of community detection algorithms on artificial networks, *Sci. Rep.*, vol. 6, Aug. 2016, Art. no. 30750.

465. Xiaopeng Yao, Hejiao Huang, Hongwei Du: Connected positive influence dominating set in k-regular graph. *Discret. Appl. Math.* 287: 65–76 (2020).

466. Xiaopeng Yao, Yunpeng Zhao, Ningtuo Gao, Hongwei Du, Hejiao Huang: Causal Related Rumors Controlling in Social Networks of Multiple Information. *IEEE/ACM Trans. Netw.* 32(3): 2085–2098 (2024).

467. Xiaopeng Yao, Yunpeng Zhao, Ningtuo Gao, Hongwei Du, Hejiao Huang: Causal Related Rumors Controlling in Social Networks of Multiple Information. *IEEE/ACM Trans. Netw.* 32(3): 2085–2098 (2024).

468. Grigory Yaroslavtsev, Samson Zhou, and Dmitrii Avdiukhin: Greedy + max: Near-optimal 1/2-approximations for submodular knapsack. In *International Conference on Artificial Intelligence and Statistics*, pages 3263–3274. PMLR, 2020.

469. Yinyu Ye: On the complexity of approximating a KKT point of quadratic programming. *Math. Program.* 80: 195–211 (1998).

470. J.-Y. Yeh, H.-R. Ke, W.-P. Yang, and I.-H. Meng: Text summarization using a trainable summarizer and latent semantic analysis. *Inf. Process. Manage.*, 41 (1): 75–95 (2005).

471. H. Peyton Young: *Individual Strategy and Social Structure: An Evolutionary Theory of Institutions*. Princeton, 1998.

472. H. Yu, M. Kaminsky, P. B. Gibbons, and A. Flaxman: Sybilguard: defending against sybil attacks via social networks. In *SIGCOMM. ACM*, 2006, pp. 267–278.

473. Ying Yu, Jinglan Jia, Deying Li, Yuqing Zhu: Fair Multi-influence Maximization in Competitive Social Networks. In *WASA* 2017: 253–265.

474. Chunyuan Yuan, Jiacheng Li, Wei Zhou, Yijun Lu, Xiaodan Zhang, and Songlin Hu: Dyhgcn: A dynamic heterogeneous graph convolutional network to learn users' dynamic preferences for information diffusion prediction. In *Joint European Conference on Machine Learning and Knowledge Discovery in Databases*, pages 347–363. Springer, 2020.

475. He Yuan, Ziwei Liang, Hongwei Du: Two-Stage Pricing Strategy with Price Discount in Online Social Networks. In *COCOA* 2020: 185–197.

476. Jing Yuan, Shaojie Tang: Adaptive Discount Allocation in Social Networks. In *MobiHoc* 2017: 22:1–22:10.

477. Jing Yuan and Shao-Jie Tang. No time to observe: Adaptive influence maximization with partial feedback. In *Twenty-Sixth International Joint Conference on Artificial Intelligence (IJCAI)*, 2017.

478. Jing Yuan, Weili Wu, Yi Li, Ding-Zhu Du: Active Friending in Online Social Networks. In *BDCAT* 2017: 139–148.

479. W. W. Zachary: An information flow model for conflict and fission in small groups. *Journal of anthropological research*, pp. 452–473, (1977).

480. H. Zhang, T. Dinh, and M. Thai: Maximizing the spread of positive influence in online social networks. In *Distributed Computing Systems (ICDCS), 2013 IEEE 33rd International Conference on*, pp. 317–326, July 2013.

481. Huiling Zhang, Yilin Shen, My T. Thai: Robustness of power-law networks: its assessment and optimization. *J. Comb. Optim.* 32(3): 696–720 (2016).

482. H.Y. Zhang, H.L. Zhang, A. Kuhnle, M.T. Thai Profit maximization for multiple products in online social networks. In *INFOCOM* (2016), pp. 1–9.

483. H. Zhang, H. Zhang, X. Li, and M. T. Thai: Limiting the spread of misinformation while effectively raising awareness in social networks. In *Prof. of the International Conference on Computational Social Networks*. Springer, 2015, pp. 35–47.

484. Xiansun Zhang et al, A combinatorial model and algorithm for globally searching community structure in complex networks. *Journal of Combinatorial Optimization* 23 (2012) 425–442.

485. Yapu Zhang, Jianxiong Guo, Wenguo Yang, Weili Wu: Mixed-case community detection problem in social networks: Algorithms and analysis. *Theor. Comput. Sci.* 854: 94–104 (2021).

486. Yapu Zhang, Jianxiong Guo, Wenguo Yang, Weili Wu: Targeted Activation Probability Maximization Problem in Online Social Networks. *IEEE Trans. Netw. Sci. Eng.* 8(1): 294–304 (2021).

487. Yapu Zhang, Jianxiong Guo, Wenguo Yang, Weili Wu: Supplementary Influence Maximization Problem in Social Networks. *IEEE Trans. Comput. Soc. Syst.* 11(1): 986–996 (2024).

488. Yapu Zhang, Wenguo Yang, Weili Wu, Yi Li: Effector Detection Problem in Social Networks. *IEEE Trans. Comput. Soc. Syst.* 7(5): 1200-1209 (2020)

489. Zhao Zhang, Wen Xu, Weili Wu, Ding-Zhu Du: A novel approach for detecting multiple rumor sources in networks with partial observations. *J. Comb. Optim.* 33(1): 132–146 (2017).

490. Qingyuan Zhao, Murat A Erdogdu, Hera Y He, Anand Rajaraman, and Jure Leskovec. Seismic: A self-exciting point process model for predicting tweet popularity. In *Proceedings of the 21th ACM SIGKDD international conference on knowledge discovery and data mining*, pages 1513–1522, 2015.

491. Y. Zhao, Y. Xie, F. Yu, Q. Ke, Y. Yu, Y. Chen, and E. Gillum, Botgraph: large scale spamming botnet detection, in *NSDI*. Berkeley,CA,USA: USENIX Association, 2009, pp. 321–334.

492. Fan Zhou, Xovee Xu, Goce Trajcevski, and Kunpeng Zhang. A survey of information cascade analysis: Models, predictions, and recent advances. *ACM Computing Surveys* , 54(2):1–36 (2021).

493. Jie Zhou, Ganqu Cui, Shengding Hu, Zhengyan Zhang, Cheng Yang, Zhiyuan Liu, Lifeng Wang, Changcheng Li, Maosong Sun: Graph neural networks: A review of methods and applications. *AI Open*, vol. 1, pp. 57–81, 2020.

494. T. Zhou, L. Lü, Y.C. Zhang Predicting missing links via local information Eur. *Phys. J. B*, 71 (2009), p. 623.

495. Jianming Zhu, Smita Ghosh, Weili Wu: Robust rumor blocking problem with uncertain rumor sources in social networks. *World Wide Web* 24(1): 229–247 (2021).

496. Jianming Zhu, Smita Ghosh, Weili Wu: Group Influence Maximization Problem in Social Networks. *IEEE Trans. Comput. Soc. Syst.* 6(6): 1156–1164 (2019).

497. Jianming Zhu, Smita Ghosh, Junlei Zhu, Weili Wu: Near-Optimal Convergent Approach for Composed Influence Maximization Problem in Social Networks. *IEEE Access* 7: 142488–142497 (2019).

498. Jianming Zhu, Ye Xing, Runzhi Li, Smita Ghosh, Priyanshi Garg, Weili Wu: Efficient algorithm for stochastic rumor blocking problem in social networks during safety accident period. *Theor. Comput. Sci.* 1023: 114898 (2025).

499. Jianming Zhu, Junlei Zhu, Smita Ghosh, Weili Wu, Jing Yuan: Social Influence Maximization in Hypergraph in Social Networks. *IEEE Trans. Netw. Sci. Eng.* 6(4): 801–811 (2019).

500. K. Zhu and L. Ying: Information source detection in the SIR model: A sample-path-based approach. *IEEE/ACM Trans. Netw.*, 24 (1): 408–421 (2016).

501. K. Zhu and L. Ying: A robust information source estimator with sparse observations. In *Proc. IEEE INFOCOM Conf. Comput. Commun.*, Apr. 2014, pp. 2211–2219.

502. Yuqing Zhu, Deying Li, Wen Xu, Weili Wu, Lidan Fan, James Willson: Mutual-Relationship-based community partitioning in social metworks, *IEEE Trans. Emerg. Top. Comput.* 2(4): 436–447 (2014).

503. Yuqing Zhu, Deying Li, Ruidong Yan, Weili Wu, Yuanjun Bi: Maximizing the Influence and Profit in Social Networks. *IEEE Trans. Comput. Soc. Syst.* 4(3): 54–64 (2017).

504. Yuqing Zhu, Zaixin Lu, Yuanjun Bi, Weili Wu, Yiwei Jiang, Deying Li: Influence and Profit: Two Sides of the Coin. In *ICDM 2013*: 1301–1306

505. Yuqing Zhu, Weili Wu, Yuanjun Bi, Lidong Wu, Yiwei Jiang, Wen Xu: Better approximation algorithms for influence maximization in online social networks. *J. Comb. Optim.* 30(1): 97–108 (2015).

506. Z. Zhu, G. Cao, S. Zhu, S. Ranjan, and A. Nucci: A social network based patching scheme for worm containment in cellular networks. In *INFOCOM 2009, IEEE*, April 2009, pp. 1476–1484.